Science's Next Steps

By

Paul Dutton

'There are more things in heaven and Earth, Horatio,

Than are dreamt of in your philosophy.'

Preface.

In no way is this book an attack on science. Far from it. Science has done many wonderful things and I am confident this will continue in the future. No. All I am attempting is to suggest that science does not leave itself sufficiently open to alternative ways of thinking.

There are equally valid belief-systems. It is these beliefs that I wish to address here. It seems to me that science has turned its back on competitors in the field of knowledge and I would like to take the opportunity to redress the balance. Scientists pride themselves on their open minds and to an extent this is justified but I do feel, and I am sure I am not alone, that scientists can come across to the general public as unaware, unfeeling and (though I hesitate to use the word) arrogant.

There is no need for this. I hope to show in these pages that scientists, with a little more understanding and application of the common touch, can regain the esteem and recognition in society which was previously theirs. That this will mean becoming more flexible in their thinking is inevitable.

The structure of the book is that I present a topic which seems ripe for full scientific investigation. Proponents of these ideas speak for themselves followed by my comments where I feel the topics are already being partly addressed by science. If this is not the case, I suggest where science may go next in order to put these ideas on a full scientific footing.

The new ideas are in **bold**. My comments are in ordinary text.

There is much that is not yet within modern science but which many people have a feeling should be.

This book is for us.

Chapter 1: Aromatherapy

The word chakra literally translates as wheel or disk and refers to a spinning sphere of bio-energetic activity emanating from the major nerve ganglia branching forward from the spinal column. Generally, six of these wheels are described as stacked in a column of energy that spans from the base of the spine to the middle of the forehead, the seventh lying beyond the physical world.

Science has already in recent times introduced a number of theoretical entities which lie beyond our normal physical world, tachyons for one. These are particles which are able to travel faster than light and so move backwards in time. Anatomy confirms that major nerves branch out in the forwards direction from the spinal column.

Modern medicine is now capable of measuring the combined electrical output of thousands of nerve cells working together in ganglia using sensitive voltmeters and these electrical discharges can be thought of as bio-energy. What remains to be confirmed is that the chakra energies are arranged stacked in a column from the base of the spine to the middle of the forehead.

A recent development in Western practices dating back to the 1940's is to associate each one of the seven chakras to a given colour and a corresponding crystal. For example, the chakra in the forehead is associated with the colour purple, so to cure a headache you would apply a purple stone to the forehead.

Science has shown that each colour of light has its own particular frequency of vibration and also that crystals do have a resonance frequency. As is well known, white light can be split into the seven colours of the spectrum. These are, in order of increasing frequency, red, orange, yellow, green, blue, indigo and violet.

As a headache cure, purple stones have obvious advantages over traditional painkillers, not least because they are unlikely to lead to long term dependency, though scientists would be prudent to confirm this in clinical trials.

Patricia Mercier, author of 'The Chakra Bible: The Definitive Guide to Chakra Energy' introduces the relation of colour energy to the science of the light spectrum. As humans, we exist within the 49th octave of vibration of the electromagnetic light spectrum.

Below this range are barely visible radiant heat, then invisible infrared, television and radio waves, sound and brain waves; above it is barely visible ultraviolet, then the invisible frequencies of chemicals and perfumes, followed by x-rays, gamma rays, radium rays and unknown cosmic rays.

Vibrations can occur in octaves as musicians know so it should come as no real surprise to learn that we humans exist in the 49th octave of vibration of the electromagnetic light spectrum. More research is needed concerning the exact frequencies of chemicals and perfumes and I suggest that science may like to concentrate some effort here. Likewise, research is lacking at present in narrowing down the exact octaves of sound and brain waves. Cosmic rays however are beginning to be investigated by science and so are not entirely unknown.

Chakra healing emphasises: Balance your chakras: There are hundreds of chakras joined together by Nadis energy lines and connected to the nervous system at the Plexuses. Emotion is directly hot-wired to health at these points.

This shows development in the field of chakras. The idea of energy lines is already widely accepted, for example electrical energy flows down such lines and they form part of the National Grid. These are probably analogous to Nadis energy lines. The idea of a plexus is a common way of describing localised collections of nerve endings, the solar plexus probably being most familiar to the public. Psychologists, working closely with anatomists, are expected to confirm shortly the hot-wiring of emotion and health at the Plexuses.

Once you have found the chakras that are disaffecting your wellbeing you can shop for the associated chakra balancing creams. Creams are available in two sizes. You only need a little, which will work within seconds to dissolve any blockages allowing the body to heal itself.

You will be familiar with blackheads being blockages of the skin which can be treated and removed by application of the correct type of cream. Chakras are no different.

'The holistic way of getting well, losing pain and disease and staying healthy is to take control of your own health with the vibrational tools of organic-aromatherapy's powerful chakra attuning and organ balancing creams. These creams are simply wonderful.' Caroline Myss PhD.

It is a huge step forward that a Doctor is willing to put her name to the theory of chakras. The fact that she does shows how close the theory is to becoming fully accepted. In fact, chakra theory has widened and deepened in recent years as the following passage shows:

Chakra is a Sanskrit word meaning wheel, or vortex. Our chakras are energy centres which function as pumps or valves, regulating the flow of energy through our energy system. How we choose to think, feel and react to the world around us is reflected through our chakras via our subconscious. All of our senses, all of our perceptions, all of our possible states of awareness, everything it is possible for us to experience, can be divided into roughly seven categories and each category can be associated with a particular chakra.

None of this should come as a surprise. After all, the heart is now understood as a mechanical pump, also containing valves, which regulate the flow of blood. In the same way, our energy system could well be regulated by the energy pumps we call chakras.

Chakras are more dense than auras but not as dense as the physical body. They are hardwired to the physical body through two major vehicles, the endocrine system and the nervous system. Each of the seven main chakras is associated with one of the seven endocrine glands. Thus, each chakra can be associated with particular parts of the body and particular functions within the body.

The density of the physical body has been measured accurately. Science should now put more effort into measuring the density of auras and chakras. Obviously, accurate density determinations are important here and the National Physical Laboratory (NPL) which leads in such measurements in the UK should become involved.

If a chakra is out of balance or blocked then an area of disease can develop. This can be caused by a build-up of unacknowledged negative emotions. If left unacknowledged the disease, or whisper, can become a greater problem as the body starts to shout to gain our attention, eventually breaking down into pain or disease.

Pain however is not well understood and blocked chakra theory offers a persuasive explanation.

In order to read the body we need to acknowledge that tensions in your body represent emotional angst. For most people their right side is their yang side, masculine, moving forward in the world, and the left side is their yin side, female, emotional.

Oriental philosophy has long recognised that everything can be divided into two and the concepts of yin and yang are well established in the literature today. The findings of quantum physics confirm that even at the most basic level there is a yin and yang structure to the Universe. For example, for every particle there is an antiparticle; if you stand on an ice rink and project a weight forward you go backward; all atoms are made of positive charge and an equal amount of negative charge and so on.

Have a look at your body shape. Are you: Pear-shaped? Look at base and sacral chakras. Apple-shaped? Look at solar plexus first, then sacral and heart chakras. Carrot-shaped larger bust/weight on chest? Look at heart chakra. Some areas may not appear to have a chakra nearby but are connected. Elbows, these relate to heart and solar plexus chakras. Knees/thighs/calves, look at base chakra, earth star and soul star chakras. Ankles/wrists, look at sacral chakra. Feet, look at earth star and soul star chakras.

The extension of chakra theory to hitherto unconnected areas of the body is similar to the development of more well-known scientific theories. Newton's laws, for example, were first discovered by considering a single apple but were later extended to orchards.

Once you have found the chakras that are disaffecting your wellbeing you can send for the associated chakra balancing cream. These are available in two sizes and work within seconds to dissolve any blockages.

Although science has yet to work out exactly why the creams work this should not dissuade people from buying them. Penicillin was shown to be effective long before the biochemistry of its action was uncovered.

The creams: There are 15 chakra balancing creams and 18 others to discover, some of which balance the energy of the vital organs and help to clear disease. All the creams have been developed to dissolve energy blockages in our systems within minutes, facilitating a natural flow of energy so that balance can be restored and our bodies can get on with the job of healing themselves. The results are amazing.

It is obviously good news that energy blockages can be dissolved within minutes. What must now be the subject of confirmation is the exact combination and ratios of creams which have the most beneficial effect. Scientific trials and assessments should be conducted to establish these dosage figures. With 15 creams already well known and a further 18 to also take into account there will be many permutations available.

With knowledge of essential oils and psychoneuroimmunology I have been spiritually guided to create these creams. They contain

natural essential oils, specially chosen for their powerful healing effects which balance the energy of the chakras and major organs.

We have used the technology of real time Kirlian imagery to watch the action of the creams as they balance the chakras. Those who have witnessed this are very impressed with both speed and quality of effect. The synergies take the form of organ support and soul harmony chakra balancing creams. The latter have been endorsed by the leading medical intuitive and author, Caroline Myss PhD, who describes them as simply wonderful.

Organ support is of major concern for post-operative patients. Soul harmony could be just as important. The technology of the creams therefore does appear to go beyond energy and enter the realm of synergy. It is significant that Dr. Caroline Myss, using her medical intuition, endorses the creams. Intuition is making a comeback as a validating tool in modern day science after taking something of a back-seat in previous years.

'I think these creams are amazing. I feel the energy hit immediately. I used the earth star first and really felt it clean up my meridian pathways. I have more of an issue with my earth star than I do the stellar gateway so I find that one really cleansing.'

Although a sample of one cannot be extrapolated with confidence anecdotal evidence suggests that proportionately more people do have an issue with their earth star compared to their stellar gateway.

Meridians are the pathways of Qi (Chi energy) and blood flow through the body. Qi flows continuously from one meridian to another. Any break in the flow is an indication of imbalance. If a person's vitality or energy is recognizably diminished it is an indication that the body's organs or tissues are functioning poorly, therefore the Qi flow is inadequate.

Modern medicine had assumed that arteries and veins are the pathways of blood flow through the body but this view now seems outdated. The wisdom of other cultures should be fully integrated into Western medical practice and a good place to begin is a reassessment of the importance of Qi flow.

The meridian healing system is based on the concept that an insufficient supply of Qi makes a person vulnerable to disease. Restoring the Qi is the ultimate goal in restoring overall health and well-being to the individual. Practitioners (acupuncturists, Chinese herbalists, massage therapists, etc.) assist clients in repairing dysfunctioning areas within their meridian systems to restore a natural balance by utilizing various healing methods.

Perhaps an exchange programme between junior doctors in Western hospitals and Chinese herbalists would make a suitable starting point.

The twelve major meridians correspond to specific human organs: kidneys, liver, spleen, heart, lungs, pericardium, bladder, gall bladder, stomach, small and large intestines and the triple burner (body temperature regulator). Yin meridians flow upwards, Yang meridians flow downwards. Pathways corresponding to the Yang organ is often used to treat disorders of its related Yin organ.

Science should be able to readily confirm that Yin meridians flow upwards and Yang meridians flow downwards and once that theoretical underpinning is established the rest follows.

'I can thoroughly recommend the gall bladder support cream by organic aromatherapy. Since using it I have had no indigestion pain or biliousness. Before discovering the curative properties of this cream I was in agony with the pain from my gallstones. I also use the solar plexus cream which has been extremely helpful whilst I have been dealing with personal stress. The creams are worth every penny and I never want to run out.'

Gall bladder problems are all too common these days and surgical intervention is often the only method of cure that Western medicine can offer. But now we have gall bladder support cream. This test subject also affirms the benefit of using more than one curative. The solar plexus cream is seen here to be just as effective dealing with a non-physical disorder as the gall bladder cream is in treating a physical problem.

'About five years ago Sarah introduced me to her organic aromatherapy creams. It soon became clear that I seemed to be benefitting from most of them. One really stood out for me, the anti-inflammatory cream which, over the months, has done so much to help my neuropathic nerve pain. I use this several times a day, and as far as I can work out it is not a placebo, sometimes I forget to gently apply it before I get up. A short while later I'm wondering why I'm starting to burn-up a lot more quickly than usual.'

Scientists say that some medicines work by the 'placebo effect'. By this they mean that the medicine has no effect at all on the chemistry of the body, instead the effect is only in the mind. So it would make no difference if water was administered instead of a chemical, the effect would be exactly the same. Yet here we see a person who has been using the creams and receiving benefit but who says that when she forgets to use them her symptoms return very quickly. This indicates that the creams are not placebo.

'**I had a huge tension at the back of my head and top of my neck which made me feel really agitated and dizzy. My vision was odd, sometimes double, and when I turned my head it seemed to take time to refocus my eyes, a horrible feeling and disorientating too. I was starting to really worry about my sight. Admittedly, I was feeling very tense about life in general, with lots of family and work stresses. I saw the alta major cream online and decided to give it a try. The tension released, my eyes felt normal again and I couldn't believe how quickly this happened once I started to apply the cream twice a day. Now I use it whenever I feel a whisper of tension at the back of my head/neck. It was such a relief to realise that I wasn't losing my eyesight. Thank you so much.**'

It is interesting here is that the correct dosage has been established.

'**I just wanted to say thank you for the healing and guidance you have given me on my visits to you for treatment. As you know, I came to you experiencing anxiety and panic attacks and I can honestly say I feel totally different now. I can drive with confidence and through your treatments and using the creams I now feel a stronger connection with something much greater which gives me real comfort. I can feel the creams working as soon as I put them on and they smell wonderful. I have also been using your base cream on my two little ones and we no longer have the big emotional scenes when I drop them at nursery. The base cream as you suggested makes them much more grounded and secure little souls. I am recommending your creams to everyone.**'

It is a major benefit to all road users that the creams can enable drivers to feel more confident. The aptly named base cream seems to ground children under a certain age. Scientific agreement on the attributes of the creams should lead to them being widely available on prescription to the benefit of parents worldwide.

'**This earth star cream really helps my restless ankles to feel normal. It works after only a few minutes. Thank you.**'

In the same way that base cream grounds us, earth star cream appears to help with body parts which are close to the Earth. Restless ankles is not a well-known symptom but that doesn't make it any less uncomfortable. Science has a duty to help those who suffer from minority complaints as well as those who suffer from more widespread ailments.

'I met you and saw your stall a couple of years ago in Brighton. I have been using the heart chakra cream and have had really powerful experiences that have astounded me. I am not otherwise tuned into such forces/energies in my life, but when I use your cream I feel a spiritual fire come over my body from the heart chakra. It is very beautiful and I sincerely thank you for bringing these gifts to the world.'

Here we have the perfect opportunity for science to demonstrate that it is not as out of touch with ordinary people as it often appears. Physicists will complain that force is not the same thing as energy but that misses the point.

'I have had a wheat intolerance for over 3 years now and it got really bad within six months where even the slightest bit of wheat would cause my body to cramp and I would be in such pain. I tried everything and tried changing my diet but it's hard when you are in a busy job and are constantly having client lunches and dinners. When I met Sarah I was slightly cynical of the creams as I've never known anything like this to work but the solar plexus chakra balancing cream is my magic wand. I don't know what I'd do without it. I don't know how it works but it does. I use the cream day and night and after every big meal and since using it, I haven't had any symptoms except when it ran out and I was waiting for my new one. I have been recommending it to all, this is amazing. Sarah is amazing.'

Intolerances to foodstuffs is an increasingly important issue in today's world. Allergies are on the increase and it is beholden to science to at least attempt to find a cure. Here we see the solar plexus balancing cream providing a positive outcome to one sufferer and yet again that symptoms returned only when the cream ran out. A pattern is beginning to develop.

'I just wanted to thank you for the soul star chakra attuning cream. It's made my life so much easier, not sure how but it seemed to improve the minute I unscrewed the cap and smelt the cream, even before putting any on I felt more relaxed somehow. More able to cope. Genius.'

With yet another report along the same lines a clear relationship can be seen.

'This cream really helps my husband lie still at night during sleep, instead of disturbing me. Thank goodness for a peaceful night. I always know when he has forgotten to apply it.'

Yet again we see confirmation of the non-placebo effect of the creams.

'I had a sore throat and a stiff neck, and thought I was going down with something. I put on some of the throat chakra balancing cream and by next day my neck felt better and the sore throat disappeared. It was truly amazing.'

This is interesting because it could lead to a cure for the common cold, the holy grail which science has been pursuing for decades. Research is urgently needed here in order to establish how the cream has this effect. Perhaps the clue is in the name of the cream. Balance is increasingly being seen as important in all medical cases.

'Sarah's healing chakra synergies and amazing depth of holistic knowledge together with her amazing touch has made me feel balanced, grounded and revitalised after each treatment. I love to use the total balance cream on my clients. I know they will be totally balanced and feeling fantastic if I do so. I am privileged to be able to take my massage to a new dimension by using her chakra synergies in my treatments and therefore helping even more people.'

Science should confirm what healers using the chakra system already know. Individual scientists should attempt to become more holistic in their outlook and not so compartmentalised. The creams should be analysed, perhaps using a mass-spectrometer, to ascertain their exact composition. Then they can be mass produced, under licence, in order to bring their healing powers to the rest of the world at a fraction of their current cost.

The following extracts refer to the end of the Mayan people's Long Cycle calendar, an event much in the news at the time.

Chakra balancing for the 2012 shift: Why is it so important to be balanced at this exciting time. A course at the New Atlantis Academy, Brighton UK Thursday 26th August 2010. Extra chakras are now being activated as we approach the planetary shift from 4th to 5th dimension at the approach of 2012. Sarah will talk about this and bring her soul harmony chakra balancing creams. She will also be giving free chakra readings. Entry is £3.

This is very reasonable. Strange as it may seem, ancient people knew a lot more than we give them credit for and science should seek to explore that knowledge. At present, string theory is undecided as to whether there are 10 or 11 dimensions so the shift in 2012 is of obvious importance.

Sarah Williams is an internationally trained holistic bodywork therapist with her roots in aromatherapy. She has studied many aspects of healing from the shamanic to the angelic and has worked professionally, full time, both in the UK and internationally over the past 15 years. Her treatments incorporate many different techniques, healing, ancient bodywork and advanced massage movements.

Shamanic healing refers to a type of medicine used by ancient peoples. It is summarised as follows by Chris Waters, a practising Shaman:

What I love about the Inca tradition of shamanism is that the letting go can be done with such ease and grace, rather than pain and struggle. That's the beauty of working within the luminous energy field. There is no need to re-experience the trauma and emotional or physical pain that's held there. We are simply dealing with the original imprint that informs the physical, mental and emotional bodies. We are working at the source of the wounding. Dipping our fingers into the rivers of time, and intervening at the energetic level so that changes can be made with ease and grace at a cellular level within the body.

Intervention at the cellular level is something which is commonplace today. It therefore does not seem as though it will be very long before shamanism is put on a fully scientific footing.

Sarah was guided to channel the soul harmony range of 15 chakra balancing creams over 10 years ago. They are powerful vibrational tools which facilitate release of energy blockages in the body. This allows the body to heal itself naturally, with no side effects. She has also developed a range of organ balancing creams, also available to buy from her website.

Medical physicists will confirm that vibrations can indeed release blockages, for example gallstones being vibrated into non-blocking dust by the application of ultrasound.

Complete set of 15 chakra balancing and attuning creams. Get everything you need at this special price. 15x 15ml rrp £225. Offer price £195 + p&p £5.50. 15 x 30ml rrp £450. Offer price £400 + p&p £8.50. Owning a complete set of chakra balancing and attuning creams makes sense. You always have a cream for whatever is going on in your body's energy system or those of your friends and family. The need for regular medication seems to diminish as balance can be maintained naturally and gently. Please feel free to join this group

meeting if you are the practitioner of any type of alternative and/or complementary therapy, healer, shaman, visionary, psychic artist, psychic, clairvoyant, coach, martial arts or yoga teacher etc.

And this list should be extended to include scientists. The reason they are missing can only be because of the inappropriate hostility science has shown so far but as we have seen, this hostility is misdirected when judged on purely scientific grounds.

Treatments: I provide a range of one-on-one treatments, all of which are holistic and include a mixture of techniques. Treatments include: raindroptherapy, aromatherapy, lymphatic, Swedish and Balinese, full body, head, back and shoulders, rejuvenating facial including non-surgical facial lift techniques and cranial massages. In addition I offer reflexology to hands, feet and ears, acupressure, injury remedial work, specialist back and neck treatments, vibrational essences, crystals and spiritual healing, but in practice, I can dip into any of these toolboxes and bring a unique blend of healing techniques to each session.

Blending, or holistic, therapy is an area of medicine which needs further scientific study. Scientific research is needed into the benefits of Swedish and Balinese lymphatic treatments as well as cranial massage and the use of vibrational essences and crystals.

Benefits of treatments: Relief from neck and shoulder tension and headaches. Reduction of all pain in body. Helps to mobilise stiff joints. Eases stress, anxiety and tension. Improves circulation and skin tone. Drainage of toxins within the body. Promotes natural sleep patterns. Revitalises the whole body. Lifts the spirit. Chakra balancing and clearing. Promotes natural healing. Natural re-alignment. Release of trapped nerves including sciatic. Energy clearance.

Draining toxins from the body is of obvious importance. The fact that natural re-alignment and energy clearance can also be achieved is an added bonus.

Prices: 60 minutes - from £40, 90 minutes - from £60, 120 minutes - from £80, initial consultation £10, prices vary depending on location and will be confirmed at time of booking.

These costs are very reasonable when compared with conventional alternatives such as a CAT scan machine which costs millions of pounds to build, operate and maintain.

Raindrop therapy: This is such a releasing and effective treatment involving a series of nine very strong and very pure essential oils being dripped onto the spine in a special sequence. I have seen huge shifts and improvement in those who have experienced it. Benefits include spinal alignment; specific massage techniques and powerful essential oils are concentrated on the spine raindrop style, freeing up stuck congested and misaligned areas, hence promoting natural energy flow and realignment. I have seen an old scoliosis 'S' shaped spine removed in one treatment. Improved general health: Clearing of emotional, viral and bacterial issues that we all hold in the spinal tissue usually only emerging at times when we are feeling low. Clarity and a sense of wellbeing: Everyone who has experienced this treatment has said wow, I feel fantastic as they get off the couch.

It is difficult to know why modern medicine hasn't rushed to embrace this therapy.

'I felt fantastic after my raindrop therapy with Sarah. I had to adjust my rear view mirror in my car, as I was taller at the end. Note: Practitioners have noticed that clients increase in height post treatment by as much as 2cm. This is achieved by the release of the effects of gravitational compression on the spine. As flow in nerves and arteries, passing between the vertebrae of the spine, are positively affected, the benefit to the entire body is obvious and often dramatic.'

Perhaps there is too much emphasis on looking for a successor to the Higgs Boson using billion pound particle accelerators when in fact science could better use its time and money on helping people be the height they could be.

Mind-body connection: Your entire body totally re-builds itself in less than two years through natural cell regeneration. 98% in less than one year; a new brain in one year; blood in four months; skeleton in three months; DNA in two months; liver in six weeks; skin in one month and stomach lining in five days. Ask yourself the question: Why am I still creating the same body? Take a close look at your body. Are there any areas of your body that are either overly large or small? Is there a layer of protective fat in places that is hard to shift? If you do, then these are areas where you are feeling challenged and vulnerable energetically. Your body is responding to the vulnerability of the area.

Science has previously been of the opinion that we do not regenerate a whole brain in a year but it seems that it may have to take a fresh look.

Everything starts in your consciousness: We should orient ourselves to the idea that the causes of symptoms are within us. While it's true that germs cause disease and accidents cause injuries, it is also true that this happens in accord with what is happening in our consciousness. Germs are everywhere. Why are some of us affected and not others? Something different is happening in our consciousness.

This is a subtle but important point and one which it seems science has either missed completely or thought unimportant. Yet it is so obviously true. Germs are everywhere, all scientists will agree with that, yet it is only some people who become affected and fall ill. Why should this be?

Our consciousness, our experience of being, who we really are, is energy. This energy does not just live in our brain; it fills our entire body. Our consciousness is connected to every cell in our body. Through our consciousness, we can communicate with every organ and every tissue.

Brain surgeons are of course aware of consciousness being energy, electrical energy to be precise, so there is no argument there. And nerves do penetrate all parts of the body carrying the electrical energy.

This energy which is our consciousness, and which reflects our emotional state, can be measured through the process known as Kirlian photography. When you take a Kirlian photograph of your hand, for example, it shows a certain pattern of energy. If you take a second photograph while imagining that you are sending love and energy to someone you know, the picture will look different. Thus, we can see that a change in our consciousness creates a change in the energy.

The images may show holes in certain parts of our energy field, aura, these are said to correspond to particular weaknesses in specific parts of our physical bodies. The interesting thing about this is that the weakness shows up in the energy field before there is ever any evidence of it on the physical level. A change of consciousness creates a change in the energy field. A change in the energy field happens before a change in the physical body.

Research into refining the images produced by Kirlian photography would lead to swifter diagnosis of the holes in the energy field.

When we make a decision that leaves us stressed, we create a blockage in the energy field. If intense it will create a symptom on the physical level. The symptom serves to communicate to us an angst about an emotional issue. Thus, instead of saying, I can't see, we are actually saying, I have been keeping myself from seeing something or I don't like what I'm seeing. This is the science of psychoneuroimmunology.

Breaking down this word to make it clearer, psycho refers to the mind, neuro refers to the network of nerves which go around the body and immunology is the science of protecting the body from harm by pathogens before they have a chance to establish themselves. This is an exciting new field and I expect to hear much more of it in the future. Schools may like to set an example here by directing their brightest students in the direction of psychoneuroimmunology as it probably won't be long before leading Universities offer undergraduate courses in this new science.

If we change our way of being, we have received the message, and the symptom has no further reason for being. It is able to be released, according to whatever we allow ourselves to believe is possible. If we created the symptom with a decision, we are also able to release it with a decision. For example, if we cannot express what we want, then we will eventually develop symptoms in the neck and throat area, see throat chakra.

And again, we see the importance of the chakra.

Raise your vibration advanced energy awareness healing journey: Advanced energy awareness course - the big shift. An entertaining and informative opportunity to learn about the 9 additional chakras including the very important out of body chakras that have reappeared in recent years. What they are, why they are here, where they are located and how they affect us and our health and ultimately, how to heal them. This will be an important further stage of development for those who are concerned with progressive healthcare, vibrational healing or have a basic understanding of the importance of the body's energy field.

It is essential that science realises the importance of the additional chakras and investigates their sudden reappearance as soon as possible. The out of body chakras will obviously be more of a challenge.

Course objectives: To activate, balance, attune and understand your expanded chakra energy system. To facilitate the maintenance of your good health and that of your family, clients/patients. To

enable you to make your practice more effective and profitable. To enable the achievement of a pain free, healthy, peaceful state. To shift your emotional body and raise your vibration as further preparation for the 2012 planetary ascension. This is the progression from the 7 chakra energy awareness courses.

Enabling the achievement of a pain-free healthy and peaceful state is of course the objective of all medicine.

Cost for the three day course is £160 to include 1 free 15ml soul harmony cream and refreshments. Deposit required. Group size limited so please book early to avoid disappointment.

This seems very reasonable.

Course leader: Sarah Williams ITEC MGPP ac regd. Holistic therapist, light-worker, healer, qualified teacher, international speaker.

And even more so when the course leader has these impressive qualifications.

Testimonials from March 2010 course: I want to say a huge thank you for hosting the course. I really enjoyed it and feel like I got a lot out of it. You did really well to maintain your energy and enthusiasm throughout. I just wanted to thank you for having us on your course this weekend. It was very enjoyable and enlightening.

We have booked our places on the reunion already. I just want to thank you for an amazing 3 days. Not sure if my aura remains as balanced now, can't be too bad as I still feel very centred and what could potentially have been an extremely difficult day today, is not. I have a real sense of coming home to myself, still lots of work to do but I am seeing the signposts more clearly. Well done. You really are doing your bit for the evolvement of the planet by increasing people's awareness and also encouraging people to help each other with the camaraderie you create.

These are the words of satisfied people

In conclusion I think it is obvious that chakra therapy and the properties of healing creams needs a full examination by mainstream science.

If science cared enough to investigate more deeply I think it would discover and be able to bring to a wider audience some amazing things.

Investigation of Aromatherapy should be one of science's next steps.

Chapter 2: Astrology

Astrology has been around for thousands of years and in all that time has never been proved to be untrue. Millions of people have personal evidence that what it tells them is accurate and they use it to guide their life. There are countless correct predictions in the public domain. The data underpinning astrology is firmly based as astronomers use exactly the same data for their own calculations.

Astrology is based on the movements of the planets. These movements are well known to science and via Newton's law of gravity the positions of the planets can be calculated with accuracy and this accuracy extends thousands of years into the past and can be extended for thousands of years into the future. A few specific examples will illustrate the level of agreement.

We have a number of detailed records of the eclipse of August 2nd, 1133 from different parts of Europe. The following report comes from a monastery in Germany: 'In the year of the incarnation of our lord 1133 on the 4th day before the Nones of August, Wednesday, when the day was declining towards the ninth hour, the Sun in a single moment became black as pitch, day was turned into night and very many stars were seen.'

As for the future, on January 19th 2903 there will be a solar eclipse lasting a full 10 minutes and 10 seconds. These types of scientific measurements and predictions are what astrology is based on.

Constellations are also important. Astrology only mentions twelve because when astrology was being formulated only those which the Sun transited were found to be important. Many more exist but astronomers still consider the twelve constellations of the zodiac identified at the beginning to be amongst the most important.

An example is Sagittarius, which is so important that the Hubble space telescope was specifically directed to study it. The constellation Sagittarius the Archer, home to the globular cluster M22, is one of the best known constellations in the sky. Sagittarius contains a large number of ell-known nebulae and star clusters due to the presence of some of the richest star fields of the Milky Way. The centre of the Milky Way lies in the direction of Sagittarius.

The other zodiacal constellations are also important and Hubble has spent time observing them too. So the discipline of astrology appears well formulated. But what about its results?

From Burger King girl to Buffy the Vampire Slayer, Sarah Michelle Gellar has come a long way. Now, after living together for some time, Sarah and Freddie Prinze Jr. are married. Let's take a look at both Sarah's and Freddie's characters through their natal charts and peek into the synastry of their relationship.

Some bio facts about Sarah: She was born April 14, 1977, birth time unknown, in New York. Some bio facts about Freddie Prinze Jr: He was born March 8, 1976, at 7:38 pm, in Albuquerque, New Mexico.

Some astrological notes about Sarah: Her Sun is in Aries, and she also has an Aries Venus. Her Moon is in Pisces, Mars is in Pisces, and Mercury is in Taurus. He has a Pisces Sun, Gemini Moon, with Mercury in Aquarius, Venus in Aquarius, and Mars in Gemini. He has a Libra ascendant with Pluto rising.

It is necessary in astrology to place the Sun, the planets and our Moon in their respective constellations and this can now be done with extreme accuracy. Given the precision of the astronomical observations of successive NASA satellite launches, astrology is working with the most exact data ever available.

Sarah is a strong Aries with a soft touch. Her Pisces Moon and Mars in Pisces add a vulnerable, dreamy side to her. She can come on strong and sweet at the same time. Sarah can be bafflingly subtle and feminine and headstrong and determined as well. Perhaps that's why she did such a standout job in her role as Erica Cane's daughter in All My Children, a character that was kind of wicked, but confused and vulnerable at the same time.

I don't think anyone can seriously differ with this character analysis.

Freddie Prinze Jr. is the son of the late Freddie Prinze, best known for his role in the '70s sitcom Chico and the Man. Freddie shares a Moon conjunct Mars with Sarah, except that his conjunction is in Gemini instead of Pisces. Both signs are mutable signs but they have very different styles of expression.

The conjunction of the emotional Moon and active Mars lends an impatient, go-getting streak to both Sarah and Freddie; however, Sarah will go about expressing herself in dreamy, even moody, ways with her conjunction in Pisces, and Freddie will tend to

intellectualize his emotions. **Both will have plenty of emotional moments, expressing their feelings in an assertive, in your face manner.**

Despite six manned landings NASA has still failed to investigate fully the emotional Moon.

Freddie has a Libra ascendant with Pluto conjunct, the ascendant giving him a persona that is pleasant but intense. He possesses huge presence and will hardly go unnoticed in life.

Pluto conjunct the Libra ascendant is something which has been confirmed by astronomical observations involving the Voyager missions.

Freddie has a t-square involving the Sun, Moon, Mars, and Neptune. This is a difficult aspect, suggesting a tendency to get let down by the people around him. This is likely due to inflated expectations. He has a dreamy quality, a strong imagination and escapist tendencies.

The difficult Sun-Neptune aspects have not only been confirmed by observations in the visible light spectrum they have also been verified by observations with ground-based radio telescopes as both the Sun and Neptune are emitters in the radio spectrum.

Freddie and Sarah have a classic cross-contact: Her Moon falls directly on his Sun creating an immediate and even primal bond between them. Interestingly, Sarah's Moon-Mars conjunction falls on Freddie's Sun and his Moon-Mars conjunction sextiles her Sun. It's no wonder they noticed each other and felt an immediate connection.

It seems that for anyone whose Moon falls on another person's Sun there is a statistical likelihood of instant attraction. It is now up to science to explain this fact.

Sarah's Sun falls on his seventh-house Jupiter - another pleasant and natural connection that builds trust and forgiveness. Furthermore, it sextiles his Venus, suggesting a mutual, easy affection between them.

The seventh house is of importance as scientists of a certain age will recognise:

'When the Moon is in the seventh house
And Jupiter aligns with Mars
Then peace will guide the planets
And love will steer the stars
This is the dawning of the Age of Aquarius…'

Their composite chart does reveal some challenges in their relationship. But first, the good stuff: their composite Sun is tightly conjunct Mercury suggesting a relationship that features plenty of talk. These two will want to share all sorts of details with each other and will thrive on their intellectual rapport. They may just talk their way out of the more difficult aspects of their composite chart.

The Sun and Mercury receive a helpful trine from Saturn, giving them staying power and an easy commitment to each other. Venus forms a wide trine to Uranus and a tight sextile to Jupiter adding some spice to their love life and a pleasant camaraderie.

The difference between helpful trines and wide trines is not yet well understood by astronomers so more work needs to be done here.

The challenges come with Venus square to Neptune, Mars opposite Uranus, and Pluto opposing the Sun and Mercury. Their composite Moon and Mars are also square Saturn. These aspects reveal some problems with too-high expectations of the relationship, a tendency to go to extremes when it comes to activities they do together and arguments that can be volatile.

This latter energy can be stimulating and sexually charged but challenging when the frustrations that Saturn brings to the relationship - namely, a tendency to criticize and blame during disagreements.

It might be an idea for NASA put such information on their website alongside their pictures of rocket launches and views of the surface of the Moon. These are all very well but knowing the squares, trines, conjuncts and which planet is in which house is of far more interest to ordinary people. Perhaps NASA could take this on board and thereby make itself more relevant to the people who finance it through their taxes.

But as science tells us, one verification does not necessarily constitute a proof. So let us take another example:

Nicolas Cage possesses versatility, talent, and personal presence that sets him apart as an actor. Let's take a look at his character through his natal chart and peek into the synastry relationship analysis of Nicolas and his new wife, Lisa Marie Presley.

Some bio facts about Nicolas Cage: He was born January 7, 1964, at 5:35 am, in Harbor City, Ca, USA. Some bio facts about Lisa Marie Presley: She was born February 1, 1968. Some astrological notes about Nicolas: His Sun, Mercury and Mars are in

Capricorn; his Moon is in Libra, he has a Sagittarius ascendant and his Venus in Aquarius is conjunct Saturn.

Although Nicolas has a fun-loving Sagittarius ascendant he's a Capricorn by and large. His Sun, Mercury, and Mars are all in Capricorn and even his Venus in Aquarius is Capricorn-tinged - it's conjunct Capricorn's ruling planet, Saturn. Hard-working, status-conscious and persevering, Nicolas certainly possesses the staying power and know-how to get where he wants to go in life.

Here we see Nicolas Cage's personality described fully with no obvious discrepancies.

Nicolas' Sun, Mercury, and Mars in Capricorn are all placed in the first house of his natal chart. This suggests a person with enormous presence and willpower. His Venus 10th house, career ruler and Saturn trine his Midheaven and Moon, endowing him with natural charm and respectability. It would be very natural for him to accept that he has to work for what he wants.

Midheaven needs more investigation. The new James Webb telescope may bring new information if pointed in that direction.

Nicolas recently married Lisa Marie Presley - the daughter of his idol Elvis. She is an original, independent Aquarius with a soft Pisces Moon.

Anyone who knows Lisa Marie will confirm this assessment of her character to be accurate.

Their composite chart is a little difficult in absolute terms. Most notably, their composite Moon is unaspected - generally a red flag - suggesting problems feeling truly comfortable together. The square between Mars and Neptune means they will need to be careful to define the goals of their relationship or they may find that they are working towards different outcomes..

This prediction was proved to be completely accurate when Nic and Lisa filed for divorce after only 108 days of marriage. The warning was there: 'Their composite Moon is unaspected'. How much clearer could a prediction be?

Tests by professional astronomers have demonstrated that the alignments of the planets were accurate to a millionth of a metre. It is worth noting that the more accurate the data the more accurate the prediction. The best data on positions of the planets is currently that produced by NASA and that data obviously has to be paid for. For that reason celebrity horoscopes are much more likely to be accurate purely because celebrities can afford to have the most accurate and up to date positions of the planets made available to their astrologers.

Similar data is available for ordinary people's horoscopes but it is not as accurate. That is why your own horoscope, for example from your daily paper, is not be as accurate as it could be.

Basically, science's rejection of astrology is based on ignorance. Science's ignorance. When the astronomer Edmond Halley asked Sir Isaac Newton, essentially the patron saint of physics, how he could possibly believe in astrology, Sir Isaac answered, 'Because I have studied the matter, Sir, you have not.' And therein lies the crux of the matter. When Shakespeare made the comment of the stars influencing our lives, he too was mis- or un-informed. Astrology is about the planets of our solar system, not the stars.

The comment from Isaac Newton is particularly telling. Here we have one of the most famous of all scientists explaining exactly why he found astrology persuasive. Simply because he had studied it. One wishes that today's scientists, his intellectual inferiors as they themselves will admit, would adopt the same attitude.

If science tries to present astrology as flaky it should look to someone who has demonstrated his scientific credentials by inventing the Bell helicopter. Arthur M. Young has a few choice words about astrology and mainstream science's inability to come to grips with it.

'Science has been the great venture of modern man but I am deeply disappointed that it has stopped short of its goal. The cult of calibration and measurement has dispensed with consideration of first principles and produced tons of facts tied together with bits of fragile string. Science, in short, is a set of fragmented special disciplines, each encrusted with its own jargon and incomprehensible to its fellows'.

Just like a modern day Newton, Arthur M. Young has considered astrology fully and found science wanting.

'When it comes to constructing a cosmology based on the nonmaterial I find that the discredited astrology, the divine science of the ancients, is founded on the same vocabulary of elements that is the basis for the measure formulae of physics. The arcs and transits of planets do precisely correlate to important events in my life'.

Science should become fully involved in determining the exact precision of this correlation. Anything within a few degrees or so would provide very persuasive evidence.

Young's essay on the value of astrology for a science of life is particularly noteworthy and includes the well-reasoned argument that the biological rhythms of life are endogenous i.e. not dependent upon known external influences and that the planets are influencing mundane worldly events by other means perhaps geometrically, geometry being used here as a verb and not a noun. Thus mainstream science's rejection of astrology is due to the inadequacies of mainstream science in not knowing the manner of the physical influence.

And here we have it in a nutshell. It is very much science's fault that it has not put into operation the research programme which will find the exact nature of the physical influence once and for all.

A chart is used to calculate the horoscope of an individual. Each chart maps an individual's personality and has 12 houses. Each house represents a division of the sky and holds a different meaning.

First house: The house of self: The first house is the most important and powerful. It reveals your likes, dislikes, mannerisms and disposition. The first house concerns how you present yourself to the world and how the world perceives you. It is also important because its beginning point establishes where all the other houses will fall.

Just as the Greenwich Mean line of longitude is vital to the time zones of the world so the first house is vital to setting the point from which other houses fall. This is where astrology and astronomy have yet another common point of understanding.

Second house: The house of money and possessions: The second house deals with your finances and property. It includes both what you have as well as what you may acquire in your lifetime. The things you will take with you through life will be found in this house. The second house may also give you some insight into your earning potential and projects that may make you money.

Here we see that astronomy, having determined the exact position of the second house, can hand over to astrology for the resulting exact determination of insight into earning potential.

Third house: The house of communication: This house deals with three areas; self-expression, family relationships and daily travel. This house governs many things including memory, speech, logic and your written words. This house influences your relationship with

family. The third house also governs daily travel such as work and school.

And so would be helpful in deciding whether to walk or catch the bus.

Fourth house: The house of home: The fourth house indicates the kind of home you had as a child, the kind of home you have now and the kind of home you can expect to have in the future. It also contains anything brought from your ancestors into this life. But the fourth house deals not only with your physical home but also with your emotional home. It is a mysterious house because it represents what you like to keep secure and protected from the rest of the world.

It will probably take many years of sustained effort to remove all of the mystery. But that is no reason why a start cannot be made now.

Fifth house: The house of creativity: The fifth house governs all your creativity and things you do for pleasure, including love affairs. It rules your artistic talents but your natural nature is a large part of it. It rules your children and the happiness they give you as it is the house of your heart and things you lavish affection upon.

Science has traditionally shied away from emotion but here it has an opportunity to crack this particular case. Proper investigation can only bring massive benefit to all of mankind just as the Apollo Moon landing missions were supposed to but didn't.

Sixth house: The house of service and health: This house rules how you feel about helping others. It rules your relationship with co-workers and bosses but because it is also the house of health it will determine what type of constitution you have. Any illnesses to which you may be susceptible to will be revealed in this house.

Altruism is a complete mystery to modern day biology but here we have the beginnings of a foundation for the scientific study of why we feel compelled to help other people. Perhaps even more importantly, the sixth house, if studied intensively, promises to reveal the illnesses which we may have. This has implications for decreased life insurance premiums.

Seventh house: The house of partnership and marriage: This house deals with your marriage or partnership and shows what type of relationship you can expect to have with your spouse. But it also deals with other partnerships like those found in work, business and politics. It shows how well you will be able to get along with others.

As seen previously, it is beneficial when the moon is in the seventh house. Astronomers should refine their observations and produce charts to be distributed through normal channels showing when the moon is in such a position in the sky. Potentially tricky negotiations with the wife could then take place at such times.

Eighth house: The house of death and regeneration: Like the fourth house, this house is also mysterious. It is difficult to understand because it rules the life forces that surround many areas such as death, birth and the afterlife. It may give some insight into your own death, but never the exact time of it. Surgery is under this house as is signifies regeneration.

Some may say that observing this house is the most important of all and despite predictions that an individual's time of death can never be known the launch of the successor to the Hubble Space Telescope (the James Webb Space Telescope) is expected to reveal the eighth house in extraordinary and unprecedented detail.

This tie-in between astrology and astronomy may well bring insight into an individual's exact death time if observations are as detailed as they are expected to be. At the very least, surgery times and outcomes are expected to be announced much further in advance than at present when data from the new telescope is properly analysed by astrologers.

Ninth house: The house of mental exploration and long distance travel: This house is a broader spectrum of the third house. The journeys are both of the mind and the body. Within this house is higher education and in depth study of profound subjects. The physical travel that is governed is travel to foreign lands, but it also governs foreigners you will meet and the study of different languages.

The ninth house is an obvious target for exploration by space vehicles as they themselves travel the furthest distances of all.

Tenth house: The house of career and public standing: The tenth house rules areas outside your home - your career, community status and public status. It reflects your image as seen through other's eyes.

Observing the tenth house, while important, has been agreed by astrologers and astronomers to be less important than observing the other houses first. This will no doubt disappoint some but a start has to be made somewhere. To be honest, both sets of scientists agree that although the influence of the tenth house is strong the information gained through telescopic observation could just as well be gained right here on Earth by asking your friends.

Eleventh house: The house of hopes and wishes: This is the house of long-term dreams and goals. It governs the clubs and groups with which you are affiliated. It shows your ability to enjoy others and represents working together.

On the other hand the eleventh house is extremely important and time on the James Webb telescope has been requested by psychologists as well as astronomers and astrologers in order that all of the information gained may be fully shared by the scientific community. Deep correlations in the eleventh house together with lines of best fit on psychological graphs of personality are hoped to coincide. If they do it will be powerful scientific proof for the accuracy of astrological and psychological predictions.

Twelfth house: The house of secrets, sorrows, and self-undoing: The twelfth house is the most mystical of all houses. It defines the limitations of your life and governs sorrows, accidents and disappointments. It is sometimes referred to as the house of karma because it contains both the good and the bad we receive from what we have done in this world.

The twelfth house, though it can be just as easily observed by space telescopes as well as radio frequency telescopes here on Earth has been subject to International Governmental Discretionary Procedure. This means that information gained by such observation cannot and will not be disseminated to the general public. Astronomical and Astrological research teams will have access to the data but it is subject to the hundred year rule governing non-disclosure.

This is necessary because observing such a mystical house could lead to personal data being placed in the public domain inadvertently should research results be published in peer-reviewed periodicals.

But after that time the freedom of information act will ensure that observations of the twelfth house will be released containing all of the relevant detail gathered concerning the limitations of individual lives and their sorrows, accidents and disappointments.

Finally in this section we look at the astrological chart prepared for Britney Spears, independently checked by astrologers and astronomers for accuracy. This chart is more detailed than the others previously considered.

The Cosmo natal report for Britney Spears born December 2, 1981, 1:30 am, Mahon, Mississippi.

Calculated for: Time zone 6 hours west. Latitude: 34 N 48 53. Longitude: 89 W 30 55. Positions of planets at birth:

Sun position is 10 deg. 02 min. of Sagittarius
Moon position is 12 deg. 23 min. of Aquarius
Mercury position is 5 deg. 25 min. of Sagittarius
Venus position is 25 deg. 07 min. of Capricorn
Mars position is 23 deg. 17 min. of Virgo
Jupiter position is 1 deg. 01 min. of Scorpio
Saturn position is 19 deg. 16 min. of Libra
Uranus position is 0 deg. 59 min. of Sagittarius
Neptune position is 24 deg. 02 min. of Sagittarius
Pluto position is 25 deg. 58 min. of Libra
Asc. Position is 3 deg. 23 min. of Libra

These positions are already very accurate and their exactitude is expected to be confirmed and perhaps surpassed (though this is not yet certain) by satellite launches in the window 2017-2019.

Chapter 1: How you approach life and how you appear to others: Libra rising: You are a natural diplomat, reasonable, tolerant, fair, always willing to listen to varying viewpoints and ready to see the other side of an issue. Even if you strongly disagree with someone you will try to find points of similarity and agreement rather than emphasizing the differences.

You often avoid taking an extreme or one-sided stance on anything. You have a strong desire for harmonious and pleasant relationships and express a spirit of cooperation, compromise, friendship and fairness. You very much want to be liked and because of your need for approval and acceptance you are easily influenced by others' opinions especially when young. You so much want to please that often you will suppress your own intense or unpleasant feelings in order not to offend others.

This seems to be Britney.

Your need to create harmony extends to your physical environment and personal appearance as well. You appreciate beauty and have a natural sense of balance, symmetry and proportion. You do everything in good taste with a sense of style and art. From your home furnishings to your choice of clothing everything must be aesthetically appealing not simply functional or utilitarian.

Again, this is the definitive Britney.

You also feel that relationships are an art, one that especially interests you and one that you are usually quite skilled at for you possess tact and acute awareness of other people. Marriage is very significant to you and finding the right person to share your life with is extremely important. Being part of a close couple seems natural to you - you are not an independent loner. Having a partner increases your self-confidence.

You do have a tendency, however, to become overly dependent on your partner and perhaps not to develop a clearly defined identity outside of the relationship. Finding the balance between being yourself and blending and uniting with another is a challenge for you.

Anyone who knows Britney will confirm this.

Others see you as an agreeable, smooth, harmonious and nice person. Though there may be much more to you this is the sort of face you show to the world. You possess personal charm and an understated, noncombative manner.

This is highly accurate.

Chapter 2: The inner you: Your real motivation: Sun in Sagittarius: You are a gambler and an adventurer at heart, one who loves to take risks, to discover and explore new worlds and to take the untried path rather than the safe, reliable one. You are an independent soul, freedom-loving and often very restless. You need a lifestyle that provides opportunities for travel, movement, change and meeting new people. A steady routine which offers much in the way of security but little in the way of space and freedom is odious to you.

This sounds correct.

To you, life is a journey, an adventure, endlessly interesting and rich with possibilities and it may be difficult for you to decide where to focus your attention and efforts. You probably travelled around and experimented with many different paths before you settled on a particular career or you may go from one project to the next, for once the challenge and vital interest is gone, you are very quick to move on.

And again, we see Britney's character precisely delineated.

An incurable optimist, you have big dreams, aspirations and hopes for the future and are usually pursuing some distant goal. You have a great deal of faith and trust in life and failures don't crush your spirit. You always bounce back from disappointments often with another bright dream or scheme.

You have a sporting, playful attitude toward life and are philosophical about your mistakes. You have the ability to sense future trends, to see the big picture and you like to theorize and speculate.

This is very persuasive.

You express yourself in a very open, direct, and straightforward manner and are often blunt and tactless as well. Because you do not take yourself too seriously you may not realize how deeply your statements can wound more sensitive souls.

And a candid Britney, if asked, may well agree with this.

You do enjoy friendship and comradery but you need freedom also and do not do well with a possessive, clinging, or emotionally demanding partner. You are quite generous yourself and heartily dislike pettiness in others. Someone who shares your ideals, your sense of fun and your zest for life would be the right companion for you.

This seems undeniably correct.

Sun in 3rd house: The urge to learn and to communicate is essential to you and you express yourself very well through writing, teaching or sharing ideas and information. You are persuasive and fluent with language. Restless and inquisitive, you enjoy being mobile and taking short trips and excursions in order to see for yourself what is going on in the world around you.

Britney's Sun in the third house appears to fully explain these aspects of her character and the Sun's position there will be confirmed by astronomers once new observations involving the IRAS satellite are made public.

Sun conjunct Mercury: You tend to see things from your perspective only and to be rather subjective. You also enjoy talking and expressing your views but you don't always listen as well. You have a clear mind, a love of learning and new experiences, and need constant mental stimulation and activity.

The Sun conjunct Mercury will be fully confirmed by the newly launched Gaia satellite which will observe 1 billion stars including our Sun. At this point it might be worth mentioning that science has not yet approached a delineation of Britney's character to anything like this accuracy.

Sun sextile Moon: You feel that you have the support of your family or others in your environment for your creative efforts and personal goals which enables you to act with confidence. You are able to satisfy both your need to be an individual and your need for caring relationships and a sense of belonging. You are in harmony with yourself and are therefore an effective individual.

And the Sun/Moon sextile is expected to be observed as an important part of the same mission. Results are expected in early 2018.

Chapter 3: Mental interests and abilities: Mercury in Sagittarius: You possess vision and foresight and your mind is often occupied with big ideas, plans and goals for the future. You are interested in what is possible and what is on the horizon rather than what has already been done.

The study of philosophy, religion, politics or education is of interest to you and you are more concerned with theories and concepts than with specific applications. You are not inclined to focus on any one practical, concrete area; you find it irksome to deal with details and particulars.

You have a gambling instinct and enjoy speculative enterprises and new ventures. Business, advertising, and promotion would be good areas for you also.

Britney's interest in the study of philosophy, religion and politics is a matter of record. Her greater concern with theories and concepts rather than specific applications has been noted on numerous occasions.

Mercury conjunct Uranus: An independent and original thinker, you are excited by new ideas, discoveries and innovations. Your mind functions in an intuitive, nonlinear fashion and sudden insights and ideas often come to you out of the blue. You grasp ideas very quickly and often become impatient with those who are slower or more cautious than yourself.

You are considered rather eccentric and unusual in your interests by more conservative minds. You do well in an open, unpredictable atmosphere where flexibility and quick responses are needed. You have an aptitude for science, mathematics, electronics or astrology.

This character assessment accurately delineates Britney's true inner nature.

Venus square Pluto: You experience powerful, compelling emotional and sexual attractions and you may feel that you have little choice or control over your desires. You have an intense need for love and may be emotionally greedy or insatiable. Your love life is passionate and often tumultuous and painful as well. Jealousy, power struggles or over-possessiveness can become areas of conflict in your relationships. Positively, you can be unusually creative and bring about beneficial and healing changes in the lives of others, motivated by your deeply felt love.

Studies of Venus square Pluto could go a long way towards confirming these obvious truths about Britney. Such observations could easily be made by one of many satellites in orbit around Earth at the moment if scientists would only remove their blinkers concerning Astrology. Nothing would excite the young people of today more in the cause of science than knowing it has proved that stars and planets have influence over the lives of celebrities.

Venus trine Mars: You are warmly romantic and openly express your appreciation and love of the opposite sex though rarely in a crude or insensitive manner. You enjoy playing match-maker and bringing people together romantically. Fulfilment and harmony in love is likely for you because you know what you want and need in a romantic sense and express those desires honestly.

Everyone who knows Britney is aware that she would only rarely express her love of the opposite sex in a crude or insensitive manner. For Astrology to be able to predict this is quite remarkable.

Work that serves only your own narrow personal interests does not satisfy you. You may lack the competitive edge, the fighting spirit, the me-first attitude that is often required for material advancement and success. You also tend to either overestimate or underestimate your own power and abilities.

This is so precise that it is bordering on the uncanny.

Jupiter in Scorpio: Your strength lies in your unwillingness to settle for superficiality or shallowness in any area of your life. You are unafraid of going to extremes and experiencing the depths. The mysterious and the unknown are very attractive to you. Your intuitive understanding of others' inner drives and motivations is highly developed also.

It is difficult to think of anyone except Britney to whom this could apply.

Jupiter in 2nd house: Wealth, material well-being and prosperity are very important to you and you would never choose a profession that didn't offer you great promise and opportunity for growth, financial and otherwise. You are very generous and sometimes careless with money; you don't want to hold on to it, you want to enjoy it and use it to enrich your life.

Whatever you do with your money, you do in a big way - making it, spending it or blowing it. You may accumulate elegant, high quality things art, jewellery, etc.

And likewise here. If just this character assessment was presented to an ordinary person who was then asked to whom it might apply, it would be surprising if Britney's name was not mentioned.

Jupiter conjunct Pluto: You aim high and strive to position yourself in a manner that will ensure success. You are not above joining organizations or rubbing elbows with others primarily for the opportunities and social status that they confer on you. You abhor pettiness and your broad perspective and vision incline you to an influential and successful position in life.

Science could well consider at this point just how much it has failed in the field of insight into human character. Here Astrology has presented paragraph after paragraph which could only possibly describe Britney yet science lags behind not even being able to approach a convincing analysis of Britney's character, then compounding the error by actively decrying the very method which indisputably comes closest of all to succeeding.

Saturn in 1st house: You have a mature, disciplined, serious attitude toward life which colours everything you do. Caution and realism are your virtues though you limit yourself at times by being too careful, shy, or fearful and not believing in yourself enough or being assertive when necessary. Others find you difficult to get to know intimately as you tend to distance yourself from them or to put forth a rather stern, adult face to the world.

This is a facet of Britney's character which is not often recognized.

Saturn conjunct Pluto: You are suspicious of groups, crowds and social organizations. You are quick to notice social games and insincerities and you avoid the limelight and glamour.

Britney is noted for avoiding the limelight and glamour.

Uranus in Sagittarius: You are part of a 7 year group of people who are extremely enterprising and forward-looking. You are optimists and explorers. Your age group shoots for the stars, figuratively and literally. Space exploration takes on new dimensions as your age group pushes fervently to the next frontier.

Without wishing to appear biased, I am sure Britney would make a fine ambassador on off-planet missions.

Neptune in Sagittarius: you are part of a 14 year group of people that are extremely idealistic and farsighted in their dreams. Your age group is very liberal and expansive in outlook, and consequently churches become much more flexible and more eclectic in their approach during your life time. Religions that do not adapt to the broad-minded attitude of your age group simply are unable to attract very much interest and involvement from you.

Britney's non-conformist religious investigations are well-documented.

Neptune in 3rd house: Daydreaming, lack of concentration and inattention to your surroundings can be problems for you. Your mind tends to wander unless you are using it in an imaginative, creative way. Dry facts and cold logic hold no appeal for you and unless a subject has a colourful or personally inspiring element you won't stick with it for long. You have a sensitive, intuitive bent.

Hardly anyone except Britney could possibly be described by this passage. As I have noted before, the individualistic nature of Astrological analysis is something which science, with its generalisations, would do well to emulate.

Neptune sextile Pluto: The entire generation to which you belong has tremendous opportunities for spiritual rebirth and awakening. This will not be forced upon you or precipitated by unavoidable events, rather it comes from an inner yearning and a natural propensity to seek the depths.

Science in its unreformed state is perceived by young people to be moribund, boring and lacking in depth. This is entirely science's fault and it would do well to engage the interest of young people by showing that it too offers opportunities for spiritual rebirth and awakening. If it could bring itself to develop an inner yearning and a propensity to seek the depths that would be a start.

Interest in psychology and sociology is high in your age group. There is a tremendous heightening of awareness of social skills. Your age group will experiment with different marriage styles, family relationships and even business relationships in an attempt to bring fair treatment and effective communication between people. Interest and appreciation for other cultures is also strong and your age group will work hard to preserve and support the cultural heritage of all ethnic groups.

It is no surprise to see Britney at the fore in these matters.

Your strong yearning for equitable and harmonious relationships is also reflected in major advancements in trade agreements, arms control, and international cooperation that are designed and implemented by your generation.

Britney's role here, though understated, is obvious enough.

Pluto in 1st house: You are an intense person, inclined to be wholeheartedly and passionately involved in whatever your current interest is. You easily go to extremes and you make more moderate, easy-going souls feel uncomfortable. You may have unusual charisma and also be profoundly influenced by powerful, charismatic figures whom you model yourself after.

Learning to use personal power appropriately is an issue for you. Negatively, you can be manipulative and obsessed with personal aggrandizement. Positively, you can be a potent force for change, growth, and healing in the world.

This is indeed prescient.

I think it only a matter of time before science admits there is more than 'just something' to astrology. Yet this is only what millions of people have known for thousands of years. And all of it from a scientific measurement of the exact positions of the planets at the moment of birth. If quantum mechanics is the most accurate theory of modern-day physics it seems likely to be the case that astrology will be the quantum mechanics of the future, such is the similarity between their predictive powers.

However, it is not the only system which is capable of predictive powers that requires investigation. A complementary technique is that of Numerology and again we can use Britney as an example.

Your soul number reveals your inner, private self, the underlying motivations that influence your decisions and actions, your subconscious desires and most deeply ingrained attitudes. It is determined by adding the values for the vowels in your full birth name.

Your personality number shows how you express yourself outwardly, your appearance and the image you present, how others see you, your power of attraction and the surroundings you enjoy most. It is determined by adding together the values for the consonants in your full birth name.

Your destiny number represents your overall aims and the path you will follow in order to accomplish your life's purpose. It is determined by adding together the values for all the letters in your full birth name.

Your career number shows your talents and gifts and what types of careers or vocations you are most suited for. It is determined by adding together the digits for your birth date.

Your missing number or numbers show your areas of weakness and what is underdeveloped in your nature. They are determined by whatever number values are not represented in your full birth name.

What could be more scientific than the use of numbers?

The first initial in your name indicates the most significant quality of your personality and the traits which make you unique in the eyes of other people. The first vowel of your name reveals your instinctive reaction to people and situations.

And what could be more scientific than the fact that Numerology Theory doesn't just end there it also offers extrapolations. A mature science offers testable predictions and if these predictions are falsified there must be a new theory. However, if the predictions are confirmed then the theory is accepted throughout the scientific community. Judge for yourself based on the mathematical/linguistic analysis below.

The Athena Numerology Report for Britney Spears:

Your numerology report is based on the following calculations:
Total for each letter:
a=1 b=1 c=0 d=0 e=2 f=0 g=0 h=0 i=1
j=0 k=0 l=0 m=0 n=1 o=0 p=1 q=0 r=2
s=2 t=1 u=0 v=0 w=0 x=0 y=1 z=0
Consonant total: 7 43
Vowel total: 2 20
Grand total: 9 63
Date total: 6 6
Missing numbers are: 3 4 6 8
First letter is B
First vowel is I

Your soul number is two. You need people and feel incomplete without someone to love and care about. Shared happiness and togetherness are more important to you than personal glory or being

in the limelight. You are most comfortable being in a supporting role and you are likely to depend a great deal on your mate and your close friends.

Considerate, tactful and sensitive to the needs of others, you instinctively know how to blend with people. You are adaptable and you understand the ebb and flow, the give and take, the compromises necessary for harmonious relationships. You are a peace maker, the one to pour oil over troubled waters but because you value peace so much, you may submit your own will to another, stronger person rather than fight.

Though you don't like to rock the boat you should learn to stand up for yourself when you need to. It is often difficult for you to decide within yourself what you want and to carry out your own plan without the approval and assistance of others. You may also have a secret competitive or envious side, constantly comparing yourself to more forceful or confident people. You are a gentle person at heart whose happiness lies in fostering peace, cooperation and harmony.'

This is compelling evidence. It is instructive to compare the present analysis with the previous. When two different theories predict similar things science acknowledges that more than coincidence is at work.

Your personality number is seven. Quiet elegance, poise and refinement characterize your outward appearance. You seem stately, reserved and aloof, with a dignified bearing that bids others to stay at a respectful distance. You are hard to get to know and meet and there is an aura of depth, secrecy, and mystery about you. You keep much to yourself. You may give the impression of being a loner for you seem self-sufficient and you keep you own counsel. Actually, you are quite sensitive and shy and have difficulty communicating in a free and friendly manner.

Though gracious and always displaying impeccable manners there is often a formal or restrained quality about your interactions with others. Frequently you will simply observe and analyse what is going on around you and say nothing. When you do voice an opinion, others listen and respect what you say. Very often you are happiest by yourself, reading or enjoying nature. Being near the ocean is especially satisfying and rejuvenating for you. In clothing, you prefer classic or simple lines and avoid bright, loud colours.

This seems to be an almost exact match for Britney's character.

Your destiny number is nine. Your concerns are broad, your sympathy is all-embracing, and your path lies in living your ideals of universal brotherhood through compassionate service. Tender-hearted and generous, you are easily swayed by emotion and by people in need. You need to learn discrimination, who you can really help and who will only drain you. Your interests are so wide, it may be difficult for you to focus and concentrate your attention on developing a single one.

Try to discover where your talents lie as shown by the other numbers in your name and birth date and develop them as fully as you can, for the benefit of all. Working as a counsellor, teacher, minister, in public or civic work, or in any area that involves making bridges between people and bringing them together would be appropriate for you. Music or the arts also suits you. Much of the work you do will be subtle, inner, and spiritual: encouraging people to love, accept, and forgive themselves, bringing hope and being an ambassador of good will wherever you go.

It will be difficult to find anyone who disputes that this is definitely Britney.

Your career number is six. You are suited to activities that call for harmonizing, adjusting, giving advice or counselling, helping people, serving, nurturing, the cultivation of beauty, a sense of harmony, artistic creativity. Careers and vocations: homemaking, kindergarten or elementary school teacher, guidance counsellor, doctor, nurse, marriage and family counsellor, volunteer work, community service, director of boy scouts or girl scouts, governess, professional hostess, interior decorating, artist, craftsperson, instructor, social worker, social services director, waiter, waitress, singer, musician, promoter of the arts and social welfare, manager of restaurant or cafe.

It is extraordinary how Numerology is able to 'home in' precisely on the facts.

Your missing numbers are: Three: You often lack a sense of humour and joy in living. You need to allow yourself to play, be spontaneous, and express yourself more freely and creatively. Try to adopt a more generous, cheerful attitude. You are also uneasy and unsure of yourself in social situations and must learn to overcome your tendency to withhold your thoughts and emotions. Singing would be an excellent way for you to free up your resistance to open communication.

An uncanny prediction.

Four: You lack stamina and are disinclined to persistent effort and hard physical labour. Laziness and lack of discipline may prevent you from realizing your goals. Develop a more down to earth, realistic approach to life, and learn to finalize and complete one project before undertaking another one. You may be a person with much abstract knowledge but little technical know-how. Develop some practical skills.

This seems rather hectoring in tone and it is doubtful whether Britney needs to be talked to quite like that.

Six: You lack domesticity and are disinclined to take on the responsibilities of a home and family. The desire to devote yourself to the care and well-being of others is not strong in you. You may neglect the needs of your family and feel minimal interest in the welfare of those in your community. Try to develop more loving concern for others and realize that the personal touch and caring is more valuable than any strictly material gift you could give.

This seems most unBritney like. Perhaps the adding up was wrong earlier.

Eight: You lack business sense and may have little interest or ambition for material achievements and success. You can be very impractical and naive about finances, investments, legal procedures and the ways of the world thereby making foolish decisions. Try to develop an interest and an education in these matters to prevent regrettable and costly mistakes of judgment in your finances.

You need to learn to manage time and resources more efficiently. You disdain commercialism and don't understand its value. Therefore you are unlikely to be able to sell either products or ideas to others, which in turn may prevent you from sharing your creative efforts with a wide audience.

Although this would explain why Britney needs a record company.

Your first vowel is I. Your instinctive response to any situation is intuitive discrimination. You act according to your intuitive sense of a situation and may seem entirely illogical and impervious to rational argument.

You are sensitive and intelligent with the ability to perceive and understand subtleties and complex issues. You are sympathetic, idealistic and have a strong humanitarian impulse. You can be overly subjective, overly sensitive and emotional. You avoid confrontation as much as possible and are sensitive to criticism and

rejection. You are also inspired intellectually or artistically and require a supportive atmosphere in which to develop these gifts.

Everyone will agree that this is the case for Britney.

Your first initial is B. You have a soft, gracious, hospitable manner and are easily influenced by kindness, consideration and emotional appeals. Though you appear stable and calm, both praise and blame from others affects you strongly and you are sensitive to criticism. You are diplomatic and you strive to please. You prefer to work in cooperation with others rather than on your own and have a strong desire for marriage and partnerships.

Close family ties give you a sense of security. Your home is very, very important to you. You love material comfort and have a talent for accumulating money. You can save and squirrel away your earnings and do not easily let go of anything once it is in your possession. You can be a hoarder who never throws anything away. You are more instinctive and emotional than intellectual. Negatively, you can be stubbornly opinionated and overly simplistic in your thinking.

Knowing Britney from magazines, newspapers and TV shows you will be able to form your own opinion as to the accuracy of this report.

The results speak for themselves; only negatively disposed scientists could possibly contest them. Perhaps the coming scientists of the 'New Age' will approach such matters with a more open mind.

Chapter 3: Bermuda Triangle

Disappearances have confounded and entranced people for thousands of years but one thing is consistent within the claims. Whether true or false, a vortex or funnel is often involved. In the great scientific age of the 20th century it was discovered that all matter is actually at a level of vortex kinesis, that is, swirling or spiralling motion. Whether it is the great galaxies, planets, Suns, or the smallest atoms, all things rotate upon an axis and revolve around a core. The natural action of energy is vortex kinesis.

Vortex kinesis is of course well known to science but this is the first time that it has been proposed as an explanation for the well documented disappearances of objects in the Bermuda triangle.

Now, in the example of speeding cars obviously they cannot go fast enough to truly bend space and lock one into a different progression of time. But would it really require huge mass and speed to do it? The planet was indeed doing it slightly but if time's pathway lay along energy, could not bending or changing its electromagnetic frequencies also bring this about? There must be other methods to bend electromagnetic wavelengths.

The idea that time's pathway lays along energy is a concept which needs further investigation. Science at present has very little idea which pathway time lays along.

Atomic clue: We must first begin with the atom, any atom. There are only a little over 100 types of atom in all the Universe that we know of. All that we see is built out of these and out of a combination of these. All atoms are made up of protons, electrons and neutrons, except hydrogen which has no neutron. No atom is different than the other except in number: helium, calcium, titanium, tin, whatever; they are all the same thing except in number.

They follow an orderly mathematical progression. Hydrogen, the simplest atom, has 1 proton in the nucleus and 1 electron in orbit. Helium has 2. 3 protons in the nucleus and 3 electrons in orbit is no longer a gas, it is lithium, a silver-while metal. It goes on and on to Seaborgium, with 106 protons in the nucleus.

All of this is standard science confirmed by multiple observations.

Therefore, the difference between all matter must be the electromagnetic energy created by the different number of these charged particles, not in the basic make up of these particles as individuals. The reason nitrogen and oxygen are such different gases cannot be that nitrogen has 7 protons and 7 electrons and that oxygen has 8 of each, it must be in the different amount of electromagnetic energy that that one proton and one electron make that changes the substance.

This is where things get really interesting.

The difference is not in the makeup of the particles but in the space between the particles. That space is the substance of everything. That space is electromagnetic energy. That space is reality. That space is the atom. Electromagnetism is the substance of everything. In essence, that space is the substance of all structure. That space is energy.

Once it has been shown that space is the atom it is a short step to the conclusion that electromagnetism is the substance of everything.

Understanding the structure of all matter is fundamental to understand the limitless potential of manipulating or warping electromagnetism, since by doing so you are capable of affecting the very energy that creates all things. This is far superior to the mere splitting of the nucleus of a complex atom. That only provides a violent form of energy. Within the electromagnetic force of the atom is the key to everything: creation, transmutation, limitless power.

Science really does need to concentrate on understanding the structure of matter. Affecting the very energy that creates all things is a far off goal but one which could be realised given sufficient funding. That this energy would be far superior to the violent energy obtained by splitting the atom is axiomatic. Science should redirect its efforts away from releasing violent forms of energy and concentrate on releasing more serene forms. This would solve the energy crisis as it would, by definition, be a source of limitless power.

Water molecule H_2O. In other words, 2 atoms of hydrogen and 1 atom of oxygen. Two gases combine to make a liquid. It is not a material difference: The protons are still protons, the electrons still electrons. Electromagnetic energy has been affected by the proximity of these charged particles making up the atoms. A new substance is created.

Chemists may well need to rethink their ideas on the properties of compounds given this new information.

To give you an idea of how universal are these substances and how limitless the potential of what they can create, here is an atomic formula: C_{3032} H_{4812} N_{780} Fe_4 O_{872} S_{12}

This formula will be instantly recognisable to any qualified chemist.

Do you know what this formula is? It is the most complex molecule known to man: The red blood cell.

And if biologists know their chemistry they will also recognise it.

It is something so tiny one needs a high powered microscope in order to see it. The speed with which the red blood cell carries out its complex function is incredible. The complex data it carries is even more mindboggling. Its communication with the miniscule receptors in the bodily tissues is equivalent to writing the Gettysburg Address on the point of a pin while passing that pin after being shot out of a cannon.

None of this has ever been refuted.

Transmutation: Radioactive decay. The means by which one element becomes another. The isotope uranium 238 with 92 protons and 146 neutrons in its nucleus is unstable, meaning it sheds some particles from its nucleus. Doing so, the vital loss of these protons and neutrons changes it to another substance, a lesser complex one. In this decay chain uranium changes eventually to common and stable lead.

This is all fully established.

The loss of these particles obviously means a change in the electromagnetic force of the atom in question. Losing some of its 92 protons requires that it shed the equal number of electrons. It becomes a less complex atom eventually such as lead, where the process stops. Transmutation is the result of the electromagnetic energy changes that result. In this we can see the power contained in the electromagnetic forces that bind the atom, created by the various particles, their systems of orbit and their number.

The electromagnetic force change of the atoms in question had previously been assumed to be zero but perhaps physicists aren't looking hard enough.

Most substances, fortunately, aren't like uranium 238. They are stable and do not decay. But the example of radioactive decay shows us the power of changing the electromagnetic energy of any atom. If we could do so, then we could effectively carry on any form of transmutation we want.

And here is the crux of the matter. He who controls the electromagnetic energy appears to control which atom is created as a result of changing that energy. This is a very powerful result.

This must be the basis behind the success of the Hutchison effect. There can be no other proposal. To render items invisible, at other times transmute them from one substance to another, make objects levitate, burn white hot but not melt the surrounding flammable material, or make things cold melt, fuse or disappear can only be because the atoms of that particular substance and the particular frequency each one has, was interrupted or warped, while the other substances, being on a different frequency due to the different number of charged particles therein were unaffected. This is limitless power if harnessed repeatedly.

The Hutchison effect is demonstrated by John Hutchison, the Canadian inventor, who can supply VHS video of all of the above effects for a nominal sum of $100 (available from his website).

Now we get to the triangle and to the reports of electromagnetic anomalies and aberrations, disappearances, disintegrations, the crux of its infamous mystery. Are these indicators that electromagnetic aberrations are occurring for some unknown reason? Underscoring this it is crucial to understand a space warp.

Space warp: The atom is often likened to a mini solar system, most of it space. It is always surprising to realize that this is true, that the structure of the atom is energy in a confined space between revolving charged particles. Since everything is made of atoms everything in this Universe is mostly energy, not materially particulate. The electrons, protons and neutrons are only a small part of the atom. It is the space between them that is, ironically, the substance. Space, in this case, filled with electromagnetic energy.

Space is substance. It is surprising how often science can appear completely counter-intuitive.

It is possible to estimate how many charged particles are in any given substance. Simply consider there are 30,000,000,000 red blood cells in the body. Mathematically, it is possible to determine how many electrons and protons and neutrons make up the blood. Add the volume of all tissues, of every atom from calcium to iron, to whatever. It is beyond number probably.

So much so that the calculation has never been attempted.

If all is mostly space why can't one substance pass freely through another? Go ahead and run to a wall. You won't go through. It's

been tried. It doesn't work. It is not, however, a collision of particles which prevents you from walking through, it is a collision of forces. The atoms of our body: hydrogen, oxygen, nitrogen, calcium, zinc, iron, whatever, are the same as those in the wall, they are in tune, on the same frequency. They are mutually interactive. It is not possible to break the forces or is it?

Now, excitingly, we begin to see where the theory is leading.

If I were to take 1,000,000 Volts of high frequency electricity at 60 cycles alternating current and pass it through my body to ground there is no question I would look like Ramses' Mummy in short order. I would be instantly killed. The trillion, trillion, countless trillion and giga trillions of electrons would destroy my whole body. Remember, 1 ampere measurement of electric energy is equal to 63.3 billion, billion electrons per second at any given point of the current. That's a lot of electrons in 1 million volts of high frequency electricity.

These figures would obviously need to be checked but there is no reason to suppose they are incorrect.

But change the frequency of that electricity. Change it to 65,000 cycles and you can channel that 1 million volts through your body and out your fingertips like lightning if metal thimbles are provided to allow a point of discharge. No? Yes, very true.

The question was put to science long ago. In 1952, Dr. Irwin A. Moon, director of the Moody Institute of Science, demonstrated the reality of it as have others. The same particles that make up any wall Dr. Moon channelled through his body. The electrons passed through harmlessly. Though I don't recommend anybody doing this test at home what he did was merely to change the frequency.

Modern science should be well capable of changing the frequency in the way suggested. It has been over half a century since the question was first put and the time for an answer is surely long overdue.

The very same particles that are key to giving a wall its atomic stability passed freely through him. If we could change the frequency of the entire atom and not just the charged electron particle what would be the potential? Could one not pass freely through a wall? What if there was another system of atoms that are not mutually interactive with ours? Could there not be another whole world? Like adjusting the position of a hologram card would another whole scene emerge if you could manipulate electromagnetism and then the atom onto frequencies we have not yet discovered?

The discovery of new frequencies should be one of the most important next steps for science. For too long we have been held back by only knowing about frequencies discovered many years ago. It would be fitting if we could discover new frequencies and it would galvanise science if they could be discovered soon. Nothing really new is ever discovered these days leading to a lack of 'wow factor' in science. New frequencies would go a long way to correcting this.

Time warp: All atoms are in vortex kinesis as was stated at the beginning. Therefore all matter is at some level of vortex kinesis. Since it is the nature of matter, anything that might cause a change in this intensity might bring about surprising changes indeed.

Science has unfortunately failed to measure the correlation between the intensity of vortex kinesis and the change in the nature of matter. Many scientists think they are proportional to each-other.

We already know that the great spinning orb of the Earth and its great mass affects time fractionally at the surface as compared to orbital heights. Time is slowed slightly. But what if we could increase the speed at which the Earth spun, fast enough to even bend space more before it? Could we then not slow time even more? It would seem so but it is impossible to do that. The Earth is too great a mass for us to manipulate. But warping frequencies into a dangerous spinning vortex may not be as difficult.

To the best of my knowledge this experiment has not yet been performed. Yet it is only a matter of time before it is. Let us hope that sensible scientists from reputable countries get there first. The consequences of unlicensed frequency warping experiments do not bear thinking about.

A vortex may be the only way because it is spinning energy, an intensified example of the natural state of any matter. Interestingly, could such a vortex bring about different progressions of time, different exchanges of space by changing the frequency of the atoms in the area? An energy vortex like a magnetic vortex would have unlimited potential on the wavelengths around this planet since it would disrupt them first. One would not have to waste time trying to bend them with a huge atomic mass like a planet.

There is such a lot of new physics in this paragraph that it will keep researchers occupied for many years. Science should start to tackle these questions now if only because of the spin-off effects.

One example of an energy vortex on the force fields of this Earth may come from the eye of a hurricane. Gravity anomalies have been reported recently. These should not surprise anyone considering the vast swirling energy of a hurricane is centred at its imploding vortex core.

No-one disputes a hurricane possesses a vast swirling energy.

But this is a massive atmospheric vortex like a tornado. No one can get close enough without being obliterated usually. How could anything experience a time or space anomaly before being destroyed?

Obliteration has been the default assumption of science but in the light of lack of actual evidence that could be incorrect.

The recent invention of Dr. Evgeny Podkletnov certainly underscores the potential of a natural magnetic vortex. His device was created at Tampere University labs, Finland, in 1996. It was nothing more than a superconductor sealed inside an outer metal casing filled with liquid nitrogen. The superconductor was set over three solenoids to magnetically levitate it and start it spinning. The liquid nitrogen in the outer casing was to keep it cool while this superconductor ring spun at several thousand rpm. The phenomenal result was accidentally discovered.

One of the assistants' pipe smoke wafted over the device while it spun at 5,000 rpm. The smoke straight away went to the top of the room where it hovered. It was discovered that gravity had been slightly shielded over the device. In essence a spinning magnetic field vortex affected the most mysterious force field of gravity. Dr. Podkletnov discovered via measurements that there was a funnel of less gravity 12 inches in diameter above the machine going through each floor of the building and possibly then out into space. When spun to 25,000 rpm, the device took off on its own.

Dr. Podkletnov probably had his work confiscated by the military, which is what tends to happen with discoveries like these.

Gravity is essential, both its creation and disruption. The Sun has gravity, but so does the Earth. Therefore, our tempo is set by the Earth's pace, not the Sun's. The Earth is a pocket of separate gravity to the Sun. To be independent of the Earth's tempo we must create a field independent of the Earth. So far, however, gravity has eluded us.

This is lucidly put.

Gravity is more than simple attraction. It is orientation. With time, with space there must be gravity of some kind. Even the solar system is tempered by the heliosphere, the gravitational field of our Sun. The void of space in our solar system is a lie. There are billions of wavelengths whose tempo may be set by the Sun's gravity, bent by planets, upset by several interactions, just like the mini solar system of the atom.

It is interesting to note that when Pioneer 10, Pioneer 11 and Ulysses left the solar system the last NASA could detect of them was an incomprehensible acceleration. They left the gravitational pull of the Sun and in the smooth ocean of wavelengths in intergalactic space they took off. How fast does light travel out there? How fast does time flow? One day we will know. But first we must understand all is energy. This is the blueprint to conquest and exploration. Nothing is impossible.

It is a shame the scientists of the day didn't think to place a speed of light assessor aboard any of their satellites. Now we shall have to wait for a future launch.

If gravity is not a curvature of space only, then it is a force within mass. As such it too is tied up in the atom. It becomes the conduit which directs the force of energy. It reflects order, not chaos; it tempers the progression of all things. It orients all things towards itself.

What is the potential in a natural magnetic vortex to affect gravity, time and space? Have they really been discovered in the triangle? Can they come from other sources? Is all this great power and spinning really needed to affect time or have other, very weak, electromagnetic tests and warpings been done to show phenomenal things can be achieved just by interlocking electromagnetic fields like in the Hutchison effect?

The above passage is for specialists only but it does give a flavour of the excitement centred on these studies.

What is time? Is time merely our concept for the span in which events unfold or is time a part of the Universe, an integral force that acts to set the tempo for all things? We still ponder at the invisible force fields around us: gravity and magnetism. Time, like gravity, remains a mystery. Is it tied in with these force fields? What is time?

These are profound questions.

Barry Setterfield of Australia did tests in which light seemed to be slowing down.

Barry deserves to be better known. Only scientists who continue to follow old dogma concerning the constant speed of light continue to deny his results.

Some years ago a man in York, England, was working in the basement of a building which ran over an old Roman road. Suddenly, a troop of Romans marched through the wall, in full armour, completely oblivious to anything around them. Where the road was not completely excavated he saw them from only the knees up. Where it was completely excavated he saw their full forms. They passed through the other wall and then were gone. Is this lunacy, or did the Earth merely replay a scene captured in it 2000 years ago?

These sorts of things are the experiences of many people yet their voices are entirely ignored by science.

Long before John Hutchison began his pioneering experiments into electromagnetism and alternative energy travellers in the Bermuda triangle had reported odd occurrences involving electromagnetism: Their ships or planes would be seized by a strange vapour then all equipment would go haywire; unexplained fogs would sit over the ocean, yet in all cases the weather was not right for creating fogs and there was certainly no reason for electromagnetic aberrations such as they reported.

There are many of these occurrences recorded in the literature.

Though they remain unexplained there is no longer the lack of a cross reference for them and this has to do with electromagnetism. And that's where John Hutchison comes into play. His experiments into electromagnetism has unintentionally produced every phenomena reported in the triangle. The term Hutchison effect has become a term applied to all the peculiar and startling effects his plethora of machinery can produce.

And 'real' science continues to doubt him.

It all began back in 1979. While studying the longitudinal wavelengths of Tesla, Hutchison, limited by space in his small apartment, crammed into one room various devices that emit electromagnetic fields and wavelengths like Tesla coils, RF generators, van de Graaf generators, etc.

He turned them on and went about his work. In some unexpected way the wavelengths these machines created interplayed to create astonishing effects. John first noticed this when an object touched his shoulder, one he wasn't expecting because it was levitating. Repeated

tests have produced a number of astounding effects. These include the continued levitation of objects: wood, Styrofoam, plastic, copper, zinc, they hover and move about, swirl around and ascend, or shoot off at fantastic speeds.

With further experimentation: fires started around the building out of non-flammable materials, like cement and rock; a mirror smashed 80 feet away, metal warped and bent and even broke, in some instances crumbling like cookies; some metal became white hot but did not burn surrounding flammable material; lights appeared in the air along with numerous other corona manifestations and water spontaneously swirled in containers to name only a few.

Hutchison is probably banned from announcing the results of further experiments. It is not difficult to work out why.

More than one observer of Hutchison's experiments and demonstrations have not only been amazed at the results but uniformly express astonishment at the weak electrical power which seems to be sufficient to empower stupefying results. Since basic outlets in the house supply sufficient power to operate his many machines, the power which unleashes all these incredible effects is believed to lie elsewhere, such as where these various fields interplay, since on their own the wavelengths or fields these machines create have never been noted to do this.

It is at this point, call it a warp or vortex, whatever you like, where things begin to happen. But how they interact is still largely a mystery. Even where they interact can be perplexing. Sometimes one must wait for days for something to happen, and 99 per cent of the time nothing happens at all. To draw an analogy, it is like trying to boil water without being able to determine the strength of the heat. In such a scenario it is not surprising it may take different amounts of time to produce the boiling effect.

Science knows about the interplay of various fields and has given it a name: Superposition. But mainstream science has never gone as far as Hutchison or if it has the results of these experiments have never been released.

Over the last 22 years Hutchison's work has been subject to broad though varying interest and approval. His laboratory was ransacked by Canadian officials who seized his equipment under the pretence of confiscating his antique gun collection which was, however, subsequently returned with no explanations or charges.

Three nations have aired his work; he was courted by scientists in Japan, accused of treason in Canada when he went to Germany for funds to continue; Los Alamos laboratory has done research on his effect; the Canadian government seized his lab equipment once again while crated and in transit to Germany in 1990 and never returned it despite a court order when he was seeking financial aid to continue his studies.

The military of Canada and the US has expressed a covert interest. He has been arrested by the Canadian government and placed in cuffs on his doorstep while his lab was ransacked and investigated under the excuse of checking on his antique gun collection again; this last incident was on March 17, 2000, and may have been initiated by a neighbour who experienced an unexpected levitation in their house and then complained to the police.

While unhelpful this is understandable!

Because of all this, which has been on-going for over 22 years, his lab has undergone numerous changes, from completely disappearing into government hands, to vital pieces having been sold off to pay for subsequent financial problems.

Nothing will stop the man himself however:

He has tinkered and adjusted his equipment and continues to get varying and remarkable results, many of which have been documented and photographed by McDonnell Douglas Aerospace and the Max Planck Institute in Germany.

Two Institutions which are beyond reproach.

Continuing results prove the unlimited potential of electromagnetism on matter. One of the most remarkable effects is the fusion of dissimilar materials. In describing this, Mark Solis notes that dissimilar substances can simply come together yet the individual substances do not disassociate. A block of wood can simply sink into a metal bar, yet neither the block of wood nor metal bar come apart. Moreover, there is no evidence of displacement, such as would occur if, for example, one were to sink a stone in a bowl of water.

Science should investigate thoroughly . There must be many applications in the real world which would benefit from such interpenetration of matter.

But perhaps one of the most remarkable aspects of his effect is spontaneous invisibility of materials in the active zone of a Pharos-type Hutchison apparatus. Historical investigations took place in the

Farnsworth fusion machine in the 1930s, where solid metal portions of the apparatus became transported during a number of tests. Philo T. Farnsworth was the inventor of the electronic TV.

And as such his credentials are impeccable. Modern day scientists should recover his work and build on it.

Hutchison has said that invisibly surrounds metallic objects. Properly adjusting his equipment, he adds, the cronons and gravitons generated by his technology could cause entire buildings to disappear. The propulsive part of the effect could get us into space faster and cheaper than any conventional fuel.

Now that the space shuttle has reached the end of its useful life this offers the ideal opportunity to replace it at a fraction of the cost. Meanwhile, cronons and gravitons are still being searched for at the Large Hadron Collider at a cost of billions of pounds. Scientists there should consider spending just a little of that on Hutchison's apparatus to save themselves years of time and effort.

Since the effects and the materials vary exceedingly it seems logical to assume there is some form of disruption of the basic atomic structure, considering that fusion of non-identical materials, fires, levitating objects, invisibility, etc., have nothing else in common but these building blocks of matter. Electromagnetism is a doorway into a world of which we are all made. As such, its potential seems unlimited.

This is logical. And Hutchison is obviously on the right track when he says his effects are due to some form of disruption of the basic atomic structure. With such a clue as their starting point physicists should be able to unlock the secrets of levitation and invisibility in a fairly short time.

Tinkering in his lab Hutchison has developed extensive forms of free energy including fuel cells and the Hutchison converter which apparently can run forever. It was first displayed in Japan in 1996 and is still running. A scientist sees fact but a philosopher sees potential. John is both.

And it is gratifying to see that John has finally toppled the second law of thermodynamics. It was on its last legs anyway but someone had to make the final breakthrough. Of course, some scientists, unwilling to give it up, still cling to the archaic law but it won't be long before even they are forced to retreat in the face of the evidence.

For that, of course, is how science differs from quackery.

Chapter 4: Better Vision

Everybody is interested in maintaining their eyesight in good working order for as long as possible. Opticians do their best but they are often operating after a person's vision has already deteriorated. Science should be trying as hard as possible to stop this deterioration before it starts but if that is not possible it should be working to endorse a programme which has impressive testimonials. The Cambridge Institute's program for Better Vision is such a case.

More than eye exercises, the program for Better Vision is a total mind/body method to improve eyesight and see clearly. From eyeglasses to perfect vision: I went from wearing glasses all the time to 20/20 vision.

Note here the difference between science and the Cambridge Institute's approach. Science concentrates narrowly on the eyes alone. The Cambridge Institute's program is a total mind/body method.

Dear friend, the program for Better Vision is the world's #1 best-selling eyesight improvement audio course for one reason: The program for Better Vision works like no other system does. Everything in the program for Better Vision is based on a primary law about vision that governs the quality of your eyesight. Learn and use this vision law and you'll get to the core of your vision problems. Without this law, you address only the symptoms but never the actual causes.

The primary law about vision is one which science has neglected so far.

Because the program for Better Vision is the only program built around this law of vision and because this program is so substantially different from others that are just eye exercises the program accomplishes two vital things for you: It gives you eye exercises that are much more effective. It takes you beyond those eye exercises so you can heal the underlying causes of your vision problems.

Just as science has laws there can be no prior argument that vision shouldn't have laws as well. Lenses have optical laws which are well known and there are equations which govern the passage of light through them.

I've carefully selected the exercises you'll find in the program for Better Vision based on my experience with more than 250,000 people since 1977, the year I founded the Cambridge Institute for Better Vision.

I would think that given the number of people involved we would have heard by now if anything was amiss with the program but I would not want to second guess the findings of an impartial scientific study.

For example, there's one exercise you'll find in the program for Better Vision that produces moments of clearer vision for nearly everyone the first time they try it. You spend less time every day with the program and see more results because the quality of the exercises is greater than other thrown together things you might find. With the program for Better Vision you're never spending more than 15-20 minutes a day devoted to your eyesight. Compare that to others that demand 30 to 40 minutes of your time each day. Some even tell you to practice for an hour.

Scientific study will be able to compare the various programs which promise to improve eyesight. Practising for a whole hour a day does seem excessive.

Maybe you do have to practice that long when the exercises are not well thought out, not as effective as they could be and not built on the vital eyesight principle you'll find in the program for Better Vision.

Science should be able to show that the vital eyesight principle makes all the difference.

The second difference you'll experience with the eye exercises in the program for Better Vision is the power of their sequence. When properly ordered as they are in the program for Better Vision the effects of the eye exercises build and multiply and you see changes faster.

This also applies to standard medical procedures such as brain surgery. It is vital that the steps are carried out in the correct sequence.

The program for Better Vision is as effective as it is partly for this reason: It tells you exactly which exercises to do each day, in the right sequence and for the right amount of time. Just follow along with the audio vision session for that day, choose the format you prefer CD, audiotape or mp3, and you can be sure that you are doing the right exercises in the right way and for the right amount of time.

This is well thought out.

'I was very impressed with the program for Better Vision. Once patients see the improvements they make with your program many are motivated to continue' . Dr. Stanley A. Applebaum, F.C.O.V.D. Bethesda MD.

It is telling that a Doctor is prepared to give his recommendation.

Seeing clearly is the last domino in the chain. Did you know that before a person has trouble seeing clearly, there are other, more basic and underlying visual skills that have already gone out of balance? Skills like how the eyes work together, how the brain coordinates the two eyes and how the brain balances the images from the central and peripheral areas of vision. Unless you improve these skills first, like you easily do with the program for Better Vision, eye exercises alone aren't going to work very well.

It is obvious from this that the Cambridge program has delved deep into the physiology of how the eye and brain work together. Science is well on the way to proving that skills such as how the brain balances the images from the central and peripheral areas of vision is of the utmost importance in seeing clearly. The Cambridge program already recognises this.

Improve these skills with what you learn in the program for Better Vision and you'll experience for yourself how much more productive eye exercises could be. Why settle for any random set of eye exercises when you can benefit most from the comprehensive, total approach that you can find only in the program for Better Vision?

Science already knows that random approaches to healthcare rarely work. That alone should make the Cambridge program a prime target for the next step in eyecare research.

That's why so many people who use the program for Better Vision say it's the easiest to follow. It's the easiest to understand. It's the easiest to do.

Another important endorsement comes from Vegetarian Times magazine:

'Unlike others, the program for Better Vision attempts to address the underlying cause of imperfect vision'. Vegetarian Times.

In the near future I hope that science will stop spending so much money on trying to build satellites to orbit the Moons of Neptune and concentrate more on helping ordinary people to improve their eyesight. That really would be money well spent. It is not as if science would be attempting this from a standing start. The eye specialists of the

Cambridge program have already put into the public domain the reasons why their exercises work so well and they are to be commended for this public spirited act.

The exercises are as powerful and effective as they are because the entire program for Better Vision is built around the fundamental truth about the eyes and about how vision works. The vision secret: Like many other truths, the eyesight principle behind the program for Better Vision can be stated simply. It might even sound plainly obvious when you hear it.

But don't let that fool you. Understand this principle and it will make a phenomenal difference in the way you think about your eyesight, in the way you care for your eyes and in the way that you go about improving your vision. Fully apply this principle to your own eyesight as the program for Better Vision shows you and you will see how much more quickly your eyesight changes and how much more profound the changes are.

Modern day science has had plenty of time to discover these truths for themselves yet scientists continue to accept money for designing and building such things as nuclear bombs and nanobots which have the potential to destroy the planet. It has taken the professionals at the Cambridge program a lot less money to confirm the essential truth about vision and it is this essential truth that science should concentrate on. Put simply:

The essential truth about eyesight is this: You can't separate your eyesight from the rest of yourself. Your eyes do not exist in isolation. How clearly you see is influenced by much more than only the muscles in and around the eyes. In fact, the muscles around your eyes are 200 times stronger than they need to be. It's a myth that your eyes are weak and that you have to make the muscles stronger in order to see better without glasses or contacts.

This is a profound statement.

Your eyes are an important part of your body: 25% of the oxygen that you take in with every breath goes directly to your visual system. Nearly 1/3 of your food intake goes to nourish your visual system. 90% of what you learn about the world around you comes through your eyes.

Dieticians and bio-energy specialists should be able to confirm these figures.

The program for Better Vision takes you beyond eye exercises alone because it has to. Your eyesight is affected by so much more

than just eye muscles. **The program shows you how you to use the full power and force of your mind and body's desire to heal itself and apply that power to your vision. With the program you get to the source of your vision. And when you get to the source the changes you make come more quickly and with greater impact.**

As science is close to realising, the source of vision is by far the most important thing in eye care and treatment.

At a price everyone can afford: Thousands of people have spent $300 to attend my vision seminars. But it would be a pity if there was a single person out there who stops him or herself from seeing better because of the cost. The truth is that at this economical price there is only one thing standing between you and your Better Vision. You. That's right. You.

$300 really is a small price to pay especially when one remembers the cost of not paying it. Guide dogs have to be sourced, trained and deployed for example.

Consider the cost of doing nothing: Vision that continues to get worse, glasses or contacts that get stronger and stronger and more expensive, deteriorating vision that makes you a potential candidate for more serious problems in the future.

And more serious problems will often occur if nothing is done.

Better results in less time and less effort for: near-sightedness, far-sightedness, astigmatism, poor night vision, middle-age sight reading glasses, crossed eye strabismus, lazy eye amblyopia, sensitivity to light, eyestrain, headaches, burning eyes.

Once science confirms these results the outcome could be nothing less than the complete elimination of these common and not so common problems. Strabismus and amblyopia are conditions which have defied modern medical science yet here we have a proven method of producing better results for these pathologies.

'In the last 15 years, I have examined many patients who have used the Cambridge Institute's program for Better Vision. For any person interested in improving their vision, this program is a must to try'. Dr. Richard Kavner, F.A.A.O. New York City NY.

'I believe that the Cambridge Institute for Better Vision offers realistic and proven alternatives to glasses and deteriorating vision.' Dr. Leslie H. Salov, Ophthalmologist, director, the Vision & Health centre, Whitewater WI.

The endorsements of these doctors adds more weight to the program. Why would a doctor be prepared to endorse a program unless he had proof that his patients were benefiting?

'**The program for Better Vision is the most comprehensive, progressive and effective I have seen. There is no question that the body/eye relaxation and visualization techniques you use will help improve vision. I plan to recommend your program to my patients.**' **Dr. Dale Freeberg, O.D., F.A.A.O. Los Angeles, Ca.**

'**I was very impressed with your home-study audiotape system, the program for Better Vision and also with 'what they never told you about your eyes'. Once patients see the improvements they make with your program, many are motivated to continue working with a doctor who is certified in vision therapy.' Dr. Stanley A. Applebaum, F.C.O.V.D. Bethesda MD.**

'**I would like to recommend to my colleagues the fine eye care products offered by the Cambridge Institute for Better Vision. Customer service is superior and I know you will also be pleased with the items offered in the Cambridge Institute for Better Vision catalogue.' Dr. Errol Rummel, O.D., F.A.A.O., F.C.O.V.D. Jackson NJ.**

Now we can take a look at the products offered in the catalogue:

Eyemax-plus ™ vision and body formula: The 33 essential vitamins, minerals & anti-oxidants. Promote healthy eyes with the one comprehensive, complete and balanced formula that is so powerful. It also takes the place of your daily multi-vitamin.

This is very good news.

Stop being confused and stop spending more than you need to keep your eyes and your body healthy. One supplement can do it all!

And once science stops its endless search for spinning black holes and other things which are of no use to us it will be able to devote more time to things which are of real immediate benefit to all of humanity.

Eyemax-plus nourishes every part of your eyes. It has been used for years to help prevent, stabilize and reverse: macular degeneration, cataracts, presbyopia control, myopia control, eye disease prevention & recovery and to promote optimal body & eye health. Eyemax-plus: The one formula for your eyes & body: 33 vitamins, minerals & anti-oxidants in one powerful formula. eg. Price: 1 month: $39.95.

Pioneering nutritional ophthalmologist Dr. Gary Price Todd developed the original formula and used it successfully for years to help his patients prevent, stabilize and reverse eye diseases such as cataracts, glaucoma and macular degeneration. Eyemax-plus is also used to control myopia, farsightedness and other vision problems. Every nutrient in Eyemax-plus - every single one of the 33 vitamins, minerals and anti-oxidants - also has benefits for the rest of your body. The people who take Eyemax-plus can sense it and research proves it too.

People sensing that their body is benefiting is becoming more widely recognised as significant in scientific studies.

Here's what Dr. Gary Price Todd, the eye doctor who developed the original Eyemax-plus formula said the nutrients in Eyemax-plus have been proven to do:

Reduce the chances of developing cancer by 90%

Reduce the rate of heart attack or stroke by 50%

Reduce the chance of developing arthritis by 82%

Reduce the risk of kidney stones by 86%

Slow the aging process

Reduce the chance of premature death by 75%

And to have a doctor say this is just more of the proof that science has been looking for.

There are three things to consider when choosing nutritional supplements for your eyes:

1. How important the visual system is to the body

2. What ingredients - vitamins, minerals, herbs - are in the supplement.

3. The amount of each ingredient contained in the supplement

People should be educated into choosing nutritional supplements for their eyes and the list above is helpful in deciding which supplement they should opt for. It is not a decision to be taken lightly.

How was Eyemax-plus created? Ophthalmologist Dr. Gary Price Todd was a genius. Years ago he discovered the truth about eye health. And he did it in a methodical scientific manner befitting one of the most forward-thinking eye doctors in the country.

So there can be little argument concerning Dr. Todd's methodology.

How did Dr. Todd pave the way to healthier eyes for thousands of people? First of all he carefully monitored the blood levels of his patients. And when he gave them different vitamins, minerals and anti-oxidants he studied their effect on the eyes and on the body. He even had other eye doctors participate in his research and he measured their blood levels too.

This adds still more credibility to the work.

After his death, Dr. Robert Titcomb, a long-time associate of Dr. Todd's, worked with Martin Sussman, president of the Cambridge Institute for Better Vision to update Dr. Todd's formula. In fact, the original formula was so perfect it didn't need to be changed at all.

Given the amount of work put in by Dr. Todd it is no surprise.

Only additional nutrients were added. Nutrients that weren't even known when Dr. Todd started his work or he certainly would have put them in his formula. Lutein, Zeaxanthin, and l-Taurine, for example.

Again, this is to be expected and is a model of good practice.

Since Dr. Todd started his pioneering work, leading universities and government agencies - including the National Eye Institute and Tufts University - have proven that certain vitamins, minerals and anti-oxidants can prevent and stabilize cataracts, macular degeneration and other eye problems yet, to this day, to my knowledge, no one - other than Dr. Todd - has studied the effects of a truly comprehensive well-balanced formula.

This is a little surprising but is possibly due to people at these establishments being upset they did not discover the formula themselves. Petty instances like this are common amongst scientists but they should really be put aside for the benefit of patients.

Here's just one example from Dr. Todd about how different nutrients have a synergistic effect: 'A person taking 200 units of vitamin E alone reduces their risk of developing a cataract by 56%. However, persons who take vitamin E and vitamin C together have approximately 86% reduced risk in developing a cataract. Vitamin E is recycled by vitamin C so you can actually reduce the amount of vitamin E necessary by increasing the amount of vitamin C.'

With detailed results such as these it should not be long before scientists can show that Dr. Todd was ahead of his time in his research.

What's the recommended amount to take? Each bottle of Eyemax-plus contains 120 tablets. It is recommended that a person take 4 tablets a day (2 in the morning and 2 in the evening). Thus, each bottle of Eyemax-plus contains a 1 month supply.

Does Eyemax-plus replace my daily multi-vitamin? We don't recommend that a person take a generic daily multi-vitamin on top of Eyemax-plus. Because you can't separate the overall health of the eyes from the overall health of the body, every nutrient in Eyemax-plus nourishes both your visual system and your body. Compare both the list of nutrients and their dosage amounts in Eyemax-plus to your current multi-vitamin. Virtually every time, Eyemax-plus is the more powerful, comprehensive and complete formula.

Is Eyemax-plus all I need? For maximum healing, conditions such as macular degeneration, cataracts or glaucoma may call for other specific nutrients in addition to Eyemax-plus. If you have any of these conditions, contact the Cambridge Institute after you've started taking Eyemax-plus.

Everyone familiar with modern medical practice will recognise the advanced methodology here. Eyemax-plus is suitable for the vast majority of conditions but for maximum healing additional medicine will be required. The Cambridge Institute has foreseen this and has a large store of additional medicine ready and available on request.

Why does Eyemax-plus contain 100%, 200%, or even more, of the minimum daily requirement of some vitamins and minerals? The minimum daily requirements were originally set by the United States government to ensure that people did not acquire a deficiency disease associated with a lack of a particular nutrient. Preventing a disease and acquiring optimal health are two different things.

I have an eye condition that you haven't mentioned. Will Eyemax-plus help me? It's important to understand how Eyemax-plus works: Eyemax-plus builds up the health and strength of all parts of your visual system and your body. It doesn't replace regular medical care. As you and your eyes get healthier and healthier, the symptoms of disease will diminish.

Many conditions are merely different symptoms of the same underlying issue: The eyes have not received the full nourishment they need and over time have become weakened, thus creating the underlying conditions for different problems to take hold. As you make your eyes - and your body - healthier, stronger and clearer by giving them all the vitamins, minerals and anti-oxidants they need for optimal health, a wide variety of these different conditions can improve.

In addition to using Eyemax-plus for macular degeneration, cataracts, glaucoma, near-sightedness and middle-age sight, people with the following conditions have been helped: Drusen, Retinitis Pigmentosa, macula holes, floaters, detached retina and more.

Nourishing the eyes is of paramount importance and it seems that this is the fundamental idea which modern day science has forgotten. Proper nourishment of the eyes will undoubtedly be shown to help with macula holes and detached retina given further research.

Where can I get Eyemax-plus? Eyemax-plus is available exclusively from the Cambridge Institute for Better Vision. That's how we can keep the price as affordable as it is, even though Eyemax-plus is a high quality product. We send Eyemax-plus directly to your door, wherever you live. We have hundreds of customers from around the world. International shipping is $17.95.

Is Eyemax-plus gluten-free, dairy free, etc.? Eyemax-plus does not contain any of the following: yeast, sugar, wheat gluten, soy protein, milk/dairy, corn, sodium, starch, preservatives, artificial colours, flavouring or other sweeteners.

How does Eyemax-plus compare to other supplements? It's amazing what some other companies will try to sell with their so-called vision supplements. For example, some might contain the 'latest' ingredients that are good for the eyes - like bilberry, l-Taurine, lutein or zeaxanthin - but in such woefully small amounts that they are nearly useless to the eyes and body.

Some supplements leave out other, even more important key ingredients, altogether. Still others pay little attention to achieving the delicate balance necessary so the right combination of vitamins, minerals and anti-oxidants work together to boost their individual effect. The quality of different supplements is not always the same either. Cheaper ingredients manufactured poorly can be sold at a lower price. But are they any good?

This question answers itself.

Are supplements really necessary? Many people find the issue of supplements so confusing they do the worst thing of all - they don't take any!

It is shocking that so many people think that they will enjoy good health by not taking supplements. Scientists should mount a campaign to stop people thinking this way once and for all. It wouldn't take much effort for science to finally confirm that taking supplements is probably the most important thing a person can do in order to improve their health.

Or, some think that if they eat right, they'll get everything they need.

This is another error. Eating right has never been proven to give a person everything they need.

But even if you eat a perfect diet (whatever that is - should it be high protein, low fat, vegetarian?) And eat all the fruits and vegetables you're supposed to, it's still nearly impossible to get all the trace minerals your body and eyes need. Why? Because modern farming methods have depleted the soil of important minerals such as selenium, manganese and vanadium. Even calcium and vitamin C are found in lower levels in fruits and vegetables now compared with 50 years ago.

You could eat all organic fruits and vegetables, which have been shown to have higher levels of these important minerals and vitamins. But organic produce is often hard to find and costs much more. Certainly, some foods are better than others for the health of your eyes and your body. We do recommend eating as well as you possibly can. Even then, it's still extremely difficult, if not impossible, to confidently get all the nutrients you need from your diet alone.

And this is the message science should be spreading with all the tools at its command.

Does Eyemax-plus have any poor quality ingredients from China? No. First of all, most of the concern centres around rice, wheat and whey none of which are in Eyemax-plus. Secondly, Douglas Laboratories, the United States manufacturer of Eyemax-plus, rigorously tests all of its ingredients and they source ingredients that are manufactured in many different countries, including China.

Recently there has been a substantial amount of negative press concerning ingredients that are manufactured in China. These reports have centred around contamination of certain proteins such as wheat and possibly rice with melamine and other related compounds. Our whey and soy proteins are not manufactured in China. Our rice protein is a high end material that is different from the material covered in the media. It is manufactured in China in a high quality facility that operates under specific good manufacturing practices. 'This material has been tested and has been shown to be free from any contamination with melamine or related compounds, ' according to Douglas Labs.

This is very reassuring.

Can children take Eyemax-plus? Not only can children take Eyemax-plus - we recommend it for everyone, regardless of their age - but Eyemax-plus is essential for preventing and controlling myopia

in children. Those 18 and older are considered adults and the daily dosage is 4 tablets (two twice a day). It is recommended that younger children take one tablet a day per 50 pounds of body weight.

Parents of short-sighted children should really be asking why Eyemax-plus has not been mentioned in their consultations with their optician.

Dr. Ben Lane, New Jersey Optometrist and pioneer in the role of nutrition in myopia and other eye diseases, believes that trace minerals are crucial in maintaining the strength of the eye. Dr. Lane has found that chromium levels in myopic children are 1/3 that of children with normal vision. (Chromium is depleted in the body by white sugar.) Calcium levels are also lower in near-sighted children. Children increasing in the degree of myopia have diets extremely deficient in calcium. Dr. Lane thinks that in the face of this dietary deficiency the body takes calcium from the eye to help support bone growth. This calcium lack then makes the eye susceptible to the forces playing on it during prolonged periods of near work and visual stress.

The thoughts of Dr. Ben Lane are of obvious importance in eye care.

Vitamin C is also important. Dr. Lane has noted that low levels of dietary intake of vitamin C are associated with increases in pressure in the eye. This increasing pressure also is associated with the visual fatigue that can result from extended periods of near work. The focusing mechanism needs adequate levels of vitamin C and chromium for efficient functioning. Adequate levels of vitamin C are also needed to ensure the strength of the eyes.

Again, vital work from Dr. Lane which science has overlooked. Science continues to completely ignore the importance of chromium when focussing the eyes.

Another vitamin that Dr. Lane thinks is of critical importance is folic acid, which helps the eyes to maintain near focus for longer periods of time as well as increase the eyes' ability to absorb nutrition from the body. All of these nutrients - and so many more - are in Eyemax-plus.

If Dr. Lane can come to these conclusions by thought imagine what eye doctors could achieve by following up with experiments.

How long do I have to take Eyemax-plus? Eyemax-plus is not like a medicine that you take for a short period of time until symptoms disappear. Eyemax-plus doesn't get rid of symptoms. Instead, it gives the eyes and the body the essential nutrients they

need for optimal health. As your eyes and body get healthier and healthier, there's less and less room for disease. Eyemax-plus is the foundation for building optimal health. We'd expect that you'd want to take Eyemax-plus for as long as you want your eyes and body to become healthier and healthier - for the rest of your life!

This seems obvious. It is like a person facing cancer who wants to remain in remission.

Is Eyemax-plus all I need? Eyemax-plus is the essential foundation for optimal eye health. Some people - when their eyes are really bad or when they want to get faster results - could use more.

This is standard medical practice. People who want to get rid of their cancer more quickly can increase the dose of their anti-cancer medicine.

Is Eyemax-plus vegetarian? Does it come in any other formulations? Eyemax-plus is not completely vegetarian as there is fish oil used in the vitamins A and D. (Shellfish is not used in the formula.) Eyemax-plus does not come in any other formulations or combinations. Eyemax-plus was developed and refined as the result of careful research and individual testing. The nutrients in Eyemax-plus work together in a balanced, synergistic way to promote optimal health of the eyes and body.

Balance and Synergy (as noted previously) are very important in health matters and it is high time that science set about proving so.

What about other eye problems such as Glaucoma, Retinitis Pigmentosa, etc.? Depending on the nature and severity of your problem, it could be necessary to add to Eyemax-plus some of the additional supplements that we offer.

Science should make it a priority to confirm that Eyemax-plus is to be recommended for any kind of visual problem. (The last such confirmation was for a new drug called 'penicillin' as an antibiotic and look where that led.)

Retina specialist agrees: 'I have been taking Eyemax-plus for about nine months. I had a big bleed behind one eye which required an injection as I had lost 90% of the vision. My retinal surgeon then insisted that I would require these injections on an on-going basis. However the Eyemax-plus arrived just in time! On my return visit one month later I did not require an injection after they tested my eyes.

I told him about taking Eyemax-plus and he shrugged and said, 'If it works for you, well done.' I have been tested recently only to

confuse them all as I have 20/20 vision. My retina specialist thinks I am doing beautifully and my vision is not deteriorating as it was prior to Eyemax-plus. I've already received the most valuable gift a person could wish for.....my eyesight'. Ann Williams, Australia.

This is powerful evidence. Retina specialists are in a good position to verify that the Eyemax-plus arrived just in time to reverse the loss of 90% of vision. That it should then be restored to 20/20 is an added bonus. No longer should a 'big bleed' behind one eye be such a problem as it is often made out to be by eye-doctors.

Makes a mistake: 'I took Eyemax-plus for approximately eight months and noticed a definite improvement in my eyesight especially being able to see with my left eye that has a cataract. Then I decided to go off Eyemax-plus and see what would happen. (I was taking other supplements that supported eye health.) Well, after a couple of months I noticed that I had made a mistake in going off the Eyemax-plus because my eyesight definitely started declining again. Needless to say I am now faithfully taking the Eyemax-plus with no plans to ever stop. My vision is improving again, and I am very grateful to the Cambridge Institute!' D. Ellebracht, Capistrano Beach, CA.

Going off it to see what would happen is a scientific experiment which has been clearly demonstrated to be a mistake. This is an example of a controlled experiment and the methodology will be familiar to scientists.

Starting off on the right foot: 'I at least feel that I am doing something to hopefully delay eye problems. I also am glad that the vitamins take care of most of my body's needs so I do not need to be as concerned when I cannot eat fresh vegetables and fruit. I usually never have the time to write a testimonial but just felt I needed to take the time. The people in your organization that I have had phone contact with have also been most helpful.' Elizabeth Summers.

Dieticians should make it more widely known that taking vitamin supplements means that one need not be as concerned about eating fresh vegetables and fruit.

Cataracts - going, going, gone? 'Yes, my eyesight is clearing as the days go by. Last night I was watching TV and I could read phone numbers at the bottom of the screen. It seems to me that the cataracts are softening. I have been on Eyemax-plus for less than 2 months.' Al Harris.

Reading phone numbers at the bottom of the screen is good evidence for cataract softening.

Glaucoma pressure normalized: 'I gave Eyemax-plus to my Mom because the doctor told her that her eye pressure was high. I'm an Acupuncturist and I was really impressed by how complete the formula is. I'm happy to report that at her last visit her pressure was normal!' A. Florez, New York NY.

In order for a new treatment to be fully accepted it is important that it is embraced by other medical professionals.

Macular degeneration is completely gone: 'Last month I went to my eye clinic for my annual examination and guess what! My doctor can't seem to find the macular degeneration that I had been diagnosed with 2 years ago. I asked him 'Doc, what happened to my macular degeneration, do I still have it?' Doctor's response was 'I'm trying to find it Maria, but I just can't seem to see it'.

Before I sat down for my eye examination I showed my Eyemax-plus to my doctor and asked him: 'Doc, I've been taking this for several months now and I know it's been helping me. I know it's been doing good for my eyes and my entire health because I could sense something good with my entire body. I no longer feel fatigue in the early morning hours and lately, I could go without my eyeglass or without my hard contact lenses the entire day, especially on weekends when I'm at home. Also, I could sit and watch television in the evening without my glasses or contacts and the pictures seem to appear clearer and clearer as the days go by. The doctor stared at the Eyemax-plus and said: 'Yumm...yumm...it's got all the good vitamins, minerals and a lot of antioxidants'.

I asked him if it was all right for me to continue taking it and his response was, 'I don't see why not, it's got all the good stuff in it. All I can say is it might be too costly for you.' I didn't tell him how little it actually costs. Thanks again for Eyemax-plus. I need to place an additional order because my husband has been taking a portion of my supply.' Maria L.G. Cruz, Guam.

This is more good evidence and a future double-blind assessment of Eyemax-plus should take account of it.

Cataracts stabilized: 'I've been on Eyemax-plus about a year and a half. My cataracts have more or less stabilized to the point that I'm not getting worse every three or four months like I used to. I firmly believe that even if cataract surgery became necessary Eyemax-plus would be essential to continue both my eye and my general health.' Liana F. Hollywood, Ca.

It is hugely impressive when a patient believes this.

Holding off surgery & feeling better: 'The optometrist told me I have very bad cataracts and have to have surgery. I decided to hold off for now. Instead, I ordered Eyemax-plus. My eyes are feeling much better and they are really starting to sparkle and aren't as tired as they were three months ago. Thank you so very much.' Luci M., Loves Park Il.

Sparkling eyes are good proof

Cataracts are going away: 'I'm 61 years old and was diagnosed with cataracts about a year ago. I started taking Eyemax-plus (4 tablets a day - 2x2) on December 21st, 2009. It's now February 6th, 2010. I also started bouncing on a rebounder (a small trampoline) about the same time for 15 minutes a day and using Similasan cataract eye drops. I had my annual eye doctor appointment last week. Here are the results: No more dry eyes. Low/normal pressure, cataracts have gotten smaller and condensed to the centre of my eyes, peripheral vision has improved, my overall vision has improved by 2 steps.'

Scientific research is urgently needed in order to show what percentage eye improvement is due to bouncing on the trampoline.

'What I find most remarkable is that my optician wasn't interested as to why my vision has improved. I am so glad that I came across Eyemax-plus! Just like you, I endorse the idea that emotions have an effect on our physical health, including our eyes.

As a healing arts practitioner I believe that our emotions can be the cause of the disease that manifest in our bodies. I have been working with several practitioners to help me release old anger that I have been holding for a long time. In Shiatsu, the liver is where we hold our anger.' Margaret Welsh, Pennsylvania.

Science must really start to pay attention to the opinions of healing arts practitioners when it comes to our eyes. Old anger held for a long time should be released as part of an ongoing eye care programme but it seems that science has overlooked Shiatsu and in particular not given the liver the status it deserves. There are implications for transplant surgery. One would not wish to transplant a liver containing old anger which might damage the eyes.

Three weeks to Better Vision: 'About a month ago I woke with serious distortion and a dark spot in my right eye. After finding your site I ordered Eyemax-plus. Now, 3 weeks later, the distortion is completely gone and the dark spot is almost gone. I love the emails, reminding me why I'm taking Eyemax-plus, so that it doesn't become one of the bottles I forget.' ConneSusan Gilbert, Henderson, Nv.

Old people can forget so it is good to see that emails are sent to remind them why they are taking Eyemax-plus.

10 keys to Better Vision. Daily habits for better eyesight with Martin Sussman: You use your eyes every minute that you're awake. How you treat your eyes and the awareness you bring to them are crucial factors in improving your vision. In this lecture with Martin Sussman you'll learn the essential techniques to care for and protect your vision to make it as sharp and as healthy as it can be. You'll also learn the facts behind the 5 myths that stop people from improving their vision. 10 keys to Better Vision ™ . CD recording of live audio event: Order mp3 download $12.95. Order audio CD: $14.95.

This seems a very reasonable price to pay considering the benefits.

Called 'A real Pro' by NBC Today and 'An Expert' by CNN News, Mr. Sussman has appeared on radio and TV nearly 100 times. His appearance on CNN made 'broadcast history': It was the first instance that a segment from 'News from the World of Medicine' was re-broadcast.

There are fewer more impressive accolades than being referred to a 'A real Pro' by a television network. Likewise, the fact that CNN News describes Mr. Sussman is 'An Expert' should only serve to indicate to scientists that here is someone whose ideas need to be taken seriously. Organisations such as this will have done thorough research before allowing opinions to be expressed by their guests. In many ways, most of the hard work of confirmation has already been done; the job remaining for science should be a matter of dotting the i's and crossing the t's.

Mr. Sussman is a natural vision care expert, not an Optometrist or an Ophthalmologist. He goes beyond the medical and mechanical models of vision, believing instead that vision is profoundly affected by the mind, body, emotions and spirit.

Natural vision care experts who go beyond medical models of vision are a glimpse of the future. Science too often appears blinkered and fails to take into account the profound effects of mind, body, emotions and spirit. This is not the first time we encounter science failing in these areas and in the future science should incorporate Mr. Sussman's analysis into the training of new Optometrists and Ophthalmologists by amending the syllabuses of their training establishments accordingly.

What your glasses reveal about you: A mind/body understanding of near-sightedness with Martin Sussman: There's a mysterious connection between how you see and who you are. Nobody understands it exactly but it's been documented for more than 100 years. Learn how to use this connection to your advantage as you improve your vision.

Science should make it a priority to remove the mystery between how you see and who you are. The fact that it has been documented for more than 100 years shows that science has been lagging behind. Science is supposed to be in the business of making things less mysterious and it seems strange that this mystery has somehow slipped through.

Discover more about yourself, how to release the inner barriers to seeing and how to change your eyesight in a total way. What your glasses reveal about you (recording of a live event): Order download: $12.95. Order audio CD: $14.95. Topics covered include: Harvard University research into myopia; The inner reasons eyesight changes; What the vision transition period reveals; Changing your near-sighted response. Never a customer complaint. See our record.

It is highly convincing when there has never been a customer complaint.

Follow along with the audio CD's as they effortlessly guide you through each vision session. Learn about everything that influences your eyesight. Get the training materials and support you need to succeed. It's all in the program for Better Vision - and in less than 20 minutes a day! The program for Better Vision®: 2 audio CD's with 6 vision sessions, 80-page program guide, 2 vision charts, Fusion String (instructions only with download), lifetime free support. Order download: $44.95. Order printed edition: $49.95.

Effortless guiding is obviously superior to the efforts of scientific vision specialists who merely confuse people with incomprehensible jargon.

The program for Better Vision is not the first thing to do for medical problems such as macular degeneration or cataracts. The first step to take is re-nourishment of the eyes with a powerful nutritional formula. As you are re-nourishing your eyes use of the program for Better Vision will help to increase the blood flow, circulation and oxygen supply to the visual system, further enhancing the value of the nutrients.

Re-nourishment of the eyes (see previously) is something which modern science seems to have placed on the back-burner in favour of spectacles.

My doctor says eye exercises don't work. What should I do? Natural vision improvement is a hotly debated topic. Ask most eye doctors about it and they'll tell you there's no proof that it works. However, what they won't tell you is that there is no proof that it doesn't work either.

Doctors are very much at fault here. Science must make it clear that where there is no proof that something doesn't work then they have no reason to doubt it.

That's why we put the program for Better Vision to the test before we released it.

An enlightened approach.

There hasn't been enough research to prove either side - one way or the other. However, to say that there hasn't been enough research doesn't mean that there hasn't been any.

Here are just two examples: In 1941 the US Navy gave vision training to its pilots so they would recognize enemy aircraft sooner. It was discovered that the training also made the pilots' vision sharper. In 1958 Charles Kelley's research demonstrated conclusively that practice of certain techniques improved eyesight. His research appeared in optometric journals as well as mainstream media including the New York Times.

That Charles Kelley's practice of certain techniques improved eyesight has never been seriously disputed.

Natural vision improvement research has continued since then. You'd think that all the research that exists would encourage further research but that hasn't been the case. Instead, most eye doctors and research are more interested in studying surgery and drugs. The sad

truth is that you'll go blind waiting for the research to catch up to the facts. And the facts are that every single day people around the world are putting the power of vision improvement to work for their own eyes - and getting results.

This is a general problem with science and one which has been highlighted elsewhere. I am sure that ordinary people would respect science more if it made an effort to catch up with the facts.

Vision therapy for refractive errors, such as myopia, hyperopia, presbyopia or astigmatism is not taught in optometry schools. One has to look to the alternative, natural, self-help arena for that kind of help.

It is strange that vision therapy for refractive errors are not taught in optometry schools. Perhaps the governing bodies in such establishments should realise how out of date they are and resign.

It's time to take the health of your eyes into your own hands. Because if you wait until your doctor says it's ok, it may be too late.

This, depressingly, has the ring of truth.

Isn't near-sightedness (myopia) caused by the shape of the eye? Sure, that's one common belief but in fact it's a misconception. This limited view of vision says that if the eye is too long that is supposed to cause near-sightedness, if it's too short that is supposed to cause farsightedness and if the shape of the eye is distorted that causes astigmatism. But the shape of the eye is only one element of the entire visual system and it is not the only one that determines how clearly you see. Besides, has anyone ever asked how the eye becomes the wrong shape?

It does seem that very few scientists have asked this.

There are three real causes of poor vision: First, the visual habit patterns that a person has developed - how the visual centres of the brain work with the eyes. Second, tension in the eyes and body. Third, mental and emotional stress and strain and patterns of not seeing.

It makes sense that bad habits can develop over time. These could easily lead to patterns of not seeing.

To get Better Vision you want to re-train the way that the brain uses the eyes and restore the proper visual habits necessary for clearer vision.

And still opticians insist on spectacles.

'Before I started the program I wore my glasses 14-15 hours every single day. I passed my driver's test without them only five weeks after starting the program. Now I never need glasses at all. I feel much more confident and free. It's a great program!' Donna Sanders, Havre, Mt.

This is very telling. It also threatens to put spectacle manufacturers out of business leading to a drop in profits for opticians.

'I have experienced major improvement in my visual attitude.' Lawrence Merkler, Morrisville Pa.

Today's opticians might like to consider if they are attempting to improve their patients visual attitude. Their own attitude could be key here.

'I've worn glasses since age 7. So at first I was frustrated without wearing them but now I enjoy not wearing glasses. The program showed me how much more beautiful we all can be without glasses.' Richard Hamre, Basel, Switzerland.

This is something else opticians don't tell you. I wonder why?

'One year after taking off my glasses (except for driving and movies) I was re-examined by my eye doctor. My vision improved 100% and I wasn't even regular with the exercises.' Harriet Russell, Lenox Ma.

This is the optical equivalent of 'over-training' as athletes do before important events such as the Olympic Games. They train for a harder experience than they will actually encounter thus making the actual event easy. Not being regular with the exercises and still having vision improve by 100% shows that the program is over-training vision in the same way making a 100% improvement almost inevitable no matter how little the program is followed.

'I was amazed to discover that my brain can command my eyes to focus. One of the sessions is a beautiful metaphysical experience for me. I find the exercises bring me new answers and insight every time I do them. Thank you for your program and all it has meant to me.' Stephanie Collins, Las Vegas Nv.

It is far from certain that opticians show people how to use their brain to command their eyes to focus. One would hope that this is not so they can sell more spectacles.

'The program for Better Vision not only can help improve one's eyesight but by helping you understand the connection between how you perceive the world and your visual condition it will make significant changes in your life for the better.' Marc Grossman, O.D.

L.Ac., Author of Magic Eye: Beyond 3D, Greater Vision, and Natural Eye Care: An Encyclopaedia.

It is particularly encouraging when an actual doctor can recommend a program and particularly one who has authored respected texts.

'For a number of years I have been teaching the Bates method. Recently I came across your program for Better Vision and I have decided to teach this program as it is the simplest and easiest program for eye care I have come across. I congratulate you for the simplicity of the program and the ease that my people will be able to recover their perfect vision.' Dr. Chris Morgan, The Anti-Aging doctor, Goderich, Ontario, Canada.

And to have the Anti-Aging doctor himself on board is particularly telling.

'There are but few informed, courageous and qualified leaders concerned with those persons seeking to improve their vision. I believe that the Cambridge Institute for Better Vision offers realistic and proven alternatives to glasses and deteriorating vision.' Dr. Leslie H. Salov, Director, The Vision & Health Centre, Whitewater Wi.

To have the Director of the Vision and Health Centre in Whitewater, Wisconsin add his own personal recommendation is proof indeed that the program is scientifically rigorous.

'We would again like to express our appreciation for the Cambridge Institute's continued efforts toward educating the American public's understanding of the care and enhancement of their vision. It is a real pleasure when your organization refers a patient to us.' Dr. O. Reynolds Young, F.A.A.O., F.I.O.S., F.C.O.V.D. Dallas Tx.

I would imagine it is.

'I have personally used the program for Better Vision with a number of my vision therapy patients as an adjunct to the usual in-office vision rehabilitation we provide. I would like to recommend to my colleagues the fine eye care products offered by the Cambridge Institute for Better Vision. Customer service is superior and I know you will also be pleased with the items offered in the Cambridge Institute for Better Vision catalogue.' Dr. Errol Rummel, O.D., F.A.A.O., F.C.O.V.D., Jackson, NJ.

Science should make it a priority to investigate the items offered in the Cambridge Institute for Better Vision catalogue. Especially when such a well-qualified doctor offers his recommendation.

The program for Better Vision® has everything you need:

1. Fusion string technique: Develops binocularity and convergence, teaching your eyes to work together and to see equally.

Equal seeing is still not taught in schools of ophthalmology despite this evidence.

2. Mind's eye visualization: Increases your ability to visualize, encouraging the mind's role in restoring vision.

Neither is visualization. Too often, lenses are used in place of encouraging the mind to restore vision. Lenses do restore vision but they cost a significant amount of money.

3. Self-massage techniques: Relieves tension of the upper body, particularly your shoulders, head, neck and eyes.

Eye tension has received little consideration up to now.

4. Spectrum visualization: Releases deep tension in your entire visual system.

Deep tension in the entire visual system is only beginning to be explored by science today. The program appears to have got there first.

5. Vision chart techniques: Increases focusing power and sharpens peripheral awareness and mental concentration.

Opticians already use visual letter charts so this extension should not be too difficult for them to assimilate.

6. Memory visualization: Helps you release the limiting images, memories and attitudes that may block clear vision.

Releasing the limits that may block clear vision seems an obvious first step yet you would be surprised how many opticians ignore it.

Lifetime support: Questions about using the program? Call or email and we'll be glad to help. You may never need to - the program is so easy to use - but our staff are on hand if you need us.

What more could be asked for?

As mentioned above, the Bates Method is also of increasing use in eye-care. We finish with a look at it.

It is now over 100 years since Dr. William H. Bates first began developing and refining his theories of the causes of poor sight. He used his method for short sight, long sight, astigmatism, old-age sight, squint, lazy eye, and even structural diseases such as macular degeneration. Nothing was left out and he found all could be benefited by learning normal and relaxed use of the eyes and mind.

His view was both radical and exciting; a proportion of the public was becoming increasingly interested in the development of humankind and found this new look at vision an enticing change.

Dr. Bates was an extremely busy man at the forefront of a new way of thinking. Bates had fully trained as an Ophthalmologist and in the late 1800s ran his practice as conventionally as any other: He put people in glasses and told them there was nothing that could be done about it. How he changed his opinion to the opposite view is not fully known but the change was absolute.

However, his radical new view was also an unpopular one. He was known by his colleagues as a brilliant man, a surgeon, who had discovered the properties of adrenalin and pioneered an ear operation for deafness that is still in use today. Over the years nothing had diminished his conviction that a revolutionary change was needed in understanding of how vision, the mind, and the whole person works. His detractors were unable to stop him: Bates was a fully trained medical man and knew all the arguments against his new model inside out.

This is all very encouraging form a scientific viewpoint. Dr. Bates, although an extremely busy man, was at the forefront of a new way of thinking. This describes most modern scientists very well. The fact that his radical views were also unpopular will find resonance with those scientists who struggle against hidebound thinking both by their peers and by the public.

The people who followed in Bates' footsteps have come to his method from a different route. A new model of vision required a new approach and almost all practitioners of the method are rarely medical people. They are educators.

While it is probably necessary that brain surgeons are medical people it seems unlikely in the modern age that eye specialists should be exclusively so. Indeed, it seems counter-intuitive given the modern revolution in medicine based on the premise that alternative approaches be fully appreciated. Education may be just what is required.

After Bates' death the road continued to be rocky. One of his students, Margaret Corbett, was taken to court on two occasions in the 1940s on a charge of practicing medicine without a license. On both occasions she was acquitted as she was able to successfully prove that what she had been doing was educating people about their eyes. One charge brought against her was that she advocated a practice that would lead to retinal burns and in response 300 witnesses appeared to her defence, including some Hollywood stars. All were tested and every one had healthy retinas.

The fact that Hollywood stars were involved is good proof that nothing was untoward in Margaret Corbett's methods. These stars have expensive lawyers and if their retinas were damaged we would have heard a lot more of it.

The years that have followed until the present day have been characterised by fragmentation and disconnection. Organised bodies for the Bates method have remained small and tucked away. In some areas of the world the misrepresentation of the method has continued - even in recent times some excellent teachers have been hounded by the authorities and made to move or cease practising.

Science itself is historically well aware of such hounding and knows very well how the proponents of a new theory can be mercilessly harassed. Galileo is but the most famous example.

We think it makes sense that eyes need no artificial aids to seeing.

This seems reasonable.

Humans are undoubtedly ingenious, but instead of devising ever more complicated ways of bypassing nature let's use that brilliance to discover what nature provided in the first place.

This is where modern eye doctors have what could be described as a 'blind spot'. All too often science will attempt to interfere with nature rather than using nature as an ally. It is difficult to imagine why scientists think they can do better than nature. There are precious few examples where it has actually occurred.

Welcome to seeing. Summary of the different categories of teachers listed. Bates Association Teachers - category: mBAVE: The teachers listed in this section comprise members of the Bates Association who have trained and qualified in approved training courses. The length of training is two years and requires considerable personal development and ability to demonstrate expertise.

All members of the Bates Association are assessed through written assessment and practical observation by an independent committee before being admitted to the Association. Teachers who belong to the Bates Association for Vision Education are listed with 'Bates mBAVE' next to their name.

This is sound scientific practice. An independent committee is the standard way of ascertaining expertise and written assessments and practical observation form part of such a committee's remit. Bates mBAVE is therefore a qualification which should command respect.

Bates teachers category: Bates Verified: There are many Bates teachers throughout the world who have trained on specialist Bates teacher training courses but do not belong to the Bates Association. Wherever possible we have tried to ascertain expertise and professionalism and where these criteria are satisfied the teacher is listed with 'Bates Verified' next to their name.

And this would correspond to the next echelon. If Bates mBAVE is likened to a Professorship then Bates Verified would be the equivalent of a doctorate.

Natural vision teachers - category: NVI Verified: Natural vision teachers are listed here as this form of vision therapy is very closely influenced by the original insights of Dr Bates. Wherever possible we have obtained confirmation of expertise and teachers who satisfy these requirements are listed as 'NVI Verified' next to their name.

And it follows that NVI Verified would be Graduate Students.

The means to Better Vision is through relaxing the eyes. Rest makes vision better, strain or effort makes vision worse.

This seems so obvious that it hardly merits mention. Yet there are many eye-specialists who would take issue. However, they should note that rest does indeed tend to make vision better. Following a nice sleep, and after, of course, rubbing away 'sleepy-dust' from one's eyes, vision is invariably most congenial.

There are several ways to rest the eyes. Close your eyes. While doing this, think of something agreeable. Cover your eyes. Called palming. If you cover your eyes so as to exclude all light the eyes will be able to achieve a greater degree of relaxation. Cover both eyes with the palms of your hands, your fingers crossed on your forehead.

Note: in order to be successful, you must be able to relax while palming. Some people cannot do this and palming becomes counterproductive. The blacker the field you see, the more relaxed you are. But if you try to see black, this may cause more strain. Don't try to see black: it is better to imagine a concrete, familiar object or scene.

Many people fall at this hurdle and it is here that the Bates method professional can help. One must not try to see black. If it happens naturally then all is well but actively trying can be counterproductive.

Bates professionals will suggest a range of shades to be attempted which could, in time, lead to black.

Observe the swing of things. As you move your gaze from one point to another things seen should move in the opposite direction. For example, if you look at the upper left corner of the letter h and then shift your gaze to the lower left corner, the h should appear to move, or swing up. If it doesn't, this is a sign of strain.

Observing the swing of things is possibly the greatest insight of the Bates method. It is surprising that opticians have not caught on and continue to persist with their old-fashioned lenses, letter-charts and the bright light they shine into your eyes. Science should make it a priority to ensure that more opticians are swingers.

There are a variety of exercises to practice the swing. You can gently swing your whole body to the left and to the right and watch a distant tree swing to the right and to the left. You can move just your head or just your eyes. The better the vision, the shorter the swing can be made to be.

Science should be capable of confirming these results fairly quickly.

Use your imagination. By seeing things with your mind's eye, and remembering them in precise detail, you increase your ability to see actual objects better. The perfect memory of any sensation can be produced only when one is free of strain. It also helps, when practicing with a test card, to imagine that the part of a letter that one is looking at is blacker than the rest of the letter, or to imagine a small letter within a small black spot of a letter. In this way you direct your mind to appreciating finer and finer detail.

Psychologists are on the point of confirming that remembering the details of objects enables you to see them more clearly.

Catch those flashes. When your eyes finally achieve a state of relaxation through swinging or palming you will see a clear flash. Paradoxically, the sight of everything in focus is such a surprise that it causes strain and the blur returns. So before the clear picture blurs out, close your eyes and remember the image in its full sharpness and clarity.

It is not entirely unexpected for those using the Bates method that the sight of everything in focus should be such a surprise.

Keep your vision centred. When you regard an object only one small part should be seen best. This is because only the centre of the retina, the fovea, has the best vision for detail. Farther away from the fovea the retinal receptors get progressively less able to pick up

fine detail. Therefore trying to catch all the detail with all of your retina at once causes strain because it cannot be done. To be able to see all the details of an image put each detail into the centre of your visual field, where it can be seen best, one at a time. Allow each detail to become less clear as you move away from it and centre in on the next detail.

After this technique is mastered speed can be worked upon as, for example, it could present some danger if while driving a patient allowed too much time between allowing each detail of the road ahead to become less clear as he or she moved away from it and centred in on the next detail. Bates professionals might advise clients to slow down a bit until they feel more confident.

Enjoy the sun. Get out into the open and enjoy every sunny day. It is especially relaxing and stimulating to the eyes if you close your eyes and let the Sun shine onto your lids as you sway back and forth.

This is obviously healthier than being confined in a dark room with an optician for a long time.

Practice with a test card. Keep an eye chart on the wall. To practice, stand from 10 to 20 feet away and read the smallest line that you can without straining. Then look at one of the letters on that line and close your eyes. Remember that letter, go over every detail in your mind; shift from part to part, from curve to corner and so on. When you open your eyes, you will see not only that letter better but also the one below it.

Of all the examples mentioned in this book the theory and practice of eye-care appears to be one of the most backward at present yet the one which could benefit most if only science could get its act together and do something. There is a mountain of information available if science would take an interest instead of mounting long-term and expensive efforts detecting gravitational waves from colliding black holes. It is surely time that science realised that gravitational waves are of little interest to ordinary people whereas seeing clearly undoubtedly is.

'Big Money' science has had a bad press for a number of years now. Confirming the Bates Method will allow science to win back hearts and minds to its cause and, if we are being honest, it wouldn't take that much time away from the search for gravitational waves if the scientists involved could be persuaded to lend a hand.

Chapter 5: Bio-Energy

Everything in this Universe is constructed of frequencies. This is not just the common frequencies of the electromagnetic spectrum, such as radio waves, microwaves, visible light and x-rays, but also the frequencies of complex bodies be they a neuron firing, a brain thinking, a planet wobbling in its orbit or the atoms that make up your DNA.

No serious scientist doubts that everything in this Universe is constructed of frequencies.

Until recently we thought our bodies were made of bone and muscle, organs and blood, and so on. With the developments in physics we now can see that the human body, or more specifically a living body, is also a complex electromagnetic system. Some frequencies are used to communicate between different parts of the organism whereas others are used to heal ourselves and repair damaged cells.

And no serious biologist would doubt any of this.

The fundamental problem with loading our planet with a huge range of electromagnetic frequencies is that we too are electromagnetic beings, and just as your mobile phone picks up a microwave signal from the nearest phone mast so do you.

Research into the real effects of electromagnetic frequencies (emf) has always been hampered by interested parties who don't want the public to know, from the military through to electric and now mobile phone companies. Seems like they all want you to be permanently sick.

Ordinary people reading this will readily acknowledge that mobile phone companies, for example, do tend to make them feel sick.

Research showed that it only took 3 generations of mice exposed to emf for the offspring to be poorly developed with stunted growth. The alarming growth in people sensitive to emf does not make headline news and the public must be kept in the dark.

Small people may consider complaining about their lack of stature to their electricity supplier.

Like many human discoveries the awareness of our body electric and its interactions with emf can be used for good or bad. On the positive side, certain frequencies can be used for healing and even

curing cancers. However, the frequencies, and especially the mobile phone pulsed frequencies, are not beneficial. Just think of the difference between having a soothing massage and being repeatedly punched: Both use hands.

A telling analogy.

What's most catching my eye is the lack of interest and lack of medical trials in the Rife and Beck therapy. If these therapies have any validity a billion dollar market will be destroyed...

As is so often the case, vested interests limit the advance of medicine.

Extremely low frequency (ELF) less than 100 Hz is exactly the frequency range your human body is working with, just think of brainwaves and the interaction of the Schumann resonance and your brain. All frequencies used in the Rife and Beck therapy are around 0-1000 Hz.

This is telling evidence in favour of the Rife and Beck therapy. But as in all the best therapies, their frequency range exceeds the body's requirements.

The SP-12 vortex energy plate is a unique energy tool designed to investigate the effects of geometrically applied frequencies on various systems. The SP-12 vortex energy plate is designed around vortex based mathematics.

Vortex based mathematics is extremely powerful. Its use in the design of the SP-12 energy plate was not foreseen by mathematicians however, as a trawl through recent papers on the subject demonstrates.

According to Konstantine Meyl in his book 'Scalar Waves', information is a type of energy which is organized into cybernetic biofeedback energy vortex patterns or 4 dimensional energy vortex matrix holograms: spatial: x,y,z and time which look like a collection of energy swirls and patterns of both uncoiled vortex energy waves and coiled quantized energy particles that our spirit, brain, and body senses, interprets, and remembers as data. These scalar waves are sometimes known as Aetheron flux monopole emanations, or Aetherons.

This is transparently clear and needs no further comment.

The SP-12 hyper-dimensional vortex antenna operates in either of two primary modes: Connect to a frequency generator, audio or electronic device: The SP-12 can be attached via 2 wires to a GB-4000 frequency generator, LIFE force 2000 advanced bio-photon

analyzer, any computer system, stereo system, CD player (the resistance is 14-17 Ohm), or other Rife machine, creating a hyper-dimensional vortex transmitter with stunning results. The SP-12 hyper dimensional vortex antenna amplifies and expands the effect of music, subliminal, affirmational and meditation CD's five to 20 times when connected to stereo equipment.

A recent goal of scientific research has been to create a hyper-dimensional vortex transmitter. Here, it seems, science has been beaten to it.

Place the SP-12 anywhere you wish to harmonize the surrounding environment: The vortex energy plate does not have to be connected to anything to work. The generated field is self-sustaining, meaning that the modulator can simply be held in the hand or placed against any desired area of the subject or substance.

Science hasn't yet been able to produce a scale of harmonization. Indeed, science has been unable to definitively demonstrate whether harmonizing has occurred a little, completely, or somewhere in between. How are people expected to take science seriously if it cannot even offer a rudimentary scale of harmonization?

Nutri-Energetics Systems – Professional: NES-Pro combines Peter Fraser's discovery and mapping of the quantum electro-dynamic QED human body-field with the simplicity of computer technology. This has led to the development of the most advanced system for whole body-field analysis and treatment currently available. The NES-Pro enables the therapist to identify the root energetic cause of illness, and prescribe a personalised corrective treatment regime of Nutri-Energetics systems infoceutical remedies.

Peter, although not yet well known, will likely become famous in the near future.

The science behind the NES-Professional: Your body has energy continuously moving through it creating a dynamic energy field. This body-field can be viewed at many levels, at its simplest it is both the internal and external aura. As a practitioner, it can be considered as the energetic field which homeopathic remedies, acupuncture needles and healers are able to effect. In biochemical terms the body-field is the supra-chemical system which acts as the master control for all metabolism and growth.

Science has unfortunately been on the wrong track up to now by failing to consider the master control.

Nutri-Energetics System NES research has suggested that the human body-field (HBF) has structure and consists of magnetic vectors or signposts which are able to direct sub-atomic particles within your body. This enables the control of molecular reactions above and beyond that of the traditional biochemical system.

Particle Physicists should make a joint effort utilising the LHC to research the HBF. They should invite NES researchers on board to provide relevant expertise. Peter could be contacted with a view to leading the current team at CERN.

The NES-Pro software carries a map of the HBF information, as devised by Peter Fraser. When the client is connected to the NES-Pro system via a simple touch pad, there is communication between the clients' HBF and the field information stored in the software. In this way, any discrepancies between the clients' current state of energetic health and the optimum state are highlighted graphically on the user friendly NES-Pro computer interface. By working with the results the practitioner can prescribe an NES infoceutical treatment regime to rectify any underlying field issues and thus allow the body to return itself to the optimal state of health.

Science has completely failed to communicate with the HBF, so this is a real step forward.

What can you screen and treat with the Nutri-Energetics Systems - Professional? Peter Fraser's research at Nutri-Energetics Systems has enabled the HBF to be divided into its constituent sub-field systems. The NES-Pro is the only system able to assess each sub-field in turn, providing an amazing amount of information in a scan which takes only a matter of seconds to perform.

This division, and its accuracy, will come as news to all scientists associated with this research. As a general point, it is a measure of how much ground scientists have already lost that they are nowhere near inventing a scan which takes only a matter of seconds to perform, let alone one which provides an amazing amount of information.

The NES-Pro scan shows detailed results in 5 main report screens:

Big Fields: Shows the body's alignment with the Earth's main energy fields; gravity, magnetic polar and equatorial. Misalignment with these fields restricts the body's ability to maintain optimal health.

This has long been suspected by Big Field scientists.

Energetic Drivers: Reports on the integrity of the individual sub-fields which energise and regulate each organ system.

If the sub-fields have no integrity then organ systems are in trouble. Transplant surgeons will readily confirm these facts.

Energetic Integrators: These energetic pathways are related to meridian theory, allowing the flow of information between systems and around the body.

Information flow around the body is vital. Meridian theory is quickly becoming accepted as research continues.

Energetic Terrains: These disturbances are linked to the field component of disease-forming microbes, viruses, bacteria, fungus etc. and provide an energetically supportive environment for microbial growth.

Energetic Terrains have, as far as I am aware, not been considered by health professionals. Reading this, they may well change their minds.

Energetic Stars: Similar to energetic integrators, this area of the body-field concerns the pathways governing energy usage of specific organ systems.

This is so new that even I am not qualified to offer informed comment.

Musculoskeletal: This reports on physical shocks recorded in the body field, or pain that has been or is currently being experienced.

Reporting on pain which has *already* been experienced is also brand new.

Environmental: Reports on the body's ability to deal with the impact of environmental factors, including insecticides, electromagnetic fields, solvents and heavy metals.

Of these, the body's ability to deal with solvents is the most pressing. One hopes that the 'glue-sniffing' craze has died a timely death but even so, forearmed is forewarned.

Intolerance: This gives indications for the need of specific infoceuticals to reduce acute reactions, improve intolerances and aid absorption of nutrients and regulate sugars.

Sugar regulation has obvious implications for diabetes sufferers.

Nutrition: This reports on the functional integrity of the body-field in relation to the absorption, regulation and metabolism of nutrients.

Nutrition scientists working in this area will want to hear more concerning functional integrity.

The NES-Pro software facilitates the correction of any issues highlighted in the above reports by suggesting the most appropriate infoceutical for treatment. The Nutri-Energetics systems infoceutical protocol: The scanning process itself is a remarkable piece of technology, but the main focus of Nutri-Energetics is to improve the health of their clients through the prescription of NES infoceuticals. A protocol has been developed in which the major issues presented in the clients' scan are addressed from the ground up.

The scanning process is indeed a remarkable piece of technology. However, the real breakthrough is the 'ground up' addressing of major issues. This contrasts sharply with current medical practice which insists on a 'top down' addressing of the issues. Given the contrast, Nutri-Energetics systems infoceutical protocol seems to offer an obvious advantage.

Typically, the first session will reveal a strong reading in many of the sub-fields outlined above. However, the holistic nature of NES therapy is reinforced by the practitioner first addressing those fields with the largest ranging influence, namely the Big Fields. In this way the initial focus may not be on addressing the specific complaint for which the client is seeking therapy, but after a number of visits, all pertinent issues will be covered.

It is, of course, the experience of most if not all medical professionals that the client will insist on having a specific complaint dealt with immediately. But clients fail to realise that the medical professional has undergone years of training and has many years of experience to call upon.

Through the foundation and certified course training the practitioner is taught how to identify priorities in treatment. They will then prescribe a maximum of 5 different infoceuticals which the client administers at home for up to a month.

Infoceuticals are administered as drops in water, 6-28 drops a day per infoceutical. Feedback from clients shows an excellent level of compliance in the majority of cases and numerous testimonials as well as my own personal experience support the effectiveness of the treatment.

The fact that foundation followed by certified course training is given should remove all doubts concerning the Nutri-Energetics system, should any exist.

Why do we believe Nutri-Energetics systems Professional is the best energetic medicine device available today? The NES-Pro is one of the fastest and most comprehensive energetic diagnostic devices available based on a revolutionary understanding of the human body.

It is not often that revolutionary medical advances are made so when one is science should embrace the fact and turn all of its efforts onto achieving an understanding of the new paradigm. In that way science advances, and when science advances, we all benefit.

Full interpretation of the results naturally requires a certain degree of skill and understanding, however the results themselves are completely operator independent. NES provide extensive training, with an in-person foundation training course followed by a comprehensive distance learning certificate course. All standard training is included in the purchase price.

A comprehensive distance learning certificate is rapidly becoming standard medical practice.

The support available to both training and certified practitioners is second to none: International and localised distributors facilitating the ordering of remedies: On-hand practitioner support to deal with specific questions or queries: A dedicated research team able to answer technical inquiries: Discussion groups and practitioner forums where we can exchange experience and ideas. In conjunction with the book, 'The Unturned Stone' NES – Professional provides an unparalleled education into the latest advance in quantum biology and energetic medicine.

Quantum Biologists complain that their field hasn't advanced much in recent years yet I wonder how many of them have read the book 'The Unturned Stone'?

On a more practical note, the NES-Pro is simple to operate, running off a Windows based computer. Priced below most energetic medicine devices yet with far superior performance. Providing a dual income for practitioners from both diagnosis and remedies sold. Continually expanding in public interest, there are currently over 300 practitioners world-wide and a rapidly increasing client base. Quite simply the future of medicine.

Modern healing can be done quite effectively by following a distance learning course and reading the book 'The Unturned Stone'. Doctors should not be slow to realise that this could well be, quite simply, the future of medicine.

Bio-Resonance Explained: Bio-resonance is an umbrella term which is traditionally used to describe the interaction between the bio-field of a living organism and frequency patterns that interact with specific parts of it. There are many different types of bio-resonance modalities which fall under this umbrella term.

No serious scientist has disputed the interaction between the bio-field of a living organism and frequency patterns.

It is the ability for an external frequency/pattern to resonate with a specific item in the bio-field of the physical body. True bio-resonance is a conscious interactive exchange between the consciousness of a life form and a bio-resonance modality such as a piece of bio-resonance equipment.

It is highly encouraging that consciousness can be picked up using equipment these days.

Bio-Resonance within World Development Systems (WDS): Our particular focus is in the research and understanding of how individual parts of the human body resonate with specific combinations of frequencies and patterns. Through our research we have seen that frequencies alone are only a partial answer in the quest for finding an interface with the bio-field. We have discovered that a key function in the movement of energy in the bio-field is to do with consciousness rather than just a traditional 'hard wired ' frequency (Hertz).

Some scientists, to be fair, have suspected that there was more than just frequencies in their quest for finding an interface with the bio-field.

Much research over the last 50 years has concentrated on finding direct frequencies, generated from an electronic frequency generator, and matching them up with the response of specific organs and structures within the body. Although this traditional frequency research has contributed a greater understanding in the workings of the human body, it is a bit of a 'scatter gun' approach that deals with a physical Newtonian dogma rather than a holistic viewpoint.

The holistic viewpoint encompasses the subtle energy anatomy including meridians, chakras, distinct subtle energy fields within the bio-field, negative emotional charges, emotional and physical trauma as well as the physical body.

As science progresses, it is becoming increasingly obvious that traditional authority figures such as Newton will inevitably begin to lose their allure. Today, a physical Newtonian dogma is recognised to be inferior to a holistic viewpoint. Viewed this way, traditional direct frequency research has been superseded by the subtle energy anatomy including meridians, chakras, distinct subtle energy fields within the bio-field, negative emotional charges and emotional and physical trauma as well as the physical body. All of this would have been news to Newton.

The consciousness of a person is autonomous (i.e. in control) in the process of responding to bio-resonance patterns. We at WDS have created an interface which communicates with the body's consciousness and allows us, via our technology, to present patterns from our Resonance Pattern Matrix to the recipient's holistic bio-field and receive bio-feedback.

This is a breakthrough of potentially enormous ramifications. Communication with the body's consciousness has long been a goal of neuroscience. The fact that an interface already exists is beyond modern day medicine and the fact that the technology can be used to present patterns from the propriety Resonance Pattern Matrix to the recipient's holistic bio-field is a real step forward. The point that bio-feedback can then be received is the icing on the cake.

This consciousness contained within the bio-field is so incredibly advanced that it is hard to imagine why anybody suffers from disease in this day and age: The answer lies in the ability to ask the body's consciousness the right question and understand the answer.

Doctors should really be practicing this rather than indulging in the rote learning of the names of different pills.

There are indigenous cultures in different countries that still possess knowledge and understanding about the bio-field and consciousness itself that have been passed down for thousands of years. Some of these cultures have developed techniques and methods of interfacing with the bio-field which are far beyond what is accepted in the West.

Not for the first time (and not for the last) we see the knowledge of indigenous cultures being proved to be far beyond what is accepted in the West. The fact that knowledge of the bio-field and consciousness itself occurs in different countries and has been passed down for thousands of years merely adds credence to the evidence and makes one wonder why science hasn't taken more of an interest in indigenous cultures before now.

Regardless of which 'camp' the researchers are in, they all strive to achieve a change in the bio-field by externally stimulating it and then observing a change in the physical. Many bio-resonance

equipment can deliver an assortment of frequencies/patterns that include everything from frequencies (electronically generated), vibrational patterns from herbs, plants, crystals, flower essences, minerals and aromatherapy oils, as well as colour, sound and many other frequency/patterns.

Vibrational patterns from herbs has been wilfully ignored by science up to now.

At WDS we seek to find imbalances in the bio-field and through our proprietary interface, correct the imbalances in the energy field.

Imbalances in energy fields however are a perennial topic in modern day science. For example, the magnetic confinement field in a fusion reactor.

In addition, our latest research recognises that bio-resonance can play a part in changing the way the epigenetic layer of the DNA expresses itself. This is done by reinstating the energetic component of the original blueprint of the DNA and by balancing the environmental/social conditioning that surrounds it.

This promises to be an entirely new field of science. Reinstating the energetic component of the original blueprint of the DNA could fairly be described as the Holy Grail of Biologists. Balancing the environmental/social conditioning that surrounds it is the icing on *their* cake.

Adaptive Behaviour: Adaptive behaviour is a safety mechanism created by the brain to block out trauma and shock as and when it occurs. The trauma and shock is encapsulated and hidden to protect the brain from continually activating the fight and flight mechanism. Failure to encapsulate the trauma and shock would cause severe damage to the nervous system in the body. The down side of encapsulating these traumas and hiding them is that the brain will adapt around these traumas as if they did not exist.

Every trauma causes a chemical imbalance in the body which may eventually lead to an imbalance in the pathology around the area of the trauma or an area connected to that trauma.

Most modern medical research today is concerned with identifying not the area of the trauma (which is obvious) but an area connected to that trauma. This is known as the 'Hard Problem' of Medicine.

The WDS bio-resonance technology can change adaptive behaviour by safely releasing traumas that have been trapped within the body by matching the exact resonance that holds the emotional charges. One trauma can be held in many places within the body and

have multiple emotional charges in different parts of the body attached to them.

Safe release of traumas that have been trapped within the body has also traditionally been seen as difficult. Traumas have been released under controlled conditions but not as safely as would be wished therefore the technique is not widespread.

Inverting Chemical Substances: Further WDS research shows that many imbalances in the bio-field come from environmental factors that includes toxic loading from chemicals, social conditioning, emotional attitudes and beliefs. Bio-resonance has a good track record in successfully detoxifying a whole range of chemicals including heavy metals, poisons and drugs.

This research is ahead of science which has only partially succeeded in detoxifying some heavy metals. Poisons and drugs have proved intractable.

Different names are given to this process such as phase cancellation of a substance, frequency inversion of a substance and others. They all respond in a similar way by using sympathetic resonance that helps the body eliminate these substances through the normal body channels.

Phase cancellation and frequency inversion are terms known to science but, crucially, in different areas of research. It is to the credit of WDS that their researchers have brought what were previously thought to be disparate fields of study together under one umbrella and provided a unified theory for the first time.

Conclusion: Many forms of practice make the energy system of a person more evolved allowing greater movement of energy during bio-resonance therapy. This refinement allows for a better connection to consciousness and better bio-feedback. The reverse is true for a person whose system is energetically blocked.

Biologists following the Neo-Darwinist paradigm might like to look again at their theory in light of the energy system of a person becoming more evolved in this way.

What is E-lybra 9? By Victor Sims, Managing Director of WDS: The E-lybra ® system is designed to bring energy balancing to the masses. Its unique technology has been developed to enable practitioners to detect and balance imbalances using a fully interactive software program incorporated into bio-resonance equipment. The E-lybra ® is easy to use and has friendly

ergonomically designed computer screens with layouts that guide the user through the analysis and balancing process.

In the absence of science being interested in providing energy balancing to the masses it is good of Victor to take it upon himself to do so.

What does E-lybra 9 do? The E-lybra ® is designed to analyse imbalances in the bio-field of the client and at the exact same time produce bio-resonance patterns that harmonises the imbalances found.

Nobody has ever analysed anything and as a result of the analysis produced something at the exact same time. This is a major breakthrough.

These balancing bio-resonance patterns are sent to the client during a session via comfortable connection cables. The range of imbalances covered includes the complete physiology of the body, psychological, emotional and subtle energy systems. There are 222 categories containing approximately 300,000 bio-resonance patterns which the E-lybra ® will use to make a comprehensive body status profile. The E-lybra ® then uses this profile to analyse imbalances in detail whilst sending balancing to the client. The items that are found in the analysis are stored on the database for the client each time they have a session.

Sending balancing to the client. With those few words a scientific revolution could be at hand.

There is no limit to the number of clients nor the number of sessions that can be saved on the system.

As if that wasn't enough the problem of infinite storage seems to have been solved.

For new users there is an automatic balancing schedule that requires only one button to be pressed and it provides a complete analysis and balancing session without the need for any further button pressing on the software or bio-resonance equipment. In addition, all the corrective bio-resonance patterns that are sent to the client during the session can automatically be programmed into the bio-resonance capsule or drops for the client to take away.

This will remove the need for prescriptions to be written and avoid time spent waiting in pharmacies. Everybody wins.

Within the 222 categories there are several hundred formulae that can be extracted and used by the practitioner taking only a few seconds to create one's own formula for specific imbalances e.g.

synovial fluids in the knee, bones in the wrist, all of the organs and complete systems of the body.

Imbalance in the synovial fluids have been noted by knee doctors but up until now they have been unable to create their own formula for specific imbalances. It seems that those days are over.

The reliability of the E-lybra ® allows WDS to offer a 3 year guarantee on the hardware base unit and also 3 years free software updates and support. 2 days basic training is included in the package and this will be carried out by a WDS certified E-lybra ® trainer. We also include a free voucher to attend an additional 2 days training at any of our refresher trainings courses which are held at various locations throughout UK and Europe. Standard training worksheets are used by the WDS certified E-lybra ® trainers and these completed worksheets are returned to WDS who then issue an appropriate training certificate direct to the trainee.

This is becoming standard medical practice.

In addition to E-lybra ® there are many other balancing products created by WDS that can be purchased and resold to clients.

And science, once it has completed its analysis and confirmation of the E-lybra system, should investigate these other balancing products as a matter of importance.

Additional possibilities. Because of the holistic viewpoint that the E-lybra ® works from, it is normal for the system to find imbalances that are present in the bio-field but not yet pathological. This is a great help as a preventative measure to help kick start the body's healing systems ahead of time.

The ability to find imbalances before they manifest themselves is a key theme of predictive medicine.

The E-lybra ® formulae can be used to create effective creams and ointments for all sorts of conditions, eczema, acne, face lift (favourite amongst women), for athletics injuries and so on.

The scope is endless. Women in particular should be extremely grateful.

The E-lybra ® can be networked so that several E-lybra ®s can send data to a main server. This is ideal in a clinic situation where the data from several E-lybra ®s needs to be collected on one computer.

This is particularly useful. Science should ensure that several E-lybra's can indeed perform in this way by purchasing a number of them without delay.

A typical balancing session protocol consists of 10 minutes initialisation/balancing program followed by 5 minutes full body scan. The body scan produces a profile automatically selected/tested from 222 categories. The profile is then used in the main analysis and balancing session which is usually run for 45 minutes. During this process the E-lybra ® produces bio-resonance patterns which are delivered to the client and also stored on the bio-resonance capsule which the client takes away and wears on their body.

This analysis is impressively quick.

At the end of the analysis and balancing session an 'after' body status is recorded showing changes that have occurred in the client's bio-field which takes a further 5 minutes. Lastly, a grounding formula is run for 5 minutes to make sure the client is fully awake before completing the session.

One can quite imagine that a client might doze off during this procedure.

Variations to the protocol include adding extra formulae for specialised processes such as addiction therapy, weight balancing, to name but a few. The use of the E-lybra ® is only limited by the creativity of the practitioner.

In contrast to MRI scans where the emphasis is on following the rules.

Practitioner's may also sample and create their own bio-resonance patterns using the built in sampling function. In this way your favourite products, remedies and test sets can be stored on the E-lybra ® for inclusion in any formula and to test against the client connected to the system and to produce balancing remedies to take away.

This sampling function makes practitioner's expertise readily available to clients without undue delay and is an obvious improvement on modern clinical practice where a doctor may need to search through the relevant literature and/or consult a database to aid his memory of what to prescribe, all of which is time consuming.

How does it work from a distance? Part of the design of E-lybra ® incorporates a bridge to consciousness of the selected client allowing it to perform powerful energy balancing at a distance, i.e.

client not present. **Although this sounds a bit far-fetched to the newcomer of energy medicine it is in fact standard practice.**

X-rays were once thought miraculous.

To make this 'bridge' as strong as possible the E-lybra ® provides a function to sample the hair or bio sample of the client which is resonated electronically and creates the signature of the hair sample.

This creates an electromagnetic field and a unique energetic connection to the client. This 'connection' is then sampled and stored on the E-lybra ® system. When the client is being balanced at a distance the stored 'connection' is used in the circuitry to receive information from the client as well as to send patterns across this quantum bridge.

Science is nowhere near approaching this level of sophistication.

Experiencing this type of balancing just once is enough for most practitioners to open the way forward to using this exciting feature when and wherever possible. Ideally used for clients in hospital and as a general support 2-3 times a week.

Anecdotally, it has been known for hospital doctors to claim the credit for patients recovery when in fact they had been distance healed!

Practitioners of E-lybra ® say that it adds extra effectiveness to their balancing sessions because clients can be helped a little each day through their difficult challenges as a regular part of the support offered by the practitioner. Many clients can be scheduled one after another each with their own needs when E-lybra ® is used at a distance. It is not uncommon to give each distance client 15-25 minutes each night when the E-lybra ® is not being used for normal day sessions.

Doctors are noted for being overworked and as a result stressed. The E-lybra, once endorsed, will enable any doctor to become much more effective. Assuming a nominal 8 hour shift pattern, it would mean three times as many patients can be treated than was previously the case. This will cut waiting times dramatically and will enable all doctors to have some much needed rest.

What is included in the standard E-lybra kit? E-lybra 9 energy balancing base unit, comprehensive E-lybra instruction manual, E-lybra 9 registration documentation, E-lybra 9 software install CD for Windows XP and Vista, 6 x e-capsule's and practitioner information leaflets, 1 x quantum star, 1 x crystal spiral, 1 x water enhancer, 1 x

silent healing CD, 1 x computer clear CD, 1 x WDS product information DVD, 1 x WDS E-lybra practitioner talks DVD, 1 x standard USB 2.0 cable type a to type b, 1.8m, E-lybra connection kit, 1 pair wrist straps and connection cables, 1 pair heel and ankle straps and connection cables, 1 x 9-volt power supply, Voucher for additional 2 day E-lybra refresher training course.

Science should move towards establishing which of the pair of wrist straps or heel and ankle straps are the most effective for healing purposes though both are included as standard which is a point in favour of the E-lybra system.

Testimonials: This elderly client had mild dementia when he was brought to see me. He cannot remember places well and had frequent urination. At the beginning, he goes to the toilet at least twice during each therapy session. After about 4 months of therapy his dementia improved significantly and he began to recognise places. Also, he can hold his bladder better and needs to go to the toilet at most once towards the end of the therapy.

Improvement in toilet problems is a significant indicator.

My first experience with Perm and her machine was mind blowing indeed The first week after my first treatment I also noticed a difference in myself as to the way I felt, not only was I pain free but I also felt ready and able to do things, I felt alive again and full of energy, like I had been cleared of all the rubbish that was inside of me that needed moving, like someone had found the delete button and pressed it and emptied it.

The subjective analysis of symptom and cure by a patient is rapidly becoming the standard by which medical intervention is assessed.

Stomach infection for 5 years, crippling pain (on left side), often on sick leave, difficulty functioning...30 doctors visited, no help. Two treatments, full of energy, functions well at work and daily life. Seven treatments: Free from pain.

The fact that 30 doctors were visited and there was no improvement is an indictment of modern medical practice. If 30 doctors between them cannot produce a cure for a 5 year long stomach infection then there is something fundamentally wrong with modern medical techniques. Yet just two treatments with the E-lybra fills the patient full of energy and seven treatments leaves him/her free from pain. One can draw one's own conclusions.

Liz and Kelvin are just fab! Their combination of caring, knowledge and expertise is powerful and very effective. They make a great team - a kinder, more caring and lovely couple you couldn't hope to meet. Liz especially is the calm voice of reason on the end of the phone when I'm having a blip or when the going gets tough and patiently and calmly puts me back on track. I'm so grateful to her for all those moments when she's come to my rescue, bless her! The E-lybra is mind-blowing and Liz is very skilled at using it. It's been a massive help. Client of Liz and Kelvin Kellond, Devon, UK.

It would be instructive to know how many medical professionals who don't use the E-lybra have been publicly praised in such a manner.

As Sarita's mother I requested remote bio-resonance sessions for her as for quite some time she had been a very poor, fussy eater and as a result underweight. If she was ill with something like a cold then she could lose more weight rapidly and was mostly quite pale. Soon after beginning the E-lybra programme her eating tastes started to change. After a few sessions on some days she mentioned that she was actually hungry and occasionally asked for foods which before she would have refused, like cherry tomatoes, and even asking for seconds of broccoli. Also she seems very thirsty and has been happy to have water.

As the weeks went on she seemed to eat with gusto, including meats and salads as her appetite is much more vigorous now. A month after her sessions ended she is still doing well and eating fairly regularly without a lot of hand holding or excessive encouragement and has surprised us by finishing her meal before her older sister eating more than or the same in some cases, which is a HUGE difference - she is growing up so fast - this will be a wonderful summer as I am not worried about her any more, she's fine - thank you, Theresa - Sarita's mother.

And if scientists were in any doubt concerning the effectiveness of distance healing then they have only to ask Sarita's mother. It must seem like a miracle but it is no miracle, it is merely the result of using the machine in the appropriate way.

Inergetix Core: The Inergetix Core system claims to enable energetic frequency balancing based on informational analysis of clients' resonant frequencies. The Inergetix Core system analyses the clients' specific frequencies and balances the body with frequencies that have the greatest resonance in context with the symptoms

entered during analysis. The practitioner can also use designated frequencies from a disease list containing 1100 conditions.

1100 conditions is a major step forward. Science should use this list to extend its own.

Informational broadcasting of frequencies by the Inergetix Core system affects the body on an informational level, for example by re-establishing control functions or mental-emotional issues. This activity will indirectly affect the energetic, biochemical and structural levels, helping to return the client to health.

This has long been suspected to be the case.

Inergetix Core vs. other biofeedback devices: Inergetix have taken the approach to absorb as much of existing bioenergetic technology as they can as without actually creating anything new. The result is an integrated delivery of frequencies over the whole spectrum of physical energies, i.e. light, sound, electricity and magnetism. In this way, the Inergetix Core system provides informational and energetic therapies in a single device.

This is the first time the 'big 4' have had their frequencies integrated in this way. It is a major advance.

The Core frequency database contains 25,000 remedies divided into 400 groups to help focus analysis, and additional items and remedies can be added to the database by the practitioner as required. The Core system also contains a catalogue of emotional concepts which reveal psychological connections.

This places the device into areas wholly new to science.

Treatment with Inergetix Core: The Core scan itself lasts only a matter of minutes and therapy can be applied by one or a selection of the available add-ons. The main Inergetix Core interface unit acts as both the symbolic connection to the client, enabling radionic interaction, and the physical connection for energetic interaction. The Core interface unit has an audio connection to external speakers for broadcast of audible frequencies; a connection to the gold-plated imprint tray for imprinting of potentised remedy information; a connection with hand and foot electrodes; optional electric frequency therapy unit and an optional plasma generator.

The optional plasma generator will be of interest to those physicists who are still trying to achieve nuclear fusion while scientists attempting to imprint potentised remedy information will doubtless want to investigate the tray. A search of the literature reveals that they don't seem to have considered gold-plating as yet.

Inergetix Core add-ons: The Inergetix Core hand and foot electrodes are easily sterilised and ergonomically shaped.

And they certainly haven't ergonomically shaped their electrodes.

They form the physical connection of the client to the system, acting as informational facilitators to increase body awareness and energy flow.

Increasing energy flow is a major goal of research. The fact that body awareness is increased as a by-product is another breakthrough.

The optional core electric frequency unit (EFU) is a high resolution arbitrary waveform generator (AWG) which measures electromagnetic changes in the client. This unit increases the therapeutic range of the Core system from the 20 kHz of the interface unit up to 1 MHz. This higher frequency is said to be more effective in treatment due to higher energy and therefore deeper penetration. The Inergetix Core EFU also allows control of amplitude and waveform, allowing high resolution measurements of the clients' response to energetic frequency therapy.

This passage is largely for specialists but I include it here to demonstrate the level of science underlying the equipment.

The Inergetix Core plasma generator emits modulated plasma light from discharge tubes to carry energies and information deep into the tissues of the client sat in front. The discharge tubes are filled with noble gases to supply the carrier wavelength. This therapeutic option can be used for clients not able to hold the hand electrodes, including animals, pants and soil. It should be noted that light therapy was the greatest success for pioneers in frequency therapy but has been neglected by other frequency therapy devices.

Science has demonstrated that energy can be carried deep into tissue but never suspected that information could. Noble gases were not thought of to supply the carrier wavelength and this now appears to be a mistake.

The Core software can also be used to run a stand-alone hologram generator, a white noise diode and analogue digital converter. The Inergetix Core hologram generator makes it possible to analyse acute cases in short-term mode by working through high speed data sampling. It also increases the resolution of the clients' picture when used in conjunction with the interface unit by introducing more white noise, integral to real-time bioenergetic and bioinformational feedback.

Again, here we have a passage which will only be understood by professors or at the very minimum holders of PhD's. If you don't understand it, don't worry. You are not alone.

Oberon – organ functionality testing. The Oberon system assesses fluctuating wave characteristics of body tissue, individual cells and chromosomes. The client is scanned by the Oberon using a pair of headphones which emit a range of frequencies of magnetic fields. The results of the Oberon scan are then compared to the resonant frequencies of all normal organs and common disease tissues in various stages of progression. The resonant frequencies of medications, allergens and foods are also scanned for comparison.

The results of the comparison are then fed into the Oberon computer software programme Metapathia where areas of significance are highlighted for the practitioner on a graphical physiological display.

How science could have missed the use of headphones emitting a range of frequencies of magnetic fields is one of the great failures of modern scientific endeavour.

How the Oberon works: Around biological systems exists an exceedingly weak low frequency magnetic field. The distinctive oscillations of every organ and cell are recorded in the Oberon software in the form of a spectral analysis of vortex magnetic fields. Electronic oscillators within the Oberon resonate at electromagnetic wavelengths and energy levels sufficient to break weak bonds, to read bio-activity of the body tissues, whilst selectively amplifying signals against background noise.

This is so advanced it is untrue. Science has never suspected that a spectral analysis of vortex magnetic fields was possible even in principle yet here we have a fully functioning device which does exactly that. Perhaps the money supposed to be ring-fenced for vortex magnetic field spectroscopy has been diverted into that pet project so beloved of Physicists, the Large Hadron Collider. If that proves to be the case then the LHC and its proponents have a lot to answer for.

In an unpublished trial of 100 clients, the Dove Clinic, UK, reported an 80% correlation between known pathology and the Oberon readout.

One can only speculate why this trial remains unpublished.

Areas of interest/concern in the client's body are identified by the Oberon in terms of entropy values of the degree of quantum

organisation, the lower the entropy, the greater the level of structural organisation and functioning; the higher the entropy, the lower the level of structural organisation and functioning.

Experts in entropy values and quantum organisation have never contested this.

The entropy values of specific structures are depicted with colour on the Metapathia computer display. Low entropy fully functioning areas are displayed in pale yellow. As the entropy increases the colour displayed on the Oberon Metapathia becomes darker, ranging from orange to red, purple, and nearly black.

Fortunately the Metapathia computer display intuitively bypasses the need for specialist interpretation in favour of full colour representation. A patient can literally see what is wrong with him.

The Oberon diagnostics system also gives a unique opportunity to record frequency oscillations of any medicinal preparations. According to spectral characteristics the Oberon can make an immediate comparison of all the preparations stored in computer memory which may be several thousand in number, with the characteristics of the pathological process and reveal which remedies may be most efficient.

Experts in spectral characteristics will confirm that frequency oscillations are of the greatest importance here. The amount of science embodied in this machine is quite astounding.

Treatment with Oberon: The information therapy delivered by the Oberon influences the patient's body with a combination of different modulated electromagnetic oscillations. Oberon therapy is aimed at correcting the deranged intrabodily balance and corresponding electromagnetic radiation via information preparations metazodes.

Metazodes are so new in medical theory that the only explanation I could find was from the developers of the Oberon itself.

Oberon metazodes are specific combinations of frequencies found to resonate with the current state of ill-health. The pathological waveforms are inverted by 180°. They can be received by the client via the Oberon itself or transferred to a matrix of water, alcohol and lactose for oral delivery over a course of treatment. This method of treatment is similar to that used in the NES-pro treatment protocol and accounts for a wide sphere of influence of this type of preparation and the absence of harmful side

effects and contraindications in cases when conventional remedies are prescribed simultaneously.

The latter of which continues to dog modern day medicine.

Delivery of therapy via the Oberon itself is achieved through a laser scanning device, similar to Mora therapy, i.e. Laser radiation to influence the pathologically injured organs and system projections.

Influencing pathologically injured organs via laser radiation has just become available to modern medicine but the use of such therapy to influence system projections has, until now, been merely a dream. Science must investigate the Oberon as a matter of urgency.

MARS III - Multiple Analytical Radionic System: Computer aided radionic system for analysing and balancing electronic homeopathy and colour tuning. The MARS III is the computer-aided radionic system of Bruce Copen laboratories. The MARS III consists of a master control unit and the software Scope 4 can be easily installed into a PC computer. The unit and the software allow the individual modules to function.

The step forward here is the modular approach.

How does the MARS III work? The MARS III uses radionics to treat the patient. The device has a large data bank of over 14,000 rates that include; Acupuncture, Allergies, Flower Essences, Organs, Toxins, Gemstones, Foodstuffs, Teeth, Symptoms, Homeopathic Miasims, Emotions and Subtle Anatomy.

Dentists have been slow to investigate tooth rates. This omission is remedied here.

The radionic analysis is carried out using samples of substances, blood, nail, hair, urine etc. With the MARS III the sample radionic witness is places in the device input channel and the built in scanner is claimed to digitally convert the sample and then stores this digital representation of the sample on the hard disk of the PC. Thus the sample only needs to be scanned once and it is claimed treatment can be done without the patient being present.

This is an obvious advantage for hard-pressed physicians who may be kept waiting by patients who forget their appointment, are too scared to turn up, are running late and so on.

Treatment is claimed to be achieved by testing against predefined tests built in to the software database. Once radionic

rates have been found they are then claimed to be projected to the patent.

Predefined tests are obviously superior to unpredefined tests.

The colour tuning module on the MARS III is claimed to be programmed to balance the patient using various ranges of naturopathy. It is claimed to be harmonical and composed to mathematical laws of music. Each sound is given a colour and in the spectrum are representations of various treatments including acupuncture, Bioresonance therapy, homeopathy and others.

This builds upon modern ideas of synthesis. Sounds being assigned to colour is part of the inclusive nature of holistic practice and is a progressive methodology of treatment. Why should patients continue to be denied balance when the technology now exists to achieve it harmonically?

The electronic homeopathy module on the MARS III is claimed to be programmed to have the rates for corresponding homeopathic tinctures. These rates are then used to treat the patient.

It is a large part of today's scientific effort to determine the rates for homeopathic tinctures. At some point in the future science has promised that these will be electronically programmed then used to treat people. It is entirely these same scientists fault that they have been beaten to it and they should acknowledge this publicly. There is no shame in coming second, but, as ever, a show of humility would not come amiss.

MARS III overview: MARS III Multiple Analytical Radionic System USB port compatible, Windows XP compatible, available in English, German, Spanish: Size: 36cm x 28cm x 10.2cm: Weight: 2.4kg: Price: $15,545.00.

This seems a very reasonable price to pay for sounds being given a colour which represents various treatments. Governments should adjust their healthcare budgets accordingly.

Quantec 5 – Transmits conscious intent between 'life' organisms. The Quantec is a regular radionic system of the old style. By placing their hand on the gold plate of the Quantec rose quartz triangle the client is scanned in a matter of minutes. The Quantec will then obtain an expression of functional state and any corrective vibrational information can be transmitted.

Obtaining an expression of functional state is a real step forward and as such, the Quantec 5 should be part of the armoury of any self-respecting medical practice.

How the Quantec works: Quantec believe that biophotons, light emissions, weak electromagnetic waves, merge into a collective global consciousness also known as the morphogenetic field. The intrinsic intelligence of the morphogenetic field interacts with the white noise diode random generator within the Quantec system, the computer and the software with the desired intention. Deficiencies of the client's morphogenetic field are identified in the Quantec scan, then removed/resolved with radionics impulses. A well instructed morphogenetic field can induce desired materialistic changes, and thereby restore the client to full health.

This may be difficult for the layman to comprehend. However, it has been demonstrated by science that electromagnetic waves can indeed merge. It is a small step from there to a collective global consciousness.

The white noise diode random generator is of particular interest. No-one has previously suspected that it would or could interact with the morphogenetic field. Science should make all possible haste to investigate this fundamental interaction. In particular, it should differentiate between a well instructed morphogenetic field and a poorly instructed morphogenetic field. Both will induce materialistic changes. The question is, just how poorly instructed does a morphogenetic field have to be in order to produce adverse health effects?

Each biophoton has a twin photon, which has been shown to maintain simultaneous contact with its opposing pair over any distance.

This is a result of quantum mechanics called entanglement. If the Quantec 5, described above as 'a regular radionic system of the old style' bases its operation on entanglement one can only guess at the principles involved in newer devices. Research should start with easier to comprehend devices such as the Quantec and ensure it understands that methodology before moving on to more advanced systems. In the well-known phrase, science should learn to walk before it can run.

This enables remedy collection and distance healing with the Quantec by simply taking a digital photograph of the target, camera included, which is saved 1:1 on the computer hard disk. The theory states that a certain percentage of the biophotons in the photograph will be twin photons so the information and corresponding vibration of any object or client can be collected or transferred by the Quantec.

The fact that only a certain percentage of the biophotons will be twin photons agrees with results obtained from Schrödinger's equation.

Heisenberg's Uncertainty Principle is also involved as it is not certain what the percentage will be.

Treatment with the Quantec: Harmonization of the morphogenetic field by the Quantec is achieved by vibration emission of a remedy on the gold plate or by radionics scalar waves.

Science should make a start at showing that both methods have similar effects.

A database of 50,000 remedies and entries is contained in the Quantec software covering spiritual, mental and physical, e.g. Homoeopathy, colours, sounds, acupuncture, affirmations, allergens, Bach flowers, psycho-kinesiology, etc. Additional remedies can also be added to the Quantec database as the practitioner desires. A therapy plan is produced by the software and distance healing can be scheduled for up to 500 designated targets at a time.

Distance healing of up to 500 designated targets at a time is an obvious advantage over modern medical practice which still insists on dealing with only one target, or patient, at any given time. This improved 'throughput' to use the industry term, will give Doctors more time to catch up with relevant medical literature, write up their case-notes, or simply relax, at home or on the golf course

The Quantec may also be beneficial for plant growth and stimulation as it reduces the need for insecticides, fertilisers etc.

This is equally good news for World Agriculture as well as the more traditional weekend gardener.

The Personal Therapy Device: What is the Personal Therapy Device? The PTD promotes the body's self-healing functions. It works by stimulating our body's frequency-based information system on an electromagnetic level. It is based on the science of how the body is influenced through communication energetically.

Science has suspected that the body's frequency-based information system is not being stimulated on an electromagnetic level nearly as much as it should.

Cell molecules generate electromagnetic frequencies. These frequencies are an important part of the cell to cell communication and guidance of the body's self-regulating mechanisms. When a health condition appears the molecular movements change, and by influencing them with correct frequencies, they are stimulated to re-tune themselves back to their normal state.

Biologists will readily confirm that cells communicate. Molecular scientists have recorded many instances of a health condition changing molecular movements but have not thought of re-tuning them with the correct frequencies. This may now change.

All programs have one or more of the following aims. Stimulate wellness: PTD has programs to support stress reduction, increase mental and physical energy, improve sleeping conditions, weight reduction, detoxification and more.

Stimulating wellness is something which modern medicine has largely ignored. Too often the emphasis is (mis)placed on drugs and surgery.

Problem solving: PTD does not heal or resolve anything directly; it can send information to trigger our own self-healing mechanisms. There are programs to stimulate the body's response systems, both generally and for specific conditions such as the common cold, infection, allergy, injuries, etc.

Even the most optimistic scientists will admit that they are far from sending information to trigger our own self-healing mechanisms.

Stimulates regeneration: Where there is reduced function or lack of vitality in a tissue or an organ the PTD has programs to support the tissue or organ strength. This requires long-term use with daily treatments. The PTD has support programs for specific organs and general programs for systems including digestion, respiration, urinary, musculo-skeletal, skin.

Modern medicine practitioners will be the first to admit that the main thing they lack are programs to support tissue or organ strength. Even the PTD protocol requires long term use with daily treatments so doctors should not be downhearted that they have failed in this respect for so long. It was never going to be easy.

Symptom reduction: Symptoms are reactions our body creates to let us know about a problem so it is important to investigate the cause of a symptom. However, there are times when we have a need for short-term relief, where we can reduce or eliminate symptoms. The PTD has programs for supporting the body response to headaches, muscles pain, inflammation, constipation, depression, hypertension etc. in both chronic and acute situations.

Better and better.

The PTD does not solve health problems, it is not like a medical treatment. It is like a friend that helps support you at times when the body is struggling to cope combined with a tonic that supports the

body's natural processes. It works only on the symptomatic level, it does not address causal factors or disease processes. The PTD makes no physical or chemical intervention and it is non-medical. The programs have been tested for several years and have been shown to work with excellent results.

It is no longer of overriding importance that intervention be made on the physical or chemical level where healing is concerned, as many scientists are coming to accept. Such demands belong to the 'dark ages' of medicine. Addressing the lack of vitality in organs is the exciting new research field.

The device is small, lightweight and elegant. It has the shape, display and rechargeable battery like a cell phone and is even easier to use. You can add up to 50 additional programs which can be variations of the existing programs or new programs with frequencies for specific conditions that may not be in the database. There are innumerable benefits to such a feature.

This is empowering as it makes everyone who uses the device their own Diagnostician.

It comes complete with a sleeve and belt clip, European and American adapters and a full, reader-friendly manual. It does not attach to the body but needs to be held in close proximity while in use. The PTD is manufactured by a leading supplier of high-tech electronic equipment to the Russian Defence Industry where top quality requirements are standard.

There can be no higher recommendation.

SCENAR: Cell repair and pain management through stimulation of the neuropeptides. In principle the SCENAR Self-Controlled Energo Neuro Adaptive Regulation device works as a catalyst on the body's immune system.

Science has been searching for a catalyst on the body's immune system but the path has been obscure.

The SCENAR reads the resistance level of the skin and relays this information to the brain via the skin itself.

It is surprising that evolution did not come up with a method of relaying the resistance level of the skin to the brain but there we are. A missed opportunity perhaps.

This accelerates the body's healing mechanism through stimulation of the neuropeptides in each damaged cell.

This is an unexpected, even startling, advance in Biology.

In addition the SCENAR could be used to treat individuals whose bodies do not repair themselves properly as a result of chronic illness. A classic example of this is suffers of fibromyalgia where SCENAR may act as a regeneration and pain management tool.

Pain management without drugs is the Holy Grail of Anaesthesia. That the same tool can be used to regenerate is an additional bonus.

How the SCENAR works: The SCENAR uses biofeedback, enabling the body to heal itself. The SCENAR sends out a series of signals through the skin and measures the response. Each signal is only sent out when a change is recorded in the electrical properties of the skin in response to the previous signal. Visible reactions include reddening of the skin, numbness and stickiness. The SCENAR will have the feeling of being magnetically dragged, a change in the numerical readout and an increase in the electronic 'clattering' of the device.

Clattering specialists will immediately recognise the principle. Magnetic dragging is only beginning to be understood but the change in the numerical readout will enable the effect to be quantified for the first time.

The body can get accustomed to a stable pathological state which may have been caused by injury, disease or toxicity. The SCENAR initiates the healing process by stimulating the production of regulatory peptides for the body to use where necessary thereby re-establishing the body's natural physiological state.

Regulatory peptides are a completely new field of medicine and whilst known to cell biologists they have been thought to be merely a curiosity, existing in the body but of obscure practical use. The SCENAR scientists however have developed a model by which these peptides can be allocated a specific function by the body to use, as they clearly demonstrate, where necessary.

Neuro and other regulatory peptides are produced by C-fibres, which comprise 85% of all nerves in the body. C-fibres react most readily to SCENAR electro-stimulation and the peptides produced last up to several hours, meaning the healing process will continue long after the treatment is over. The large quantity of neuropeptides and C-fibres in the central nervous system can also result in the treatment on one area aiding with other chemical imbalances, correcting sleeplessness, appetite and behavioural problems.

I have circulated these claims and found that biologists were not aware of the results but found them fascinating. Several intimated that they may alter the direction of their research as a result.

Results from the SCENAR: In Russia, some 600 practitioners currently use the SCENAR as their principal treatment instrument with over 50,000 reported cases of individual use.

These are impressive statistics.

A vast wealth of information on the SCENAR is available from research papers, clinical reports and training manuals. The SCENAR can be used on most types of disease or injury: circulatory, sensory, respiratory, neurological, genito-urinary, musculoskeletal, gastro-intestinal, endocrine, immune and psychological disorders. The SCENAR is also credited with vastly reducing recovery times. Russian athletes have been known not only to compete after serious injuries but even to break world-records post SCENAR therapy.

And this is the clincher. How can we in the West possibly have been in ignorance of this device for so long? The only feasible answer is that Western scientists have been embarrassed by their lack of progress and have agreed to ignore its existence. It is not good enough to simply discount original research and they should be ashamed of themselves.

In Russian accident and emergency wards the SCENAR is used to aid recovery from cardiac arrest, massive trauma and coma. The SCENAR has recently been shown to aid in improving learning ability, memory, sexual function and improved physical health. Finally, trials in Russia have also realised SCENAR's usage for pain management. Both cancer and fracture patients have found more pain relief from the release of natural opioids after SCENAR treatment than from those externally administered.

There have been rumours that Western scientists have been using SCENAR devices to aid their own sexual function. That they have been denying the treatment to their patients is therefore doubly reprehensible.

SCENAR operation: The SCENAR is run over the spine and abdomen or infected area. The software records the resistive response to its signals and returns a fresh signal causing a gentle tingling/stroking sensation. The practitioner is looking for anomalies on the skin surface which may be highlighted by redness, numbness, stickiness or a change in numerical display or sound. Although these areas may not seem to directly relate to the obvious symptoms by treating these 'asymmetries' as the Russians call them with the SCENAR, the healing process will commence.

People who claim that the areas are linked to the symptoms merely demonstrate their ignorance of asymmetries.

Experienced SCENAR practitioners state that though the SCENAR device may be used after the preliminary training it takes up to four years to master. A chronic problem may require up to 6 weeks SCENAR treatment for long-lasting effectiveness. Acute problems may just take one or two treatments. It is reported that SCENAR proves effective in 80% of all cases, of which full recovery occurs in 2/3rds of them, and significant healing in the remainder.

No Western doctor can claim such percentages. Sometimes, trials of new drugs are halted because it is recognised that they are so effective that it is morally wrong to deny them to the control group. It seems to be the case here. Given the undisputed effectiveness statistics it is morally wrong to deny the treatment to patients in Western hospitals. Hundreds, if not thousands of these devices should be purchased without delay and distributed to clinics throughout the Western Hemisphere.

SCENAR overview: Of all the devices reviewed here few demonstrate the rapid treatment capabilities of the SCENAR. Not only is the SCENAR versatile in its range of treatment but it also displays an amazing ability to accelerate the healing process. As a stand-alone treatment system the SCENAR is excellent for alleviation of acute symptoms. SCENAR can also be used in conjunction with other treatment systems such as the NES-pro for the management of more chronic disorders.

And, of course, the NES-pro should be bought as well.

ACUSCEN: Adaptive Compact Universal Self-Correcting Electro Neuro-Stimulator. ACUSCEN starts to generate electric impulses as soon as it touches the skin. These pulses closely resemble the natural impulses of the neural cells. The device continuously analyses the skin conductivity/resistance impendence between the central and peripheral electrodes and adapts its signal to the body's response. This bio-feedback is what distinguishes the ACUSCEN from other electro-therapy devices such as TENS.

Bio-Feedback is increasingly being recognised as widespread in biological systems. Pulses closely resembling the natural impulses of the neural cells is a good start but with extra funding science could make the pulses virtually indistinguishable. The improvement in treatment can easily be imagined.

The body effectively controls the work of the device through changes in the impendence so no two pulses are the same. This prevents development of any tolerance to the influence.

This is sound scientific practice. Tolerance development has long bedevilled antibiotics, to give just one example.

ACUSCEN weighs around 200g and its ergonomic shape makes it comfortable to hold. It operates from 2 re-chargeable AA batteries and has 2 built-in electrodes.

The benefits of miniaturisation are ubiquitous in today's electronic devices.

ACUSCEN effect: ACUSCEN is a fantastic tool for treating acute pain or other disorders or symptoms of recent origin. Fast and permanent pain-relief is observed often especially so in acute cases. Chronic diseases, developing from internal causes, are usually slower to respond and disappear gradually. The treatment of such conditions generally requires more skill from the user.

It is a commonplace in medicine that symptoms of recent origin are the most easily treated.

The ACUSCEN aims to sedate/reinforce/harmonize while decreasing the local variations in the skin impendence which is in close correlation with the state of the relevant tissues and organs. The ACUSCEN signal stimulates the body's own electrical currents, which in turn triggers chemical reactions at the cellular level. The ACUSCEN acts on body's self-healing mechanisms and that is why it is effective for a wide variety of disorders.

Impedance meters should be on hand in health centres so the state of organs can be determined without costly referrals of patients for expensive MRI scans.

Preparing for treatment: Before the treatment the output energy is always adjusted while placing the ACUSCEN device on the skin. The tingling sensation is felt clearly but remain comfortable. It is possible to change the device output over a wide range. Besides the energy other signal parameters such as modulation, frequency, damping and depth can be adjusted.

Science should make it a priority to fully investigate these signal parameters. Depth is a real mystery.

Dose: To calculate the dose the device has to be held stationary at the given point, the simplest way of application of the ACUSCEN. Once the body stops reacting to the stimulus at the given spot the device signals the dose or the end of treatment. The device is moved

to the adjacent area until the next dose signal and so on until all problems or painful areas are covered.

A far simpler technique than the over-complicated CAT scan.

Subjective method: The user may choose to override the automatic dose calculation and to move the ACUSCEN device over the area looking for other signs of treatment progress. One of the signs to look out for is the stickiness of the ACUSCEN device to the skin in places where the problem area is. An almost magnetic pull points to the area to concentrate the treatment upon. Once such an area is identified it can be treated stationary in auto mode where the user is waiting for the dose or in double Var mode, another highly effective mode which allows deep-tissue penetrating.

It has been noted by clinicians that skin can tend to become sticky where problem areas exist. Science has been familiar with auto mode for some time but the innovation of double Var mode promises to revolutionise treatment and should be investigated immediately.

Objective method: The ACUSCEN scan mode available on Professional model ACUSCEN pro+ allows for the precise locating of the problem area and the monitoring of the treatment progress.

Precisely locating the problem area is to be expected from a professional instrument. Purchasers of less accurate versions may wish to upgrade.

The handling of the ACUSCEN device: The handling of the ACUSCEN device is very simple and the operating manual DVD also supplied is often enough to get great results especially for quick pain relief in acute cases. However in the hands of the experienced user ACUSCEN can be a very powerful tool indeed. It is used professionally by trained practitioners for the treatment of many conditions. Therefore we always recommend at the very least 1 or 2 days training. If the device is to be used professionally we recommend regular training updates and lots of hands-on practice. Contact us for more information on training.

This goes without saying. The power of the tool alone makes such investment by any responsible medical provider a given. One or two days training may turn out to be ambitious and it is possible that double that or perhaps more would be appropriate.

The Vega devices – Diagnostic and therapy systems designed to facilitate Naturopaths. The Vega devices have been developed to assist practitioners of Naturopathy in the diagnosis and treatment of

their clients. The general assessment procedure consists of recording electrical parameters of the skin at harness points based on EAV principles and analysis of the results by the specific software systems. The Vega devices currently available are outlined below.

Naturopathy has been largely overlooked by conventional medicine but that is hopefully a thing of the past.

Diagnostic devices: Vega check: The Vega check performs an assessment of the inherent metabolic rhythms of the body and corresponding regulatory processes giving an indication of the energetic state of organs and organ groups, their function and influence.

The energetic state of organs is of obvious importance but the real breakthrough here is the progression to organ groups.

The Vega check measures electrical parameters of 7 segment derivations of the body via 6 electrodes transmitting pulse currents to the feet, hands and head. A disturbance is detected by deviation of electrical parameters from the normal range for each segment. In this way the Vega check gives hints of acute, inflammatory, degenerative or allergic ailments and displays graphically which organ is affected. The Vega check also displays the clients' over all vitality factor, which can be used to trace improvements during a course of therapy.

Modern surgeons now recognise body segment derivation theory to be fundamental. The concept of vitality factor is completely novel and promises to be the new paradigm for medical intervention in the forthcoming decades.

The Vega check is very simple to use and the assessment takes just a few minutes to complete. Vega also provide client management software where the results of screening and recommended treatments can be recorded for future reference.

It is of course imprudent that modern medicine continues to waste valuable time and money on lengthy CAT and MRI scans when the Vega check only takes a few minutes to complete. One can only assume that the manufactures of CAT and MRI machines have somehow conspired to force medical practitioners to use their expensive and time wasting machines in favour of the Vega. Governments should step in at this point and order such scans to be phased out.

VegaTest: The Vega Test Expert is a self-contained computerised electro dermal screening CEDS instrument which can identify a wide range of nutritional and health issues and evaluate specific remedy

requirements. The Vega test can access 1400 digital test ampoules, measure amalgam/tooth voltage, assess the energy balance of body quadrants, energise acupuncture points and produce magnetic strip treatment cards (SI cards). If you do not want to get involved with computers and can live with the limited number of digital tests and manual testing of remedies then the Vega Test Expert represents a good choice.

Medicine should stop diverting money into stem cell research, to pick one obvious white elephant, and instead develop better ways of measuring tooth voltage.

To perform an assessment with the Vega test the client is stimulated with energetic signature patterns stored in the system while a reading is taken with the probe. If a specific stimulus causes a deviation in the readings there may be an 'issue' between the substance being tested and the subject.

Scientists have been slow to catalogue energetic signature patterns. Fortunately, the work has now been done for them.

Therapy devices: VegaMed Matrix: The VegaMed Matrix is designed to alleviate and activate the connective tissue network matrix of the body which supports all cells. It is believed that damage to the matrix is conducive to chronic ailments, allergies, reduced resistance to pathogens, depression and many other complaints.

The connective tissue matrix has been largely overlooked by transplant surgeons who tend to focus to an disturbing degree on organs. But they are missing an important point. The matrix supports all cells, and organs are composed of cells. Therefore organs are supported by the matrix. Transplant surgeons should concentrate on activating the connective tissue network and support all cells, not just the ones which happen to be in the organ they are interested in transplanting.

The VegaMed Matrix integrates a number of matrix regeneration therapies including Petechial suction massage, rhythmic direct-current therapy and Systems Information Therapy (SIT). Via this combination of techniques, the VegaMed Matrix breaks up and disposes of metabolic waste, renews structures worn out by chronic inflammation and reinvigorates the activity of the immune system and connective tissue cells. This detoxification process then enables the body to initiate the healing processes with renewed vigour.

This is impressive enough but science must ensure that rhythmic direct-current therapy is indeed superior to rhythmic alternating-current therapy. The advantages of Systems Information Therapy (SIT) already seem obvious to all except conventional medical practitioners.

The VegaMed Matrix is operated via a large touch screen display which proceeds through the treatment sequence. The therapy is administered by a hand held suction probe which simultaneously delivers the rhythmic direct-current and systems information therapies. VegaMed Matrix therapy programs can follow the basic manufacturers' guidelines or they can be specified for each client.

This is another advance on CAT scans. So far modern science has only found one way of doing these and applies it to everyone.

VegaSelect: The VegaSelect makes use of the principles of Systems Information Therapy (SIT) in the treatment of immune system dysfunctions including allergies, toxification, metabolic disorders and physiological stress. The VegaSelect system also includes a pulsating magnetic field therapy device for the treatment of pain.

Science has long sought a magnetic means of treating pain but has so far concentrated research solely on steady (non-varying) fields. The breakthrough with the introduction of varying magnetic fields was never suspected.

SIT is a vibrational therapy in which specific information is fed to the immune system to stimulate the regulatory process. This is based on the theory that the field associated with pathologically altered tissue exhibits a reduced vibrational fluctuation compared to healthy tissues. A reduced vibrational fluctuation impairs the ability of damaged tissue to adapt to external stimuli which in turn increases the disease potential. The body may come to regard these vibrational changes as 'normal' resulting in a chronic state which is not addressed by the natural immune processes. The impulse emitted by the VegaSelect provides an attention signal to reinitiate the immune system. Combined with the pain management capabilities of pulsed magnetic therapy the VegaSelect is designed to provide rapid alleviation of symptoms as well as treatment of underlying maladies.

Yet again we see vibrations and magnetism used to good effect in treatment. It is surprising that science hasn't devoted more resources to the healing effect of vibrations and magnetism before now.

The VegaSelect can also be used to reproduce and administer the vibrational signals of other holistic treatments e.g. Homeopathics, Bach flowers, precious stones, etc. as desired by the practitioner.

These treatment variations are for advanced practitioners only. Science, once it has finished investigating vibrations and magnetism may feel confident enough to move into these more sophisticated theraputical areas. VegaSelect operators should be on hand to team lead such efforts until scientists have become fully accustomed to the new technology.

Vega Audiocolour: The Vega Audiocolour is designed to strengthen the overall constitution by exposing the client to harmonizing colours and audio tones. This strengthening is said to reduce susceptibility to environmental pollution and increase the clients' capability to deal with life's daily stresses.

Colours have been shone on people and sounds directed at them and their overall constitution appears to have been strengthened. So far this is mainstream, but…

Using the Vega Audiocolour, twelve colours based on Goethe's colour cycle can be mixed across four octaves, either individually or in combination as the practitioner sees fit. The Vega Audiocolour can also be combined effectively with the VegaSelect see above.

This is something completely new. The fact that they can be mixed across four octaves was never suspected by science though there were theoretical grounds for believing that they could be mixed across three.

Vega SI card: The Vega SI card is a storage medium for complex magnetic field information. It is worn on the skin, like a plaster, and delivers system information, remedies etc. to the client between treatments as a supplement to therapies.

This is a further innovation. Science has been concerning itself with trying to develop a storage medium for complex magnetic field information and it seems as if the Vega SI card may well be the answer. Presumably, they can now stop wasting taxpayers money and join forces with Vega to develop the technology jointly.

Vega SI Trans. The Vega SI Trans is used to potentise substances for delivery to the patient via individualised Vega SI cards. The desired potency of each substance can be generated by placing it in the honeycomb of the Vega SI Trans, alternatively test ampules stored in the VegaTest can be duplicated onto a Vega SI card.

This is a technical passage. Suffice to say that science has been attempting to potentise substances for delivery for many decades now without success. Honeycomb technology was never suspected to hold the answer and science went off in many other directions, none being fruitful.

Vega overview: The techniques employed in the Vega devices have been developed through years of dedicated research and undoubtedly have their place in energetic diagnosis and treatment. Vega provide an excellent range of tools for a number of well-established therapies which will undoubtedly be of assistance to any practitioner already familiar with them. However, advancements in the understanding of quantum biology have led to the invention of a new class of device which assesses and treats patients on the most basic quantum level.

While true, this should not negate the powerful influence the Vega devices will have once scientists stop their pointless fascination with neutrinos, for example. The amount of money spent researching these particles, which have next to no effect on us, is disproportionate to say the least. I urge readers to look up the facts on neutrinos and write to their elected representative if they are at all dissatisfied.

Biomeridian Vantage MSAS Professional. What are the Biomeridian devices and Meridian Stress Management MSA? Meridian stress assessment measures the energy at the Meridian points, which directs the practitioner to probable areas of imbalance. The biomeridian technology is used in various health care practices around the world. It gives practitioners more information to consider while trying to bring their client's energetic disturbances back into balance.

Energy measurement is of obvious importance in therapy but only now is its measurement at the Meridian points becoming fully acknowledged.

Biomeridian devices: Vantage brings it all together for the healthcare professional seeking an affordable measurement platform optimized for BioRep. Vantage was developed for practices seeking to incorporate BioRep protocols with leading professional supplement providers as quickly and inexpensively as possible.

BioRep protocols are now widely regarded as the industry standard.

MSAS Professional : Highly flexible MSA platform. Can be used for quick first assessments or for in-depth testing. As your practice moves forward you can add devices like the Focus and the new Epic Probe. Ideal for practices seeking to fully integrate MSA testing into

their office with the revolutionary benefit of the automated point reading system - the Epic Probe.

Science has fallen behind recently because of concentration on the successor to the Hubble Space Telescope. This has left a funding gap as far as medicine is concerned and it is into this gap that MSAS has stepped with their development of the Epic Probe.

Energetic balancing with the MSAS-Professional system is done using bioenergetic evaluation. Meridian Stress Assessment (MSA) is used to conduct a comprehensive evaluation of a person's energetic health and balance. This process involves measuring electrical conductivity at responsive meridian points on the skin, typically on the hands and feet. The locations of the test points generally correspond to those of acupuncture points. These measurements are recorded using the biomeridian to help provide a profile of a patient's present condition.

It is impossible to overstate the importance of a comprehensive evaluation of a person's energetic health. The explanation above mentions electrical conductivity and this is an important scientific term.

According to European medical research, acupuncture points are related to the body's organs and organ systems. Major groups of points are connected through channels, or meridians. Twenty of these meridians begin or end on the hands and feet.

In this respect, European research is well ahead of research in, for example, the United States.

As a patient moves toward or away from health the condition of any particular organ or system can be sensed along the meridians at representative points. As a result, stress associated with the corresponding organs can be surveyed using the indicated points.

Stress Surveyance Point Theory is key here.

After the initial measurements with the biomeridian have been taken and recorded the results can be reviewed. If stress values are above or below equilibrium the system's extensive computer database will allow consideration of a wide range of possibilities that might help the patient regain a healthy balance. The MSAS allows consideration of thousands of herbal, homeopathic, and nutritional products.

Science has not considered whether stress values below equilibrium are significant and has concentrated far too much on those values which are above. Here we see the MSAS protocol leading the way in diagnosis.

Overall, an biomeridian MSAS provides a completely non-invasive method for gaining valuable information about the body's vital functions. The primary objective of this procedure is to disclose patterns of stress and to provide feedback for use in a program to help restore each system and meridian to an appropriate balance.

And again, unsurprisingly, we see the importance of balance.

Rife machine: The beginning of frequency treatment in energetic medicine. The Rife machine was developed by Dr. Royal R. Rife in the 1930s. The Rife machine uses a variable frequency, pulsed radio transmitter to produce mechanical resonance within the cells of the physical body. The Rife machine was, in its time, a pioneering front-runner for what today is the basis of energetic medicine.

Royal Rife discovered he could use specific electro-magnetic frequencies to kill bacteria or viruses without causing damage to the surrounding tissue. The Rife machine utilizes the law of resonance and produces possible health benefits for varied diseases both chronic and infectious. Though the first Rife machines were used on diseases such as tuberculosis, arthritis, and ulcers it is more commonly known for its use on cancer, described by authors such as Barry Lynes as the cancer cure that worked.

Barry Lynes, author and authority on such matters, should know what he is talking about. Scientists, to their shame, appeared to have ignored him.

How Rife works: Rife machines work on the principle of sympathetic resonance which states that if there are two similar objects and one of them is vibrating the other will begin to vibrate as well even if they are not touching. In the same way that a sound wave can induce resonance in a crystal glass and ultra-sound can be used to destroy gall-stones, Dr. Rife's instrument uses sympathetic resonance to physically vibrate the cells of the parasite resulting in possible elimination.

From these examples, it is a very small step to imagine sympathetic resonance vibrating the cells of a parasite to destruction. One wonders why the technique, scientific in its very essence, was never adopted and developed further.

Vibration between two objects can be seen in everyday life, from a tuning fork to a guitar string. The destructive capabilities of resonance have been widely demonstrated, for example when an opera singer hits a particular note and breaks a glass. In this

instance the musical tone sets the glass in motion and as the motion builds the glass shatters. The pulsed wave used in the Rife system produces a mechanical vibration whereby the low amplitude input leads to a large amplitude vibration in the target. If the induced resonant vibration is intense enough, the target cell, tissue, or molecule will be destroyed.

The ability of Opera singers to induce glass shattering via resonance is sufficiently widespread to need no further comment.

Rife treatment: Please note that the Rife machine used today is not the Rife in its original form and no medical claims can be made for this technology. Modified versions such as the Rife-bare device are today's modern incarnations of the original Rife machine. However it should be highlighted that the original machine was not given FDA approval which is partly why it has been modified for today's market.

And, as usual, one can only suspect but of course not prove the possible influence of the invisible hand of pharmaceutical companies.

The Rife machine and related devices claim to help people with a whole range of conditions, the most common being arthritis, rheumatoid arthritis, chronic fatigue and more surprisingly emotional or mental problems. The beneficial results reported are promising and interesting but at this stage may also be considered 'anecdotal'. On the other hand no harmful effects have been reported from the use of this device.

Interestingly, though again not conclusively as I am continually forced to point out, these very same conditions are the subject of massive product placement by guess who. They claim to spend billions of dollars producing drugs which are supposed to benefit people. The only certainty is that they do make literally billions selling such 'cures'.

Rife overview: Rife based devices may help with a variety of diseases. Despite all its uses the Rife machine has little diagnostic ability compared with systems such as the NES-PRO. Its treatment programs may be effective, as case studies have suggested, but Rife machines does not provide the versatility of other, newer devices. Systems such as NES-PRO may enable the recipient to target a wider range of maladies, providing a greater degree of adaptability for the practitioner.

And this is the good news. The NES-PRO designers have built on the good work of Royal Raymond Rife and produced a worthy successor.

Bicom 2000 Bio Resonance Vibration Therapy: The Bicom 2000 biological computer is a sophisticated bio-resonance vibration therapy treatment device incorporating over 400 pre-set therapeutic programmes. Electromagnetic information is taken from the body through any of a number of electrodes, modified and fed back through a magnetic modulation mat in the Bicom, with the effect of counteracting energy imbalances. The Bicom 2000 process enhances regulation and detoxification and stimulates the immune system.

The idea of a magnetic modulation mat has been floated in theory by scientists for a while now but they have never produced a working model.

The Bicom 2000 has a built-in electrode dermal screening EDS device which is used to assess the clients energy balance, nutritional status, toxic load etc. and can also be used to modify the pre-set treatment programmes. The Bicom has an input cup and remedy honeycomb to allow the incorporation of clients tissues and remedies flower, homoeopathic, herbal etc. into the treatment signal. The Bicom 2000 also has a chip imprinting device allowing the treatment frequencies to be recorded on a chip and worn by the client.

As science stands still others move on. The novel EDS machine is an important step forward but that pales into insignificance when entirely new devices such as the input cup and remedy honeycomb are taken into account. These are major advances, unforeseen by science. Toxic load has been discussed before in the literature but never assessed. The chip imprinting device which allows treatment frequencies to be recorded is probably several decades ahead of what science can manage at the moment.

Bicom 2000 is a comprehensive bio-resonance/ EAV device that can be used both in a straightforward manner for standard therapeutic treatments or for very advanced applications. The Bicom allows extensive modifications to treatment programmes and even the development of new treatment programmes.

Modification and development of new treatment programmes is at the heart of medical endeavour.

How the Bicom 2000 device works: A toxin, for example, entering the body, has disturbed frequency patterns which interfere with the body's own regulatory powers and in turn impair the body's functions. Using the Bicom device, frequency patterns, which cause illness, can be transformed into therapeutically effective frequency patterns.

It is common for doctors see toxins disturbing frequency patterns as well as poisoning people. Now, instead of administering an antidote to the toxin he may prefer to give you a therapeutically effective frequency pattern instead.

Biopulsar-Reflexograph - Biofeedback Imaging System: The Biopulsar-Reflexograph is a biomedical measuring device, based on Eastern and Western alternative energy and medical science. The presentation of the body reflex-zones that are read by the hand sensor are displayed in different formats: Dynamic biofeedback graphs of different organs in real time, dynamic total body aura, the organ aura, the chakra activity, etc.

One of the main problems with medicine is that it concentrates on either Eastern or Western modalities. At last we have a device which combines the best of both traditions. Suffice it to say that science has a real problem displaying dynamic biofeedback graphs of different organs in real time and that alone makes investigation of the Biopulsar-Reflexograph an urgent next step.

The Biopulsar's method of measuring is fast, precise, uncomplicated, repeatable, cost effective, and there is no need for the patient to disrobe. Once the client's hand is placed on the sensor it only takes 1 to 2 minutes for the electronic measuring sensors to simultaneously receive the exact vitality including energetic disturbance and sickness structures in more than 49 organ zones.

The measurement of vitality has long been a problem for modern medical science with MRI scans taking up to an hour. Sickness structure scans can take up to twice that time and if all 49 organ zones have to be processed the procedure can take up to a day. No-one likes lying on their back for that long.

In the whole/total holistic therapeutic practice the BioPulsar-Reflexograph® is preferably deployed for: Holistic diagnosis, pulse and constitution analysis, therapy control and medicine testing, sickness prevention, mental and consciousness training, body and intellectual performance enhancement, psychotherapy, as well as holistic dental science etc.

All dentists should consider making their practice holistic. Far too few do and patients are beginning to notice.

The various formats make the BioPulsar-Reflexograph system easy to use. Once the system is connected to your computer all you have to do is place the test subject's left hand on the hand sensor

plate. It only takes a few minutes to acquire an accurate bioenergy scan from the reflex zones which makes it ideal for organ, chakra and biofield response comparison when testing products or therapies.

Information can be recorded and reviewed. Once the screen is frozen the user can switch format modes to access the various organ, colour energetic or chakra information. Biopulsar equipment is easily transported for versatility and convenience.

Science may be able to demonstrate that these therapies are also available when placing the right hand on the hand sensor plate. This may require modification to the analysis protocol but will benefit people who, for one reason or another, cannot use or do not have their left hand. Having missed out on inventing the BioPulsar, the least scientists can do is to offer its advantages to a wider population.

The BioPulsar-Reflexograph system accurately identifies hundreds of colour possibilities giving it the maximum possible usefulness for detailed and exact biofield analysis. The BioPulsar is the first biofeedback imaging technology that can create full colour bioenergy of the full human body electromagnetic energy field utilizing the information from 143 biomedical sensors. In turn, the system then displays the vitality of 50 body parts which include 8 brain parameters. Formats illustrated on screen include organ graphs, colour analysis of the reflex zones, full & half body biofield pictures and chakra analysis. As well, additional supporting software programs are now available.

Measuring the vitality of the brain is of importance to neurosurgeons, amongst others. The main advantage here is that 8 brain parameters can be displayed at once. Even the best MRI scans can only display one parameter at a time.

Mitosan therapy: Re-energizing the body's cells with natural radiation. Mitosan suggest that over 60% of our required daily energy comes from natural radiation with the remainder coming from nutrition. Modern life has meant that we are spending increasing amounts of time indoors, obstructed from the Sun, leaving our cells depleted of energy so their natural regulatory and repair capability is undermined. Mitosan claim that their system can provide the natural radiation to restore full cellular function.

Mitosan may have underestimated the ratios here. Up to date research is tending towards a 70%-30% split between natural radiation and nutrition when it comes to our daily energy requirement.

The Sun produces radiation over the full spectrum of electromagnetic waves. The Mitosan therapy system is able to simulate this natural radiation and direct it at targeted cells. Cells may have been damaged through injury, toxins, disease, radiation etc. causing reductions in physiological energy and deficient cellular functions thereby reducing the body's ability to heal. Mitosan therapy will increase the cell's energy, natural function and metabolism.

This will be welcomed by almost everyone but especially by scientists who spend most of their time in laboratories isolated from the Sun. Modern architecture has tried to compensate with increased use of glass but it is sometimes forgotten that glass screens out ultra-violet radiation, for example, and so does not allow the full spectrum of electromagnetic waves to penetrate and be of use to the body's cells. Scientists who take annual holidays in sunny regions compensate to an extent but those who travel expecting sun but ending up with a cloudy fortnight could benefit from the Mitosan.

How the Mitosan works: The Mitosan therapy system applies measurable energy (not electric current) which penetrates right into the cells. This replenishment of natural radiation can restore physiological energy to the body to help combat ailments. The energy supplied by the Mitosan system consists of complex signals, made up of millions of individual frequencies extending over an extremely wide effective range of 1Hz to over 4 billion Hz.

Science has made a start trying to apply measurable energy to cells but their system consists of simple signals rather than complex ones and the frequency range has been limited. The figure of 4 billion Hz was expected to be reached by scientists only towards the middle of this century.

Mitosan natural radiation therapy: Mitosan therapy is administered through a series of biotrodes placed on the body, outside of light, natural clothing. A typical Mitosan treatment session lasts ten to twenty minutes and results are often seen from anywhere between the first to the tenth treatment. The effectiveness of the Mitosan therapy in these sessions is enhanced by the client's willingness to support the treatment with lifestyle changes which

include drinking at least 2 litres of water a day and taking homeopathic drainage remedies where appropriate.

Scientists should work towards establishing a table of results which show, in increments, just how increasing clothing from light to heavy and from natural to unnatural attenuates the effect of the natural radiation therapy.

Mitosan therapy is reported to be effective in treating the following conditions: Acute and chronic pain, joint and vertebral complaints, wound healing following injury/surgery, acute and chronic inflammations, susceptibility to infection, skin diseases, sleep disorders, chronic fatigue syndrome (CFS) and burnout syndrome. When used for the removal of toxins from the body Mitosan therapy can cause loose bowel movements, increased perspiration and other similar detoxification activities in the excretory organs.

Sporting organisations should ask their own specialists to verify its effectiveness in removing toxins. However, a note of caution should be advised. One imagines the International Olympic Committee, for example, would not wish the therapy to be used directly before events given that detoxification activities in the excretory organs may occur shortly afterwards.

Mitosan overview: In regards to other energetic medicine devices the Mitosan therapy system emulates effects similar to the Q2, insofar as to provide the body with direct 'energy boosts'. However the means by which energy is transferred is completely different. Mitosan therapy uses radiation whilst the Q2 uses an electrical charge conducted through water.

This is an important distinction.

Zappers: F-scan Q2 Syncrometer Perkl light: The Ultimate Zapper - the Ultimate Zapper is one of the most powerful electronic detoxifiers in the world, it uses FDA-approved electroporation therapy to deliver nutrients, herbs and medicine to all cells in the body more efficiently.

Zappers are high on the list for science to investigate and the fact that the Ultimate Zapper has now been made available make that investigation all the more urgent.

Why is it the best Zapper? 100% positive pulses: Twice as long as other zappers. Improvement # 1: Stabilized true square wave. Improvement # 2: Footpads improve zapping results. Adapter advantage # 1: 10.5 Volts dc guaranteed. Adapter advantage # 2:

Undistorted square wave. Low frequency: The best frequency at 2,000 Hertz. Electroporation is FDA-approved: This powerful therapy is also produced by the Ultimate Zapper. Harmonics: Superior to any other Zapper. Tooth Zappicator: Very effective.

These specifications are impressive and some are beyond science at present. For example, science cannot guarantee exactly 10.5 Volts dc, though it can get very close.

How does it work? The Zapper is an amazing electronic device that was invented by Dr. Hulda Clark in 1993. It creates a special kind of wave called a positive offset square wave that is set to one specific frequency only. Even though it is set to only one frequency this unique square wave is able vibrates at all of the frequencies that parasites, bacteria, viruses and fungus vibrate at thereby destroying them.

This is known as the principle of resonance. The square wave is, therefore, a broad-spectrum resonating wave. You don't change the frequency of the Zapper. It is set to one frequency to kill parasites, bacteria, viruses and fungus. The square wave hits all of the frequencies at the same time because of the very fact that it is a square wave. The new improved Ultimate Zapper does this job better at killing pathogens than any other Zapper.

Biologists will confirm that parasites, bacteria, viruses and fungus can vibrate and they are well on the way to confirming the exact frequencies of vibration. So far, their results have shown no significant deviation to those used by the Ultimate Zapper.

Dr. Hulda Clark endorses the use of footpads. She has stated using foot electrodes helps greatly.

This statement is good enough in itself for scientists as it was made by a Doctor.

The Ultimate Zapper is being sold in the USA as a plant growth stimulator, not for use on humans or animals, because the FDA has not yet given approval for the Zapper as a medical device. The Ultimate Zapper is not intended to diagnose, treat, cure or prevent any disease, disorder, pain, injury, deformity or physical or mental condition.

But given its ability to vibrate at all the frequencies necessary, science should make it a matter of serious research to bring the Ultimate Zapper in front of the FDA with a view to approving its use. One wonders why they have not done so already. Perhaps they are in the pay or otherwise under the influence of powerful pharmaceutical companies who would obviously have much to lose.

Scientists should take a long, hard look at themselves before they accept any more hand-outs and hospitality from drug companies. The day will soon come when the science behind the Ultimate Zapper will become more persuasive than the hush-money of large Pharma companies and at that point scientists who have taken backhanders will find themselves running for cover.

F-scan/F-scan2: The new Rife? The F-scan is often called 'The contact pad device with a difference'. Like the Rife, Hulda Clark Zapper, qxci and others, it is a device that uses resonating frequencies to investigate the state of the human body. The F-scan2 detects the frequency at which infections, viruses and other forms of illness resonate then 'zaps' the target at an exact frequency. The greatest advantage of the F-scan is its ability to combine Rife and Hulda Clark technology, enabling the user to target and treat cells without damaging any surrounding tissue.

This is a significant advance and science should not be slow to acknowledge it.

The new F-scan2 version 2005 offers a variety of improvements and options compared to earlier models. The F-scan2 USB is a small, easy-to use and sophisticated frequency synthesizer made and designed in Switzerland. The manufacturer has specifically programmed the F-scan2 with an extensive permanent internal library of Hulda Clark and Royal Rife frequencies. These built-in frequency tables make it very easy and quick to scan using a wide range of user-selectable frequencies in order to detect identical bio-resonant frequency matches.

Miniaturisation is one of the signs of a mature technology and science should recognise this when validating the F-scan. One thinks of clunky (yet workable) early mobile phones and the sleek miniaturised versions available today. Exactly the same development process can be seen with the F-scan.

The F-scan2 USB uses an LCD touch-screen system to operate in a stand-alone mode. No computer is required so it's very easy to use almost anywhere, any time. It only weighs 2 pounds and can be set up and ready to run in just seconds so it is a truly portable, easy-to-take-anywhere device.

Which, again, confirms the above analysis.

How does the F-scan work? As mentioned above the F-scan uses a resonating frequency. A sine wave is digitally created, producing a frequency accurate to 0.01 Hz. This wave does not suffer from any interfering harmonics or disturbances as the F-scan uses a TNOC procedure for frequency synthesis as opposed to the phase locked loop used in other devices. According to the manufacturers this is the only way to ensure a stable signal with regards to the frequency, shape and amplitude of the waves used.

Science has traditionally used a phase locked loop for frequency synthesis so the TNOC procedure is another advance.

As with the NES-pro and qxci, the F-scan uses a standard desktop or laptop pc to receive and process data through a Windows interface. The F-scan can apply the frequency directly to the body or can be used to drive a Rife/bare, thus giving it an advantage over many resonating frequency based devices. The F-scan is able to reproduce the effects of the Rife machine, which works in the frequency range of 1-10,000 Hz, but it can also produce frequencies up to 2,999,999.99 Hz. For frequencies in the Hulda Clark range 60,000 – 80,000 Hz and above, the F-scan is usually used as a contact pad device.

These frequency ranges are almost unheard of, as will be readily confirmed by any scientist competent in the field.

The F-scan works through a touch pad system. The client has two contacts attached to the middle finger of their right hand. In the left hand, the patient holds a cylinder. The F-scan first takes a baseline reading, recording the conductivity and personal standard resonance level. All readings are compared with this and are relayed to the pc where the individual's information can be logged.

For those without such digits science could research attachment to the middle toe. A bigger challenge is the protocol to be followed for people lacking even this. Science should produce experimental results for other appendages which may be used in these circumstances.

The F-scan's functions allow you to automatically scan through a wide range of frequencies. The F-scan recognizes and records resonant frequencies which are then stored in your PC. Stored frequencies correlate to areas of illness on the body and are used to target and eliminate the malady. To determine the frequency band to be tested a personal testing method, such as electro-acupuncture according to Voll, VegaTest, etc. can be used.

The F-scan recognizes and stores up to 80 frequencies in order to zap them at a later date, demonstrating its superior ability compared with machines such as the Rife. However the F-scan is overshadowed by the BodyScan and qxci which are able to record patients' reactions to over 6,000 and 8,000 different compounds respectively in a matter of minutes by interfacing with the body.

As usual in frequency medicine science is playing catch-up. I recommend scientists familiarise themselves with the working of the F-scan before trying the BodyScan and qxci technology. It is a steep learning curve but one which serious scientists should be willing and ready to undertake.

F-scan overview: The F-scan2 USB generates very precise digitally-synthesized frequencies with sine or rectangular wave forms using the most advanced technology. This allows the manufacturer to make the F-scan2 smaller and much less expensive than comparable devices while including an extensive list of features.

All functions are micro-processor controlled. Can generate individual frequencies from 0.01 Hz to 15 million Hz. Pre-programmed for all 235 pathogens and their frequencies identified by Dr. Clark. Pre-programmed with 350 sequences of frequencies based on the research of Royal Rife. Generates very accurate and selectable waveforms. Voltage output from 0 - 25 Volts, depending on output waveform selected. Jogger wheel that allows voltage output adjustment of all square wave signals.

Frequencies can be set from 0.1 - 3,000,000Hz for square wave signals. Frequencies can be set from 0.1 - 15,000,000 Hz for sine wave signals. Individual timer for each frequency in a sequence. Small, light-weight only 2 pounds and easily transportable. Windows PC ready with supplied USB cable. F-scant software included for use with most versions of Windows. The F-scan2 USB has a permanent internal memory that can be used to store application frequencies for future use. It has a memory capacity of 50 data blocks of up to 50 frequency values with assigned parameters for each, plus the results of a DIRP analysis if wanted.

This passage is highly technical and for experts only. Suffice to say that science has failed to approach such specifications despite years of trying. In particular, a memory capacity of 50 data blocks has until now seemed only a distant dream. When allied with up to 50 frequency values with assigned parameters for each (science has only been able to produce unassigned parameters) plus the results of a DIRP analysis, the implications are overwhelming.

F-scan 2 specifications: High-grade aluminium case 8.5 x 6.1 x 5.5 (215 mm x 155 mm x 140 mm). Touch-screen monochrome display - 3.23 x 2.44 (82 mm x 62 mm) - 320 x 240 pixels. 0.1 Hz to 15,000,000 Hz frequency range. Frequency stability: 20ppm. Operating memory stores up to 50 frequencies with individual parameters. Permanent memory bank stores about 6000 values in tables, time functions and limits for sweep and DIRP. Universal ac adapter - input 100-240 vac, 50-60 Hz. Output - 15 V dc, 800mA.

One scientist doubted that a frequency stability of 20ppm would ever be achieved in his lifetime until shown this specification.

Output power port - square wave, positive dc-offset, 14 VPP / 200 mA. Multi-signal output port: Sine wave - positive dc-offset, voltage preset to 10 Volts peak-to-peak. Square full wave - voltage adjustable from 0 to 25 Volts peak-to-peak. Square wave positive dc-offset - voltage adjustable from 0 to 12.5 Volts peak-to-peak. Rect port: square wave positive dc-offset - voltage preset to 5 Volts peak-to-peak. European Union device classification: Medical device class 1, type b, en60601, regulation 93/42 EEC. F-scan2 USB package price - $3,995. Includes everything needed to get started.

This price is reasonable in the extreme.

The F-scan combines frequency ranges from both the Rife and the Hulda Clark to produce an effective hybrid capable of much more than its combined predecessors 0.1Hz - 2,999,999.99Hz. The additional software interface with Windows may allow the practitioner and subject to visually identify potential troubled areas on the body and target them. The F-scan can store and recognize up to eighty frequencies meaning that it can identify up to eighty different conditions which the client may suffer from.

Perhaps a prize similar to the Nobel could be established where the winner identified the most frequencies which could be added to the F-scan and thereby identified the maximum number of different conditions from which a client may suffer.

Q/BEFE/Q2 water energy system: The Q/Q2/BEFE hydro-therapy spa, or as it's more commonly known the Q2 water energy system, is a bio-energetic device designed for home use to promote relaxation and good-health. The Q2 water energy system converts a standard electrical charge into a 'bio-charge' which is passed

through water from the Q2 to the body. The Q2 water module is attached to power unit and placed in a bath or foot bath. The bio-charge supplied by the Q2 water energy system helps re-energise the body's electrical system, rebalance the meridians and increase overall energy.

Science should investigate the mechanics of the transformation of ordinary electric charge into bio-charge. .

How the Q2 water energy system works: The Q2 water module is placed in water with the client for 20 to 35 minutes. The Q2 water energy system feeds 24 Volts of direct current to a series of copper plates, creating the bio-energetic signature which is transmitted to the client.

Early scientific attempts used 240 volts of alternating current but the results weren't good.

Inventor Terry Skrinjar claims that as the body is made up of over 80% water the necessary electrical patterns to adapt a conventional electrical charge into a bio-charge are already present in and can be beneficially received by the client. According to Skrinjar a low bio-charge causes the body to be operating at a low energy level, making the whole body tired and slow.

This seems intuitively correct.

It has obvious implications if the body in question is in a poor state of health where the immune system may be working inefficiently resulting in symptoms of chronic fatigue. Fatigue due to a low bio-charge can manifest in various symptoms such as aches, pains and ineffectual organs resulting in susceptibility to disease.

The bio-charge supplied by the Q2 water energy system causes an increase in the potential of the cell membranes in the body, helping to maintain cellular vitality thus enhancing every cellular function in the body - from brain waves to the immune system.

Cell scientists will confirm that the potential of cell membranes in the body is of high importance. The burgeoning sub-field of cellular vitality has become much more prominent in recent times.

Treatment with the Q2 water energy system: Though the treatment protocol for the Q2 water energy system is aimed at increasing the individual's energy levels, an increased bio-energy has enhancing effects on specific areas of the body as observed through continued use of the Q2 water energy system.

This is important. One-off treatments such as amputations have their place but the continued application of a given therapy is the most effective use of remedial intervention.

Specific improvements through use of the Q2 water energy system include: Faster recovery time from illness or injury, reduced inflammation, revitalized blood, improved endocrine and metabolic function, detoxification and neutralization of toxins, pain and stress relief, improved sleep, reduced fluid retention, elimination of menstrual pain, dermal rejuvenation and improved kidney/liver function.

Most of these symptoms can be alleviated by modern medicine but usually only one at a time. The most impressive achievement, however, is revitalized blood, something which has eluded modern medicine despite decades of research.

Q2 water energy system overview: A number of research studies have been conducted examining the beneficial effects of the Q2 water energy system with some promising outcomes. One of the most interesting results has been significant reductions in negative emotions after the Q2 water energy system foot bath therapy, such as conscious hostility and total depression, when compared to a placebo control group.

Scientists who display conscious hostility to the idea of foot baths improving liver function (and there are still one or two!) now have the answer to diminishing that negative emotion.

Syncrometer – Hulda Clark's frequency analyser: Dr. Hulda Clark developed the Syncrometer as a means for rapid and accurate detection of various substances in the human body. Everything in the Universe vibrates at a particular frequency usually multiple frequencies or a range of frequencies that can be used to both identify a particular substance and, in the case of a biological organism, to kill or devitalize it.

Scientists will confirm that everything in the Universe does indeed vibrate at a particular frequency so from that it seems obvious that frequencies can be used to identify a particular substance.

Based on her research into the causes of human diseases, Dr. Clark has postulated that all diseases are caused by only two things: Parasites or toxins. If you eliminate either or both of these in the human or animal body, for example by using the Syncrometer device, disease can be eliminated.

Viruses and bacteria are old hat these days and in any case, if diseases were caused by anything other than parasites or toxins we would undoubtedly have heard about it by now. But the silence is deafening, and from this we can conclude that Dr. Clark was correct in her analysis.

How does the Syncrometer work? The Syncrometer works on the principle of matching resonance frequencies within the body. The electrical circuit made when the client is connected to the Syncrometer has three parts: 1 The audio oscillator Syncrometer unit, 2 The Syncrometer test plate apparatus, and 3 The client. The test plates form platforms for testing compounds such as flu virus.

When a sample is on the test plate it will be emitting its own resonance frequency and that specific frequency becomes part of the circuit. The practitioner then listens for a resonance sound generated by the Syncrometer audio oscillator when the client is connected. If a resonance frequency is detected from the Syncrometer unit the tested virus is present in the client. If there is no resonance then the client is either not infected by that particular virus or it exists in only a very small quantity.

This appears uncontroversial.

The Syncrometer can detect where in the body the tested compound is concentrated by placing samples of various organs, tissues, etc. on the Syncrometer test plates one at a time to adjust the resonance wave. With a tissue sample on a Syncrometer test plate an audio oscillation confirms the location of the test compound.

One must of course be careful about which sample of organs or tissues are placed. It would be unfortunate if a vital part of the organ was used though tissues are less of a problem as they are distributed in many places around the body.

What the Syncrometer may be used for? The Syncrometer whole body test can be used to detect specific entities such as aflatoxin, streptococcus pneumonia, Epstein-Barre virus, orthophosphotyrosine, mercury, benzene, etc. Such a test is not as sensitive as the Syncrometer organ test but allows you to select those entities most abundant in the body and therefore of special significance.

Science should welcome these results. Though modern medicine is competent in testing for mercury, Epstein-Barre virus detection poses problems so this is a real breakthrough.

The Syncrometer organ test enables you to identify which organs contain a particular entity. For example, the mercury may be in the

kidney, the streptococcus in the joints and so on. This allows you to embark on a clean-up program for your body in a focused way such as improving kidneys or liver, etc. To do this, tissue samples are placed on the Syncrometer test plates and combined into the circuit.

And combining tissue samples into the circuit is a completely novel procedure previously unsuspected by science. This alone makes the Syncrometer a prime target for validation by science and a program should be initiated without delay. A clean-up program for the body focussed on improving kidneys or liver etc. promises to do away with costly transplant surgery at a stroke.

The search for entities can also be pushed to the sub-cellular level using the Syncrometer. For example, heavy metals in the microcosms, lanthanides in the lysosomes or ferritin on the cell surface. Latent forms of viruses can also be detected within chromosomes. This allows monitoring of a virus' presence after experimenting with different kinds of antiviral treatment.

Such treatments have long been a dream of medical science yet despite the billions poured into research results have been patchy. The sub-cellular level is largely a mystery and certainly no results have been published in any reputable journals that I am aware of concerning heavy metals in the microcosms.

Lanthanides in the lysosomes have only begun to be touched upon by researchers though ferritin on cell surfaces have been reported in isolated cases. The real step forward here is the detection of latent forms of viruses within chromosomes. This, it is fair to say, was completely unforeseen by medical professionals.

Using the Syncrometer you may identify and analyze a particular skin site and what is directly under it, for example what is happening inside and under a mole, blemish, painful spot, swelling or discolouration. The above refinements of Syncrometer testing can be applied to client saliva samples. The Syncrometer can also be used to detect entities in products, for example, lead in household water, thulium in reverse osmosis water or asbestos in sugar, if so desired.

Some scientists claim to have detected what is happening inside and under a blemish but these results are isolated and unconfirmed. No-one has ever maintained they know what is happening under a painful spot.

Perkl-light: The Perkl-light is a little different from most other devices in that its operation is more intuitively-based than mentally based. Rather than looking up frequencies in a book or going

through some analytical process to determine a setting, the Perkl-light is best used by simply observing which settings feel good at any particular time.

Science's obsession with analysis has long been seen as its major weakness. In the near future it is expected to become the norm that a patient attached to a machine will be allowed to fine-tune the settings until they feel good. Patient empowerment is predicted to be a major part of the new holistic paradigm of healing. Doctors today could make a start by allowing patients to adjust the controls of heart-lung machines, for example, until they feel comfortable.

The fact that this works at all is a testament to the power of the light-based energy mechanism in the Perkl-light. In an informal study in 2006 at the Tucson Gem and Mineral show it was determined that 83% of non-energy practitioners could feel the energy of the Perkl-light directly with their hands, and that 100% of energy practitioners could also feel the energy directly.

This is powerful and incontrovertible evidence.

So how exactly does one use the Perkl-light and what sort of results should one expect? There are a variety of methods and uses. If one wants a quick energy tune-up one can simply set the device on one of the many groups of preprogramed settings having it balance chakras, organs, meridians, etc. in a matter of a few minutes while you do other things.

In fact, one of the advantages of the Perkl-light is that it is not necessary to take time out of your schedule to use it. It is perfectly acceptable to determine an appropriate setting for yourself and then work on the computer, watch TV, read, wash the dishes, etc. while the Perkl-light works on your energy field. The electromagnetic setting is good to use when you are working on your computer.

I have heard that many scientists are themselves using the Perkl-light so that they can have a quick energy tune-up without taking time out of their schedule. They are reluctant to have me name names but they know who they are.

In addition to giving yourself an energy tune-up you can also work directly on energy blocks related to pain or disease. Several techniques are described in the included DVD for determining settings for just about any type of energy imbalance a person might have. In particular a new form of applied kinesiology called the blink technique is presented. Many people would pay more than the price of the Perkl-light to go to a workshop and learn a new form of

applied kinesiology. With the Perkl-light you get this thrown in as a freebie.

The blink technique is currently the subject of research in some of the most prestigious laboratories in the world and as a result I am prevented from commenting further.

While the Perkl-light is powerful it is also portable and affordable. With the transparent emitter removed it easily fits into a briefcase or woman's purse. There is no complicated set-up, just plug it in and start using it. This makes it an ideal travelling companion. You can use it to energetically clear out a hotel room upon arrival or to help you balance out many of the stressors associated with travel.

There are few things more annoying than finding your hotel room not energetically cleared out upon arrival.

A fraction of the cost of many of the other devices, the Perkl-light is powerful and effective. The technology it uses is not available in any other product at any price. The Perkl-light is a friend of anyone who wants to maintain their bioenergetic health.

And this is of course why further comment cannot be made at this time. The technology is that advanced. However, once science fully validates this technology and a suitable royalty is agreed then the Perkl-light protocol should be available for all interested parties worldwide.

All of the devices mentioned in this section promise to revolutionise medical care. That their inventors continue to be ignored by mainstream science is a travesty. Their urgent investigation should therefore be one of Science's Next Steps.

Chapter 6: Biology

How you rot and rust. It all begins with pH. The pH balance of the human bloodstream is recognized by all medical physiology texts as one of the most important biochemical balances in all of human body chemistry.

This is undoubtedly the case.

Our body pH is very important because pH controls the speed of our body's biochemical reactions. It does this by controlling the speed of enzyme activity as well as the speed that electricity moves through our body. The higher, more alkaline the pH of a substance or solution, the more electrical resistance that substance or solution holds. Therefore, electricity travels slower with higher pH.

It is a failure of communication of modern day science that this isn't more widely known.

All biochemical reactions and electrical life energy are under pH control. If we say something has an acid pH we are saying it is hot and fast. As an example, look at the battery of your car. It's an acid battery. On cold days you want it to be hot and ready and you want your car to start fast. Alkaline pH on the other hand, biochemically speaking, is slow and cool. Compare it to an alkaline battery in a flashlight. You want that battery to be cool and to burn out slowly.

Nothing could be more intuitive than this description.

Here is an example of how pH can control. Look around you at society in general. Do you see people getting exhausted, burned out and quick to anger? Do you see a rise in violence? In part it could be due to the fact that people today lean to an acid pH. As a society we are running hot and fast. How did we get there?

We guzzle coffee for breakfast acid, burgers for lunch acid, wash it down with king size cola acid and have a pizza acid for dinner. pH is under the direct control of what we put into our mouths. Kind of makes sense doesn't it?

It certainly does.

Hippocrates said, let food be your medicine. Let medicine be your food.

But his wisdom should not be taken to extremes. Drug addicts could not be described as having a healthy diet.

Modern medicine has gotten to where it is today in part through a scientific and philosophical debate that culminated in the 19th century. On one side of the debate was French microbiologist Antoine Bechamp. On the other side was French microbiologist Louis Pasteur. Bechamp and Pasteur strongly disagreed in their bacteriological theories. They argued heatedly about who was correct. It was the argument that changed the course of medicine.

And such was their prestige that whoever won the argument would set the course of medicine from then on.

Pasteur promoted a theory of disease that described non-changeable microbes as the primary cause of disease. This is the theory of monomorphism. This theory says that a microorganism is static and unchangeable. It is what it is. Disease is solely caused by microbes or bacteria that invade the body from the outside. This is the germ theory.

Bechamp held the view that microorganisms can go through different stages of development and they can evolve into various growth forms within their life cycle. This is the theory of pleomorphism. He observed microbe like particles in the blood which he called microzymas. These microbes would change shape as individuals became diseased, and for Bechamp, this was the cause of disease; hence disease comes from inside the body. Another scientist of the day, Claude Bernard, entered into the argument and said that it was actually the milieu or the environment that is all important to the disease process. Microbes do change and evolve, but how they do so is a result of the environment or terrain to which they are exposed.

Hence, for Bechamp, microbes, being pleomorphic, will change according to the environment to which they are exposed. Therefore, disease in the body, as a biological process, will develop and manifest dependent upon the state of the internal biological terrain. At the core of that terrain is pH.

Both men acknowledged certain aspects of each other's research, but Pasteur was the stronger, more flamboyant, and more vocal opponent when compared to the quiet Bechamp. Pasteur also came from wealth and had the right family connections. He went to great lengths to disprove Bechamp's view. Pasteur eventually managed to convince the scientific community that his view alone was correct. Bechamp felt that this diverted science down a deplorable road, a road that held only half the truth.

It was very wrong of Pasteur to behave like this and he should have apologised. Using the right family connections and being stronger, more flamboyant and more vocal than your opponent, especially when he is reserved and well-behaved, has no place in science.

On his deathbed Pasteur finally acknowledged Bechamp's work and said Bernard was correct: The microbe is nothing: The terrain is everything.

It was shameful that Pasteur waited until literally the last moment before admitting he was wrong. Presumably his conscience wouldn't allow him to take his falsehood to the grave. It must have been little consolation to Bechamp and Bernard.

It was a 180 degree turnaround. With his death imminently at hand, he as much as admitted his germ theory had flaws. But his admission fell on deaf ears. It was far too late. It could not reverse the inertia of ideas that had already been accepted by mainstream science at that time. Allopathic drug based medicine was firmly entrenched on the road that was paved by Pasteur.

Science should immediately take steps to downgrade Pasteur and reinstate Bechamp and Bernard. The inertia of ideas, which persists to the present day, should be stopped in its tracks by science right now.

The result of that road is what you see today practiced as medicine. When a body is out of balance doctors attempt to put it back into balance, first through drugs, then through surgery. The general effect is to remove the symptoms, not to deal with the ultimate cause of the ailment.

It is impossible to balance a body with unbalanced medicine.

Fortunately there have been and are today scientists who have continued along the other road, the road ignored by Pasteur. They have continued the pleomorphic line of research and think much more about the terrain.

We are fortunate to have these alternative doctors available.

For example, the American Medical Establishment rarely looks at live blood. Their practice of staining blood with chemicals kills it. It also kills the ability to really see what is going on. But in looking at live blood, you can clearly see that there are forms that look like bacteria, microorganisms and parasites that not only are in the blood, but that over time can grow and can change their shape.

Some researchers suggest these forms are markers for pathogenic disease producing states. This ability of microorganisms to change is the concept of pleomorphism we've been discussing.

Understanding this concept is essential to the understanding of cancer and its cure, and the cure of many other diseases.

Obviously, not looking at live blood is a serious mistake and should be corrected forthwith. It is surprising that the practice continues but this is probably due to the inertia of ideas mentioned previously. There are few more conservative bodies of people reluctant to change their ways than the American Medical Establishment (AME).

Looking at live blood under a microscope is an incredible learning tool and begins an incredible journey whereby we come to understand that there are dynamic life processes going on every second in our bodies. It is an environment that is an ever changing canvas of life that holds forms that develop and grow and illustrates what some call the fungus among us.

These forms have been completely missed by the modern scientists so a priority must be to remedy this omission and start to investigate thoroughly the fungus among us.

Darkfield microscopy: Using this kind of microscope technology, German bacteriologist Guenther Enderlein, a student of Bechamp, observed tiny microorganism-like elements which he called Protits. He stated that these tiny elements flourished in the blood cells, in the plasma body fluids, and in the tissues, living in harmony with the body in a symbiotic or mutually beneficial relationship. He considered the protit as one of the body's smallest, organized, biological units.

The most interesting thing about this microorganism is its ability to change and adapt to its environment. It was observed that when there was severe change or deterioration in the body's internal environment mostly noted by changes in pH, these elements would pass through several different stages of cyclic development, advancing from harmless agents to disease producing pathological bacteria or fungi. His book 'The Life Cycle of Bacteria' (Bakterian Cyclogenie) presented his theory. From his research he was able to produce natural biological answers to many of the degenerative disease processes plaguing western civilization today.

If modern-day Biologists had looked properly they would doubtless found the Protits turning into fungi in the blood much earlier. This seems to be a case of seeing only what you expect to see.

Other researchers have continued along a similar path to Enderlein and have promoted their own ideas of these things in the

blood. Gaston Naessens observed the elemental particle which Enderlein called the protit and he described it's life cycle. He called Enderlein's protit a somatid. Naessens believes this protit/somatid predates DNA and carries on genetic activity.

Virginia Livingston-Wheeler also researched these elements and called one supposedly developmental form of it Progenitor Cryptocides, Progenitor meaning it existed through millennia, and Cryptocides being a cellular killer, essentially the ancestral hidden killer, cancer. Like Naessens, Livingston also did cancer research. Some of her research was done along with two other women, Eleanor Alexander-Jackson and Irene Diller. They referred to this microbe as the cancer microbe.

This type of thing makes one wonder what Biologists do with their time. Missing a microbe is like Astronomers missing the Moon.

Here we have similar ideas from different sources, all doing private research and not publishing in known journals. It is unfortunate that many scientists work in isolation and for one reason or another a lot of information known by one is unknown by the others. Because information is not shared, or given hierarchical credit, many who follow are left in the dark and without the full picture.

Yet this is precisely the problem. Science laboratories should 'welcome all' and researchers should not have to work in isolation. Modern laboratories should introduce Open Days where private researchers can use the most up to date tools, perform their own experiments, and then the people in charge of the laboratories should listen while the private researchers outline their results.

Remember that blood is under pH control. Ideally it has a pH in a narrow range around 7.3, which is slightly alkaline. In Enderlein's theory, a pH around 7.3 is the perfect environment in which the element he called the protit lives in harmony with the body. But when blood pH is disturbed and is shifted out of that narrow range, these tiny elements which he thought of as living microorganisms can no longer survive. In order to survive, he suggested they would change to a form which can survive.

It is these new forms that he stated can become aggressive, parasitic and pathogenic agents within the blood. Dr. Enderlein contended there are thousands of forms and many of these are able to overcome the body's defence mechanisms, causing multiple disease situations.

This is standard evolutionary theory. When an organism's environment changes it must change in response if it is to survive.

Darkfield microscopic studies conducted by Dr. Rudolph Alsleben and Dr. Kurt Donsbach of the Santa Monica Hospital clearly illustrated the proliferation of many diverse elemental forms in the blood of their sick patients. What they observed was the dance of these microbial looking forms in an expansive state and increasing with the pathology of their patients. They called it the kleptic microbe.

Examining their patients live blood revealed many of these microbial looking forms darting to and fro in the blood plasma. The more ill the patient, the more forms were observed. The sickest patients had swarming hordes of these forms within the blood, said to be causing great stress to their immune systems. The doctors learned that cleaning the blood of these forms allowed the rejuvenation of the immune system to progress in an orderly and rapid fashion.

This is an important result and one which modern medicine should be capable of building on. Again, it is surprising that the forms have not been noticed by other researchers but perhaps they should go back and look again. Investigating the sickest patients should reveal the swarming hordes which even the most inattentive of observers should not fail to see.

Curious scientists who spend a lot of time in the laboratory looking at live blood under the microscope often start to wonder about the pleomorphic concept. When they see the changes in the blood taking place and correlate it with the progression of the disease process, many begin to see a pattern unfolding. This has prompted some to state that the over-acidification of the body, caused by an inverted way of eating and living, causes a proliferation of the fungus among us which debilitates the body and, if not corrected, will ultimately cause our demise.

Mycologists confirm that fungus will indeed debilitate the body and if not corrected will ultimately cause our demise. Acidification experts are happy to state that over-acidification of the body could lead to fungus proliferation. So acid-fungus theory is supported by sound science.

Looked at in this light it could be said that all illness is but this one constitutional disease, the result is mycotoxicoses, toxicity caused by mycotic infection, or in other words, by a yeast and fungus infection. These are the great decomposers of living and dead bodies.

From ashes to ashes and dust to dust, this is nature's decomposing mechanism at work.

Just as scientists seek to unify all forces in the so-called GUT, or Grand Unification Theory, which is the holy grail of Physics, so do Biologists seek to unify all diseases in the so-called GIT, or Grand Illness Theory. This research is an important step forward in GIT theory as it promises to reduce all disease to one infection via yeast and fungus. It will also place mushroom scientists in the vanguard of pre-emptive medicine.

If you leave a bowl of milk out on the kitchen table for a few days without refrigeration, it will turn sour fairly quickly. Did it turn sour because there was an outside germ that got into the milk? Probably not. It turned sour because tiny microbes already in the milk changed their form to adapt to a changed environment.

Germ theory is becoming increasingly redundant today. Many clinicians are coming round to the idea that microbes already in milk are changing their form to adapt to the changed environment, in this case being out of the fridge.

The disease paradigm shift: One school of thought in modern medicine says most disease is caused by germs or some form of static, disease-causing microbe, the germ theory. In order to get well you should kill the germs. Kill the microbes. Kill whatever is making you sick. Drugs, antibiotics, chemotherapy, radiation, surgery.

These techniques are coming under increasing pressure to justify themselves.

The other school of thought which encompasses most other forms of the healing arts unrelated to mainstream medicine says most disease is caused by some unbalance in the body. The unbalance occurs in some nutritional, electrical, structural, toxicological or biological equation. In order to get well you need to re-establish balance in your body by working with your body, not against it.

Balance is being recognised as important above all else in medicine and instead of killing microbes medical science is moving towards rebalancing the conditions in which microbes live. This makes them far happier with their environment and therefore unlikely to cause disease.

The biology of disease: Colloids are particles that measure .01 to .0001 microns in diameter, that's about 4 hundred thousandths to 4 millionths of an inch. It has been hypothesized by some that there is

some point in space and time where the colloids of life, the smallest of particles capable of expressing biological life in the physical realm, were formed from colloids of light .

To this day no scientist has disproved the hypothesis.

The first individual to maybe catch a glimpse of this occurrence was Anton Leeuwenhoek, who lived in the 17th Century. He had ground glass to create the first microscope lens. In observing some rainwater he collected, he made note that there were tiny creatures moving about. Wondering where they came from he did an experiment. He collected clean fresh rainwater and sealed it in pipettes. At first nothing was in the water. A few days later still nothing was in the water. But on the fourth day, all of a sudden, little creatures appeared. Where did they come from? It was spontaneous generation. Life out of light.

And this has also never been disproved.

To have life you must have procreation, a mother-father union. Since there was no mother or father that created Leeuwenhoek's tiny creatures, his observations were surely flawed and they were dismissed. What could not be dismissed, however, was the observation of a newly discovered microscopic world. It was a foundation for developing the beginning ideas of the germ theory.

But what the germ theory failed to explain then, and fails to explain to this day, is the answer to the question, from where exactly do germs come? Where is the mother-father microbe? In any textbook of science, medicine, or biology, there is no explanation.

It is the chicken and egg paradox. Just as today science has no explanation for the paradox of which came first, the chicken or the egg, so modern day germ theorists fail to answer the question where is the mother-father microbe? It is the elephant in the room for modern biology yet all we hear is a deafening silence. Why is it so difficult for modern biology to say exactly which came first, the chicken or the egg?

It is simply not good enough that a so-called 'mature science' should be incapable of answering a question which an intelligent child could pose. If it cannot answer the chicken/egg question then why should we believe it when it comes to germ theory or anything else? This is very much to the detriment of modern Biologists and they should be ashamed of themselves. If money is the problem a team of Biologists should form and apply for a grant but my suspicion is that they know that the problem is intractable under their current paradigm.

Germs and microbes are physical life forms. Life forms which have evolved from something. Since that something is not physically measurable, then it must be something that is on a higher vibrational level.

This is a sound scientific argument. Life forms must have evolved from something.

Hence, colloids of light. For an empirical scientist, speaking about colloids of light is akin to speaking mumbo-jumbo. How can there possibly be a higher vibrational particle which is unseen and unmeasurable? And how can a supposed colloid of light become a colloid of life?

Yet modern day science has an exact analogy. The Higgs Boson was unseen and thought to be unmeasurable yet it has been seen and measured at CERN. From that example it is hardly mumbo-jumbo to propose a higher vibrational particle of existence such as the colloid of light. Given the progress of particle physics and the money poured into it one might reasonably expect the colloid of light to be discovered any day now.

Regarding the second question, how can colloids of light become colloids of life, doing an experiment can help find an answer. From the writings of Dr. Kurt Donsbach, he calls this experiment making protozoa. The protozoa is among the most primitive and simplest of life forms. In any biology textbook you'll never find a description of where protozoa come from, yet you can create them in a test tube.

Having searched the relevant literature I also have never found a description of where protozoa come from. This raises the suspicion that biologists do not in fact know where protozoa come from.

If you take sterile water and put in some fresh hay or other grasses, then mix it up, you will have a solution that, upon microscopic examination, has nothing in it. You can scrape the blades of grass with a knife and observe the scrapings and you will still find nothing. But cork the tube, wait a few days and come back. Your mix will be teeming with bacteria, amoebas and protozoa. Where did they come from? Under time-lapse photography, you would observe an amazing transformation.

The grass blades would lose their striations and become more vesicular filled with little bubbles or vacuoles. The vacuoles would begin to merge and gradually form a common membrane. After a few days the little mass begins to move with a rhythmic, pulsing

motion. **Eventually the pulsing motion becomes more pronounced and the glob appears to gather more energy.**

Soon it breaks away from the grassy shaft and is a living mass, classified in biology texts as protozoa. From this point it can differentiate itself and other microorganisms appear. Fascinating isn't it? Just what was that pre-protozoan mass pulsing with? Is it the beginnings of life? Could it be what's called the life force, or prana, chi, or eck.

The central mystery remains. Just what was the pre-protozoan mass pulsing with? The pulsing has been observed many times and the most popular explanation has been that it is indeed the life force which is responsible. Prana has been discounted but eck has yet to be investigated. Chi is a still possibility but the odds are stacked against it.

It is a recurring suggestion by some that just as in our protozoan experiment, when turning to the blood, we find colloids of life and those colloids of life have an urge to merge. How they merge, what they turn into, their developmental function, all will be dependent upon the terrain or the environment to which they are exposed. Voila, we've just uncovered the pleomorphic theory. Microbes change based upon the environment in which they live.

Given the evidence presented, it has claims to be much more than just a theory.

The human body strives to maintain the pH of the blood at around 7.3. Above or below this level, the colloids in your blood merge into forms that can look kind of scary when viewed under a microscope. One contention is that these forms constitute pathogenic microbes.

Another more likely scenario based on the latest DNA studies discussed later is that these forms seen in the blood are not really life forms with their own DNA, but are forms that develop in synchronicity with pathogenic situations that develop as a result of internal terrain modifiers which also affect immune function.

The fact that it is possible to have things in your blood that can look kind of scary should certainly tell you something. The second paragraph belongs in the specialist literature but is worrying nonetheless.

In essence, the vast majority of scary forms seen in the blood are re-combinations of cellular breakdown by-product which in itself is not necessarily a good thing because if you are seeing it in the blood, what does it mean about your internal and overall health?

What indeed.

An interesting case history: This story involves how a patient came about getting her cancer into remission. She was diagnosed as inoperable and her medical doctor told her because of her very poor state of health there was nothing more they could do and she had maybe 30 days to live.

This heartless attitude is, unfortunately, typical of medical doctors.

At the time her husband had been seeing an alternative care practitioner for a problem of his own. In his practice, this particular doctor happened to work with a microscope for patient education. As a last ditch effort the man brought his wife in, hoping that maybe something, anything, could be done. As they sat in the office she had such low energy she could barely keep her head held up. Her face was completely white.

Blood was taken from her finger and put under the microscope, then the doctor peered into the eyepiece. What he saw, or more to the point, what he didn't see, astonished him. The woman had practically no red blood cells, and those she did have didn't look very good. There were microbial looking forms all over the place.

It is just as well her husband took her along to an alternative care practitioner. Anecdotally, far too many traditional doctors, when presented with a patient whose face is completely white, fail to diagnose that he/she has practically no red blood cells. Even the minority who do diagnose properly frequently fail to see that those red blood cells he/she does have often don't look very good.

Well, the doctor said he didn't know if it would help but he pulled four key nutritional substances off his shelf and gave them to her. These were: A total vitamin and mineral complex, digestive enzymes, a proanthocyanadin pycnogenol antioxidant compound and heavy duty metabolic body enzymes. After a little bit of talking about these substances and nutritional concepts, the couple left the office. When the doctor peered into the microscope after they left, which was about 30 to 40 minutes later, he couldn't believe his eyes. The parasitic looking creature elements were everywhere, consuming everything left in the blood.

The prescribing of a proanthocyanadin pycnogenol antioxidant compound is key here and though medical science has tried this it has never before allied it to heavy duty metabolic body enzymes.

One week after taking the nutritional supplements the woman had renewed energy and was actually feeling better. She went back

to her traditional doctor and wanted to have a blood test. The test was performed and her red blood cell count had shot up dramatically. The doctor was rubbing his hands together thinking that now they could re-start her on something like chemotherapy. She said something to the effect of no thanks doc, I just wanted another opinion, adios. Since the hospital had essentially given her up for dead, she didn't want anything to do with them or with their killer medicines.

A wise decision.

With her renewed energy and new hope she started seriously looking at alternative treatment options. She settled on a course of action and found what she wanted south of the border in Mexico. 90 days later she was doing fantastically well, with her cancer in remission.

Mexico is fast becoming the world leader in alternative treatment options and this is something which traditional doctors are keen to hide. Why this should be the case is obviously a moot question though some may say that the success of alternative medicine south of the border is having an impact on medical treatment north of the border in the United States. A financial impact.

Human blood stays in a very narrow pH range right around 7.3. Below or above this range means symptoms and disease. When pH goes off, microbial looking forms in the blood can change shape, mutate, mirror pathogenicity and grow. When pH goes off, enzymes that are constructive can become destructive. When pH goes off, oxygen delivery to cells suffers.

A word about oxygen: More and more research is showing that low oxygen delivery to cells is a major factor in most if not all degenerative conditions. To recall how important oxygen is to your life, just stop breathing for a minute. Get the idea?

This is an illuminating exercise. I urge you to try it.

Each cell in your body can breathe fully or not. Which it is depends upon having an optimum pH balance. Do you think keeping an eye on your body pH might be important in your life? pH controls the things you can't live without. Like your brain. Your brain needs fuel to run and the fuel it uses is glucose. But unlike other cells, your brain can't store glucose. It depends on the second to second supply from the bloodstream, a bloodstream that is affected by pH, which controls the efficiency of insulin, which allows sugar to enter into cells which in turn controls blood sugar levels.

Cell biologists are finally becoming aware that each cell in the body can breathe fully or not. From this much else follows.

Your heart. William Philpott MD in his 'Biomagnetic Handbook' made an important body pH/electrical connection. As the pH of the blood goes more acid, fatty acids which are normally electro-magnetically charged on the negative side switch to positive and automatically are attracted to and begin to stick to the walls of arteries which are electro-magnetically charged on the negative side. As science states, opposites attract. It should start to make sense that a society which over-emphasizes food that could push blood to be more acid will have a high rate of heart disease. And so it goes.

And so it goes indeed. The statistics speak for themselves.

Mineral assimilation is affected by pH. Minerals have different pH levels at which they can be assimilated into the body. Minerals on the lower end of the atomic scale can be assimilated in a wider pH range and minerals higher up on the scale require a narrower and narrower pH range in order to be assimilated by the body. For example, sodium and magnesium have wide pH assimilation ranges. It narrows somewhat for calcium and potassium. Narrows more for manganese and iron. More for zinc and copper. More for iodine.

Mineral assimilation scientists have never disputed these findings. Fortunately, at the highest end of the atomic scale, Uranium requires an incredibly narrow pH range in order to be assimilated by the body. This explains why Uranium miners are at very low risk of assimilating Uranium into their bodies because their pH levels are unlikely to lie within such a narrow range.

Iodine, which is high up on the atomic scale, requires near perfect pH for its assimilation into the body. Iodine, you may know, is one of the most important minerals for proper functioning of the thyroid. But the thyroid doesn't get access to iodine unless the body pH is near perfect.

Uranium rejection is a boon; iodine rejection is not optimal.

With a society in a largely pH unbalanced state one would suspect a lot of thyroid problems. Malfunctioning thyroids have been connected to arthritis, heart attacks, diabetes, cancer, depression, overweight, fatigue and more. Are you starting to see the basic metabolic picture evolving here?

Everyone reading this can probably think of someone who has a malfunctioning thyroid or if not almost certainly knows someone who has arthritis, cancer, depression, is overweight or fatigued and so on. That adds up to a lot of thyroid problems.

Body mineral content and balances control the quantity of electricity in our bodies. The speed at which the electricity flows is controlled by the body's pH balance.

Any competent electrician will confirm this.

Minerals and their deficiencies have been implicated in a wide range of off-balance health conditions. Here are some examples: Supplementing a diet with sufficient chromium and vanadium can help prevent diabetes and has been seen to reverse diabetes in those already diabetic, as vanadium is reportedly able to replace insulin in some cases.

Diabetes is a growing problem in the modern world and science should move immediately to confirm that vanadium can replace insulin. Vanadium can be taken in tablet form which would be an immediate improvement from diabetic patients having to inject themselves with insulin.

Copper deficiency is implicated in aneurysms, brain, aortic, etc. Magnesium is quite possibly the most important mineral for the reduction of coronary heart disease. The latest cutting edge research shows that heart disease is really a function of heart muscle acidosis.

Copper could be added to the diet simply by using copper pans to cook food. Magnesium is a little more tricky but given further research there is no reason why copper pans could not have a trace layer of magnesium added as well. Cardiac surgeons worldwide will be interested in the acidosis elucidation of heart disease.

Boron helps keep calcium in the bones, helps women preserve and make oestrogen and helps men keep testosterone. Boron affects alertness. Boron can help eliminate arthritis.

It is well known that people without arthritis are more alert so this makes good sense.

Potassium and magnesium along with organic sodium are some of the most important minerals for rebalancing the electrical properties of the cell, for eliminating excess acidity and for helping to balance calcium.

Science has determined that calcium may well need balancing and whilst sodium has been used, clinical trials have proved inconclusive. This is probably due to inorganic sodium being used. Science should now move forward and use organic sodium instead.

Magnesium helps conduct electrical messages between all the neurons of the body. People get irrational when potassium levels are low.

Zinc is involved in over 200 brain enzyme interactions. Drinking zinc mixed with distilled water can stop anorexia nervosa in a day. Zinc deficiency symptoms include loss of taste and smell. Zinc deficiency in children results in moodiness, depression, irritability, photophobia, light sensitivity, antagonism, temper tantrums and learning problems.

Anorexia nervosa has proved particularly intractable to scientific research but this is largely because scientists appear to have been looking in the wrong place. In my search of the literature, nowhere does it say that drinking zinc mixed with distilled water is the cure. Science should move to correct this omission immediately. Also, every parent will be familiar with the symptoms noted above in their children. For those unenlightened enough not to be dosing their children with zinc, science should immediately move to reassure them.

Zinc is needed to balance cadmium. Too much copper is an irritant to the brain. A story is told by Dr. Alex Schauss, a noted author, researcher, and nutritional mineral expert. It is about his experience with a 9 year old boy brought into his clinic some years ago. The boy had been charged with attempted murder.

His criminal record began at age 6. He burned animals, shot at people's houses and beat up mothers pushing baby strollers. The police all said he would be a lifetime criminal, a Charles Manson type of psychotic. He was on six psychiatric drugs and was kicked out of school after he tried to kill a 10 year old girl.

Dr. Schauss did a hair mineral analysis and discovered his copper levels were off the charts. He added supplemental zinc to the boy's diet to chelate out the excess copper and within two weeks the boys urinalysis showed all the excess copper had been eliminated. He went off all medication, returned to school and became a model student. Years later the boy returned to see Dr. Schauss. He was a junior in college, on the varsity basketball team and had a heart of gold.

Obviously, his heart was not literally made of gold!

High manganese levels show statistically high correlation with violent behaviour, while lithium balances and helps control manganese. The cities of the world with the highest lithium concentration in their water show the lowest homicide rates.

Readers are invited to confirm these statistics for themselves.

The trace element rubidium cures manic depression.

This has long been suspected.

The right ratio of copper to zinc in the cell acts as an antioxidant. This information shows just a tiny fraction of how minerals and mineral imbalances can affect your health. Much of this information is buried in professional journals, there for the taking. It appears that due to politics and the influence and strength that the medical/drug industrial complex has over the suppression of information, these things stay buried.

We have come across this type of behaviour before so it should come as no surprise. It is of course the M-question. Who, exactly, is making Money from keeping this information scarce? I shall leave it to the readers imagination to supply the answer.

If this type of information, along with the other things we know, could be assimilated into our society, whether through the efforts of individuals or that of our government, and if people like doctors, psychiatrists, and dieticians were to act on it, we could lessen violence in our society, close jails, raise academic achievement and greatly reduce outlays of public money for Medicare and Medicaid. We could see our health insurance premiums drop to about $50 dollars a month for a family of four because we could eliminate our need for expensive hospital visits and treatments excepting emergency care for accidents.

Exactly. And I would expect one of science's next steps to be to validate and disseminate this type of information with urgency. If scientists do not start to do this with alacrity one could well suspect them of being a large part of the answer to the M-question. Scientists should look at themselves in the mirror and ask: Do I really want that question asked of me? Then they should take steps to reallocate their research grant in favour of showing that rubidium cures manic depression.

Without a doubt, the single most important thing you can do for your health is to supplement your diet with broad spectrum trace minerals. They are that important.

To be on the safe side, before science sluggishly gets around to confirming these findings, anyone sensible should visit their local health shop or order mineral supplement pills from the Internet or invest in copper pans with boron/magnesium/rubidium coatings where available.

Peering into the microscope and looking at live blood you can see a reflection of cause and effect. When you're not feeling well, your blood doesn't look good. Often, the worse you feel, the worse it looks. When you get better the blood also looks better.

This seems very reasonable.

I recall someone asking at a seminar if anyone had ever had a blood transfusion and then felt different afterwards. One gentleman said he had had a few transfusions and he definitely knew it when he had received his sister's blood because while in the hospital he had the urge to get up and start cleaning. He also knew it when he got his brother-in-law's blood because all he felt like doing was sitting around and watching TV.

A humorous story maybe, but it holds some truth. In Australia, stories are told of the aboriginal people donating blood as urban life encroaches upon their territory. When normal white folk would get this blood during a transfusion some have been known to wake up in the middle of the night sweating and grabbing the sides of their beds as they experienced night dreams beyond any they've ever had before.

Almost everyone will be familiar with evidence such as this. Science has no need to confirm these facts but it should note that the majority of the reports are for normal white folk. Science should investigate reverse transfusions to see if the findings are confirmed. Do aboriginal recipients have the urge to wax their surfboards, for example.

Similar stories are told of individuals who have had organ transplants. Having never liked certain foods, or doing certain things, or being proficient at specific tasks, they would suddenly find themselves after the transplant with cravings for food they hate and abilities they never before possessed.

Again, this evidence is so well-attested there seems little room for debate.

Why? The tissues of the body are fed by blood which contains the vibrational imprint of who we are. Get someone else's blood or other organ tissue and you have to assimilate their imprint and remake it your own. Like a giant human organic tape recorder, over time you erase their message with your own. When people have problems with transfusions and organ transplants, part of the problem lies in the vibrational makeup of both the donor and donee.

This is an attractive theory and certainly fits the evidence.

In time, modern medicine may get around to understanding this and they will discover how to electromagnetically and/or in other ways erase the donor's imprint on blood and organ tissue. About that time maybe they'll also understand more about the microbial forms in the blood and how blood may need to be assessed and treated at our blood banks. As you get deeper into this work and research its topics, a new way of looking at health will undoubtedly begin to unfold for you.

As it should unfold for scientists if only they could be motivated to stop finding out how spherical an electron is and spend their research money on something useful. Perhaps a petition could be mounted to shake scientists out of their inertia and get them to investigate how to erase the donor's imprint on blood and organ tissue, something of far more relevance and importance to millions of people than knowing how many varieties of neutrino there are and how they manage to change into each-other so often, something which exercises the minds of far too many scientists.

The biological aspect is the pleomorphic behaviour of the wiggly things in the blood. These are the microbial looking elemental forms that exist in your blood and will shape themselves according to your metabolic balance. Their forms, and ultimately their function, are going to be driven and decided by the environment in which they live, and which you have provided through your eating, thinking, and living.

Your eating and living has long been recognised by science as precursors to the behaviour of the wiggly things. What's new is that your thinking can also affect their behaviour. This could open up an entirely new field of medicine.

Differentiation between the internal and external parasites: According to the theories of Enderlein, the colloids of life in your blood i.e. Protits develop according to the terrain of the blood. At some stages of their development he said that they are outright pathogenic and parasitic and constitute the real fungus among us. He believed these to be our true, forever with us, internal parasites.

These are to be contrasted with the external parasites which also are, alas, forever with us. The external fungus among us would include such infections as athletes foot or verrucae.

Professor Enderlein called these parasite looking forms Endobionts from the Greek endon = internal and bios = life. He said we can never separate ourselves from them. We co-exist in a

mutually symbiotic relationship. We give them a vehicle for life, they give us blood forms like platelets without which we couldn't exist. The endobiont appears in all mammalian species and has shown evidence through some of its developmental forms to be of a plant nature. Our symbiotic union with them evidently occurred millions of years ago as our species grew into existence. Without some blood clotting mechanism in place, mammals could have never evolved.

Science has yet to fully confirm that platelets are produced by parasite looking forms that show evidence of being a plant, but that is because science hasn't been looking hard enough. Further research is expected to confirm the findings of Professor Enderlein in the near future.

Wow. Now that is some theory.

It certainly does appear to be ahead of its time. In fact, more than one researcher has gone on record as stating that the theory is on a par with Darwin's Theory of Evolution but has the advantage of greater predictive power.

Chapter 7: Diode

White noise: The interface to the consciousness. There is no doubt that, scientifically, a way to breakthrough into consciousness must come in the near future. But before that it is much more likely that we will be able to access the unconscious. This is because the unconscious is unlikely to try to block interference which is how it would see direct probing from the outside. Also the conscious mind is the one we have access to. It is structured and contains definite thoughts about specific things.

The unconscious, on the other hand, by its very definition is likely to be a place of unstructured thoughts, a region to which we have no direct access but which nevertheless swirls with abundant incoherent thoughts just waiting for consciousness to grasp hold of them and bring them to view in our mind's eye.

So it makes sense that the first electrical device to make contact with the unconscious should be a diode producing at first white noise. This is how it would interface at first with the unconscious mind. Following that we would expect to be able to tune it in just like a radio tunes into a station once radio waves have been identified as coming from it.

This makes so much sense that no scientist has ever opposed it.

The diode that generates white noise can be visualized as the television or radio quartz valves used earlier. The noise is called white because it appears on a black screen as flickering white dots, rather like what you see on a TV screen after the end of transmission time. The noise of the Quantec diode itself can be compared to a radio that is not tuned into any station: It alternates between deep rumbling, high whistles and then medium-frequency hissing, completely at random and with no recognizable pattern.

However, when this noise is connected with the consciousness, it changes. These changes can be processed and interpreted by a computer. Quantec uses this interaction with the consciousness by working with white noise to search its databases for relevant entries e.g. Affirmations, acupuncture, Bach blossoms, homeopathy, colours, etc.

Connection with the consciousness is very much an on-going research programme for science and it should not be put off that someone else has got there first. Now the challenge for science is to extend the database. Affirmations and Homeopathy have already been covered as have Bach Blossoms but science should be capable of providing at least a few more items for inclusion.

The diode as interface between consciousness and machine has been studied for over thirty years - with unambiguous results. Countless series of tests on people and animals have proved that diodes with white noise can be used as an interface between different types of consciousness and computers.

No scientist seriously disputes these tests.

The access logic of the diode: White noise has quantum-physical properties. For this reason, an explanation of the access logic with which Quantec works that is simple for ordinary people to understand is only possible here in the form of an analogy. For experts too, the linking of white noise to a computer is so complicated that Quantec is the only device on the world market that can handle this technology.

So the challenge for science is to train more experts of their own, and the sooner they do that the sooner we can all have the benefits of Bach Blossoms, to choose one example.

The analogy: Let us suppose we are trying to find the ideal colour for energizing and harmonizing a client. A computer database containing all test colours is linked to white noise. If five colours are to be tested the noise spectrum of the diode is divided into five equal-sized frequency bands; these bands are then automatically numbered by the Quantec software and assigned to the five colours in the database.

Colour harmonizing and its use for energizing a client is an area of research which has been sadly neglected by modern day science. Budget cuts haven't helped but it really is too bad that science puts all of its efforts into sequencing the human genome, for example, while fundamental research into the ideal colour for energizing and harmonizing clients remains stalled.

When the consciousness is now connected to the noise, the noise which was previously distributed across all frequency bands is now in one range more than in all others. The Quantec software now just has to find out which colour was assigned to the corresponding frequency band and, lo and behold, the unconsciousness has picked out the required colour all by itself...

This software must be amongst the most sophisticated in the world at present so cutting-edge code writers and programmers might like to become involved at this point.

Meanwhile, the great success of Quantec in instrumental bio-communication has also interested our competitors. But since Quantec has the only equipment worldwide capable of handling the technology of the diode with white noise other manufacturers are trying to make do with so-called deterministic pseudo random generators as found at no extra price on every computer. At the University of Princeton two studies have been running for years to find out whether these pseudo random generators can produce the same results as white noise. The result of both studies leaves no room for doubt: Without the diode, nothing is possible.

Princeton is one of the most prestigious Universities in the world.

Study 1: 1997, duration 12 years: The final sentence states that deterministic pseudo random generators are incapable of interaction with consciousness.

Study 2: 2000: A central paragraph states that the possibilities of interaction between consciousness and machine shown in research can only be achieved with genuine noise sources non-deterministic, but in no way with pseudo random generators.

This amounts to convincing confirmation. Princeton was so impressed it went on to set up a special laboratory:

The power of consciousness: Anomalies are events that occur in practice but according to scientific theory simply cannot occur. In the USA, a special laboratory has been founded for investigating such anomalies, the Princeton Engineering Anomalies Research laboratory or PEAR.

This yielded immediate results:

After Princeton researchers had showed that it is possible to divert falling balls by the power of thought...

Not a widely published result for the obvious reasons...

...they turned to conducting research into white noise.

This was considered to be less controversial. An extract from their final report is below:

'After proving the power of thought over matter, they wanted to find out if and to what extent people can use white noise as an interface to a computer. To test this hypothesis students were placed in front of a computer that, controlled by white noise, constantly made plus or minus decisions. The result of the years-long study: If

people think plus while a computer controlled by the white noise of a diode has to decide between plus and minus, the proportion of plus decisions rises with statistical significance.'

An impressively noteworthy result.

Global consciousness: In Princeton, Roger Nelson discovered that the white noise of the diode also reacts to group consciousness. In one test he took his diodes to a concert hall and noticed remarkable changes there during the concert. He then set up about 75 diodes around the world and proved that events that affect people emotionally have an influence on white noise. The first event, during which he received simultaneous, synchronous measurement results in Europe and the USA, was the day Lady Diana was buried. Since then he has taken measurements for many events, of which September 11, 2001 returned probably the most astounding values. The data of all diodes, placed all over the world, was transferred daily by internet to the central computer of the University of Princeton and processed there.

Science should extend this impressive research by setting up more diodes, say 1,000, in other parts of the world to see if there was any variation in the white noise caused by other emotional events. The money involved would be a fraction of that spent on 'blue sky' research such as studying the Compton Effect. Who can forget the day that Princess Diana was buried? It is no surprise that remarkable changes occurred in the diodes.

What's special about Quantec? Quantec has properties that no other device has, either by itself or in this combination. The white noise diode. Quantec is the most modern equipment with the greatest experience in using white noise diodes. All other devices on the market work with so-called pseudo random generators which, as the name suggests, cannot interact with consciousness and morphic fields in the same way as the real generators. There are even scientific studies that show that the results obtained with Quantec cannot be obtained at all without a white noise diode.

It is surprising that more people cannot see through the obvious fallacy that mere pseudo random generators might be able to interact with morphic fields.

Automated radionic treatment. Quantec is currently the only device that can actually be proved to radiate target objects. Every therapy appointment is displayed on your screen and demonstrated

before your very eyes. All other manufacturers merely save the therapy appointments on the hard disk - without carrying them out.

Other manufacturers may wish to respond to this bold statement.

Miscellaneous. Quantec is now available in 13 languages, over 100,000 substances are available, it is the most widely sold computer-aided system in the area of instrumental bio-communication and it is the first system of its kind to be considered by a German University as worthy of testing in a scientific study. Quantec is quite simply something special.

There is little doubt it is something special. And it doesn't stop there.

Quantec tooth module: The tooth module for the ExpertScan takes account of the fact that in the tooth model according to Dr. Kramer, all meridians in all four quadrants of the bite run through certain teeth. The tooth module can be selected after import in the ExpertScan screen. The pairs of meridians are drawn in schematically for the individual teeth. Passing the mouse over a tooth opens a pop-up menu showing the organs and joints allocated to this tooth.

Dr. Kramer has opened up a completely new field for science to investigate. Instead of wasting money on satellites observing the over-familiar moons in our solar-system, science would do better to investigate the hitherto unsuspected connection between teeth and the organs of the body.

If the screen is set to the tab Findings, the diseases of the teeth can be entered tooth by tooth. For example, the disease paradontitis is set for the appropriate tooth, while in the right screen all findings are set which can be summarised under this general categorisation.

For example, acute toothache is allocated to tooth 35. Clicking on the next tab then displays an overview of all dental diseases. Switching to the tab Stresses initiates a generator run which helps to identify those organs which are most adversely affecting the teeth or which are being most adversely affected by the diseased teeth. The sequence of the diseases in this screen is arranged in descending order with the most severe stresses at the top.

Very little research has been done in this area. But it would be helpful for doctors to offer a more focussed diagnosis of organ failure based on patterns of tooth decay.

In addition to the diode with the white noise, an important role is also played here by how many teeth of a meridian are diseased and whether these diseases are minor, medium or severe. Quantec

therefore determines an organ status, which is then fine-tuned by the generator run.

Science could determine whether a tooth filling cures an organ disease and if so whether the type of filling affects the cure. One might empirically expect a gold filling to offer a longer lasting cure than amalgam.

If our example patient has stomach symptoms we double-click on the line with the entry 'Stomach – allocated tooth 35'. This then opens the screen from the Psychosomatics module. Here the symptoms of the patient are selected. The list is then examined anamnestically for further matches between diseased teeth and organ symptoms, double-clicking on every entry for which the patient displays corresponding symptoms.

It is most important that the list is examined anamnestically. The reason is beyond the scope of this book.

Let us assume in this case that the patient also has kidney-stones so select this disease by double-clicking the line 'Kidney – allocated tooth 11'. If we then click on the tab Summary, a screen opens in which all identified diseases are displayed. All that is needed now is to click on Start and the evaluation is now running as already known from the Psychosomatics module. In this way teeth findings can be easily weighted and evaluated with regard to their relevance to organ diseases.

People missing tooth 11 may have donated a kidney.

Dr. (Med.) Dahlke recommends the Quantec-ExpertScan: Imagine that you were responsible for a particular area, and that you were assessed on how well you kept this area under control. The result would then depend on your knowledge, your technical ability, your care and diligence. This would only apply however if your area cannot be adversely influenced by outside factors which could sabotage your area. This is exactly the dilemma in dental medicine. The dentist is responsible for the health of the jaws and teeth but has to cope with many influences which are outside his official area of responsibility.

Dentists worldwide will sympathise with this dilemma.

This also makes logical sense: An energy flow on a straight line would soon find itself at a dead end while a closed circuit presents no obstacle to the energy flow. Through empirical research in electro-acupuncture it has now become apparent that all pairs of meridians

also run through the teeth. **It has also long been known in general medical practice that suppurating wisdom teeth can cause heart complaints up to the point of heart attacks and the connection between incisors and the urogenital area has in the meantime also been confirmed empirically. It is therefore hardly surprising that the canine teeth are also known as the 'liver teeth' because it is exactly here where the inner liver-bladder meridian runs.**

Scientific research concerning the connection between incisors and the urogenital area is at long last underway. And science is on the verge of confirming the exact route of the inner liver-bladder meridian. The fact that people without wisdom teeth suffer fewer heart attacks has been unexplained… until now.

If the dentist has to construct a bridge from the canine to the first molar he must be able to rely on the fact that these two teeth will be able to support the bridge. However well he makes the bridge, if the teeth supporting it are weakened due to organ failings via the corresponding meridians, he is powerless and can only watch in frustration if his bridge collapses.

How many bridges must collapse before this is appreciated? In a perfect world transplant surgeons would be on hand to replace failing organs before dental work commenced.

The teeth cannot be considered separately from the organs and the different organs cannot be divided into groups as (medical) specialisation is (dentist, ENT, cardiologist, endocrinologist etc.) Everything is connected together. This is also nothing new. Even Goethe complained about the reductionist procedure in the natural sciences. Does the tooth put stress on the organ or vice versa? Things become even more complicated in case of chronic findings.

Goethe is widely recognised as a leading opponent of reductionist techniques in science. His opinions carry weight.

The dentist can sometimes discover from the condition of the teeth stresses and illnesses which may not even be indicated to the general practitioner by the screening measures available to him. Since the teeth are exposed they provide good starting points even on cursory examination and if necessary by additional x-ray examination. In fact, it frequently happens that the diseased tooth is the first indicator of symptoms, even before one of the relevant organs also shows symptoms.

Yet again we see the importance of dentists working together with doctors to come to a complete diagnosis. The important point is that the teeth are visible whereas the organs are not. So any surgeon is well advised not to start an operation without having a dentist examine their patients teeth first.

Instrumental Biocommunication: Psychosomatics and the responsibility principle. If the psyche loses its balance and harmony in a particular aspect it psychosomatizes it, it shows on the body where the problem lies. This is true for all people, all languages and cultures, with the same illnesses. If you have dental problems you may be lacking in bite; if you have a problem with your back you may not be able to assert yourself (to have backbone, to break the back of something); if you have digestion problems you cannot stomach something; if you have a throat disorder you cannot swallow something; if someone has problems in giving he may have asthma or constipation, and if he gives too much, diarrhoea. This list can be continues to include every type of illness.

The list is not exhaustive nor is it meant to be. But it is instructive. Of particular relevance is the familiar case of someone having problems with giving. This can occur at Christmas, when it is also commonplace to experience constipation. As we see, the two are linked.

Take gastritis as an example. To obtain an evaluation, you just click the checklist under 'stomach' and select an illness. After you click 'Start' Quantec needs just a couple of minutes to generate an affirmation, the text for the psychosomatic meaning, the medical ExpertInfo and finally, as before, to display the remedy found using the white noise diode.

Taking a couple of minutes to do all of this is a real improvement on human-based diagnostics.

After the affirmation the psychosomatic meaning is displayed. Quantec takes into consideration the name, age and sex of the patient – in this case a 50-year-old woman. We show the text just as it is displayed in Quantec: Stomach subjects: Anger, stress and aggression, overcoming conflicts, love. On anger, stress and aggression: Colloquially one says ...to be acerbic... (stomach acid), ...to swallow ones anger.... This is a matter of a level of feeling which is not adequately handled. The problem is not resolved in the head but is literally swallowed to the stomach. The stomach then has to bear the consequences and then somatises this in the form of illnesses.

This is an example of an expert diagnosis by an expert system.

On overcoming conflicts: A smouldering conflict cannot be resolved. ...To have to swallow something... means that the conflict, e.g. at the workplace or in the family is not being addressed. The opposing party is apparently stronger and so I, as a ...poor sucker..., simply have to swallow everything. On love: In the event of a lack the stomach assumes a representative position. The compensation usually takes the form of snacking on sweet things. ...I could eat that little sweetie up with a spoon... Clearly shows the connection between love and the stomach. Love however should not go through the stomach but should be expressed through the heart and the genital organs.

This cannot be stressed too often. .

Do you want to help your patients to understand their illness? You can use the gastritis example demonstrated here and select other organs and then from the checklist select the most common illnesses of these organs. The software then automatically assembles all the facts in a few minutes. If you are interested in showing your patients the connections between their actions, thoughts and illnesses and you want more information, we invite you to one of our info-evenings. And then, if you are really interested, we can arrange a demonstration in your practice.

It is unlikely that any doctor could fail to be impressed. Following this, funding comes into consideration but as billions are spent on medical care in every civilised country the following price-list can only be viewed as extremely competitive:

Update Quantec 6.0: Price € 745.00
Module reflexology: Price € 360.00
Vortex module pillar: Price € 360.00
Module of tooth status: Price € 360.00
Expertscan incl. Psychosomatic Medicine: Price € 2,600.00
Vacation engine: Price € 360.00
(Extension to the expertscan € 428.40)
Extensions for the Quantec Database extension – memory:
1-4 Gigabyte: Price € 360.00
4-8 GB: Price € 720.00
Scan categories with sending options: Price € 75.00
Homeopathy for Samuel (Key symptoms): € 100.00
Mineral library: € 100.00 or € 428.40 or € 856.80
Database (Copyright RADAR © ARCHIBEL): Price (€ 119.00)

Quantec transmitter:
Price € 3,600.00 plus 19% VAT (€ 4,284.00)

Quantec Synchro software:
Standard Quantec's Price € 360.00 plus 19% VAT
(For 2 or more € 428.40)

Quantec: New diode with matching gold clip: Price € 800.00 plus
19% VAT (€ 952.00)

'Travel triangle ' acrylic (€ 428.40)
Rose Quartz triangle (€ 952.00)

In summary, one of science's next steps must surely be to verify the findings of the Quantec team and investigate more thoroughly the properties of the white noise diode.

I fully expect that within the next decade we will be hearing much more of the properties of this device and it will only be to the benefit of humanity.

Chapter 8: Dowsing

The ability to find people, artefacts or substances by use of maps, pictures or physically being in a place are currently the most popular applications of Dowsing. One rod is held in each hand and you have located what it is you are looking for when the rods cross. When using a pendulum most people weight the line with a crystal or heavy weight. However, what does seem a mystery is how diviners can dowse over a map to find people or substances when the focus of the search can be many kilometres away.

Everybody has heard of the remarkable successes of dowsers. Some are employed by water companies to find water reserves where no geologist or hydrologist ever expected water to be found. Others are employed by oil companies and have had success in finding hidden reservoirs of oil under unpromising terrain. Even more impressive are those dowsers who have found clues in murder investigations or have located suspects simply by dowsing over maps. However they do it science must really take an interest and show once and for all that the abilities of dowsers should be taken seriously.

Dowsing, scientifically known as radioesthesia, is the interaction of the mind of the dowser and the energy of the object of interest. It can be used to dowse the subtle energy field surrounding objects, plants, animals, and people. If we are doing energy healing, we can dowse the body of a patient before and after treatment and observe any energy changes. If we quantify these results we can reach conclusions about the efficiency of our techniques.

The subtle energy fields should be measured by scientists using the most accurate instruments available. These days we have instruments such as nano-joulemeters which should be up to the job.

If you want to measure the energy field of a person, make sure they are not holding or wearing any crystals or jewellery. Step back about 3 paces, turn and face them. Hold the l-rods parallel to the ground and pointing toward your subject. In your mind, or out loud, tell yourself what the l-rods are measuring. I am measuring the reserve bio-energy field of this body. Walk toward the person, slowly, keeping the l-rods straight and level. When you enter the energy field, the wires will open wide, the left wire going to the left

and the right hand wire swinging to the right. **Measure the distance between the wires and the body.**

It would be interesting to see if the point indicated by the l-rods was the same place at which the nano-joulemeter began to respond.

Create the healing change. Once again, dowse the body. Step back at least 8 paces, turn and walk toward the body, l-rods parallel and pointing toward the subject. When you reach the energy field of the body the wires will swing open. Measure the distance and compare between the first effort. The difference is the change you made by the healing.

It would be of interest to correlate such measurements with the opinions of energy healers as to how much healing had occurred.

You can use this same method to see if your crystals or jewellery have any effect on a human energy field. Measure the subject with nothing on and then add your object and measure again.

This is standard 'fair-test' scientific procedure.

We all know that everything is created of energy and energy is vibration. If so, then everything has its own frequency and wavelength and is measurable and detectable. How? To simplify this process we can say that when you ask a question an electrical impulse is being sent to your brain. In response, compatible wavelength is created and another electrical impulse is sent back. The impulses create a micro-constriction of muscles, which push energy down along your arm and underarm to finally reach the string of your pendulum.

If the pendulum is held in proper manner, which means on the proper distance called personal wavelength, it creates leverage which is able to move the pendulum. Since the positive or negative state of our energetic system is represented by different amounts of energy we observe two totally different movements of the pendulum.

This is all perfectly straightforward. Einstein proved that everything is created out of energy. Matter is always vibrating, even the atoms in a solid. Neuroscientists will confirm that the brain is basically a compatible wavelength generator. The new concept for science is the idea of personal wavelength.

The principle of recognizing different levels of energy in regards to one's state of mind or attitude is nowadays quite popular, well-known and used by such techniques as kinesiology muscle testing and polarity to mention just two. Originally, however, it was known and used in dowsing for millennia.

Polarity refers to there being two very distinct states. A magnet has a north pole and a south pole, to pick one popular example.

Over 90% of the population today has a high enough sensitivity to become a successful dowser. In the 1970's only 75% of population had this ability. This rapid increase has a lot to do with changes in Earth's own energetic field.

This is an impressive statistic and one which science is only beginning to come to terms with. But the Earth's own energetic field has being measured and the most up to date results do indeed confirm that it is not constant. Why this should affect so dramatically the dowsing ability of large numbers of people remains mysterious.

The majority of today's science of dowsing came from many dedicated European physicists and Egyptologists. Their outstanding work on vibrations, colours and frequencies proved that by using simple tools one can achieve amazing effects. Andre DeBelizal and Leo Chaumery spent time in Egypt researching the phenomenon of pyramids, sphinx, thumbs of Pharos and other ancient structures to find out about the emission of energy from those forms.

To do that they developed simple, but extremely precise instruments based on sacred geometry. Along with shapes found inside the great pyramid and king's valley they established a set of tools known today as Egyptian pendulums or healing pendulums due to their extraordinary power generation ability.

Yet again this shows the knowledge that ancient people had. Sacred geometry has still not yet been fully explored nor understood by today's mathematicians. Like scientists, mathematicians should not suppose that they know everything just because they are living in the 21st century and should pay the proper respect due to older civilisations.

In this same time, in Europe, a substantial amount of work was being done on remote dowsing and new tools were developed to increase the accuracy of readings. Along with research done on reshaping and broadening the variety of tools, European dowsers enhanced health assessment methodology, which enables us today to measure not only the state of health, but also the probability of successful treatments. We also have more and more pendulums which possess healing abilities. Kirlian photography supports this thesis.

A number of doctors will not admit to using such techniques in public but privately it is almost certain they do.

It proved that during any type of activity the brain functions on one level only, either alpha, beta, theta or delta. Only during the process of dowsing does the brain use all 4 levels at the same time. In another words it means that when we dowse we search for the answer on all levels of our existence from conscious level to deeply subconscious levels. Due to this advanced process our pendulum can simply recognize the energy of the source in us.

Empirical evidence is the bedrock of science. It is the case that the normal working brain can be at only one level. Dowsing is the only process discovered so far which uses all 4 levels simultaneously. This is a remarkable result.

Dowsing can enrich people's lives in so many ways. To mention just a few it can: Answer the questions; clear the environment; assess health; evaluate relationships; remove emotional trauma; change non-beneficial energies into beneficial ones; find lost objects, minerals, water etc.

Answering the questions is of obvious importance.

Dowsing changes people's lives by teaching them to think straight, formulate their thoughts, be truthful with themselves. Will you dare to open the door?

Thinking straight is one of the goals of education providers worldwide. Dowsing promises much in that area.

Author's bio: Rev. Alicja Aratyn M.Eng. is the founder and owner of the Alicja Centre of Well-Being. She is vice-president of the Canadian Society of Dowsers and for the last seven years a teacher and lecturer for the American Society of Dowsers. Alicja will change your life forever as she has for thousands of others around the world. Many difficulties in her personal life before and after emigrating to Canada in 1991 led her to create the Alicja Centre.

Despite her master's degree in environmental engineering from Poland she has dedicated her life to raising consciousness on a global scale. Therefore, over last 15 years, she has travelled extensively to teach, lead and inspire as well as heal and consult, helping people bring their lives to the next level of awareness.

As a natural born healer she shares her gift generously. Like all great teachers Alicja started with transforming her own life and so now bases her classes and one-on-one practice not only on the ancient Egyptian school of vibrations but also on her first-hand experience with many alternative modalities such as: Medical Qigong, Dowsing, Reiki, EFT, as well as many European techniques transforming family patterns, releasing entities, Numerology, etc. She believes that if she was able to turn her life around for the better, everyone can. It is just a matter of understanding the Universe's rules.

Modern science can assist here by investigating more fully the ancient Egyptian school of vibrations. Reiki is already under investigation while mathematicians have long been interested in numerology.

Alicja developed her own, unique system of teaching which is known now as Science of Dowsing. The system encompasses all levels and applications of dowsing with an incredible amount of spiritual and scientific knowledge blended together to create a comprehensive and easy to follow course. Recently Alicja began teaching the leadership program which includes teacher-training for Science of Dowsing.

This is the opposite of today's mainstream scientists whose goal seems to be obscuring science from the general public by using technical jargon when there is no need for it. Alicja seeks to blend her knowledge to create a comprehensive and easy to follow course. Science could learn a lot from her approach.

In seminars, Alicja uses the latest accelerated learning techniques which allow people to increase how much knowledge they remember. Her classes are always well attended due to her unique blend of knowledge, wisdom and humour.

And she is also on top of her game when it comes to modern educational theory. Accelerated learning is expected to become the norm in most forms of education in years to come.

Welcome to the intuitivedowsing.com shop: Alicja's extensive experience in dowsing is a great help and insight when we choose new products. Many of them, such as healing pendulums, Atlantis ring or DNA spiral, have been checked with Kirlian photography to prove the power and quality of their energy.

To make your shopping even more fascinating, we offer a new special every month for the whole month so bookmark us and visit on a regular basis. You will have opportunities to buy the best products at a discounted price. We hope you will enjoy shopping with us and will find and enjoy our dowsing products and vibrational tools for you, your family, and your friends.

There are many types of rods plus some variations even with the same type which makes them more applicable to certain energy fields. They all have one common purpose however - they are used to detect or search for vibrations emitted from an outside source. As such, they are a receiver of vibrations as opposed to the ability to generate vibrations. Some rods have a witness chamber which increases their sensitivity when a sample of what is being searched for is placed on it.

L rod: The most popular rod, used to find objects, water and minerals, as well as answers to 'yes-no' questions, assessing the Chakra system and much more. Sold in pairs. $28.50.

Telescopic l-rod: for those who travel with their rods or wish to have them handy at all times, these rods completely fold up making carrying them very easy and convenient. Adjustable from short to very long. Very sensitive. Sold in pairs. $44.78.

Adjustable l- rod: A very sensitive adjustable l-rod recommended for more skilled dowsers. Sold in pairs $80.75.

Two-element Biotensor with wires and box: Price per unit piece: $106.50. The two-element Biotensor set contains: Two gold-plated rods, one with a wooden ball at the end and the other with a ring; one wooden handle; one brass handle and connecting wires; a wood storage box. This version of the two-element Biotensor allows you to make a more precise connection with your client or with your own energetic system.

To do this, connect one end of the connecting wires to the wooden handle and the other end to the brass handle. Either give the brass handle to your client to hold while you dowse with the connecting rod, or if dowsing for yourself, hold the brass handle in one hand while you dowse with the other hand. It will greatly increase sensitivity and will make your work a lot easier.

It will also allow you more freedom to concentrate on the problem you are working with while ensuring you remain connected with your clients energy field. If precision is an issue always use the rod with the golden ring. If you wish to work for a longer period of time use the one with the wooden ball, since it is lighter for holding. Made of gold plated brass with wooden handle. Price per unit: $106.50.

Aurameter: A rod well known for its sensitivity. Used to assess the condition of the Aura or Energy Field, as well as being very good for accurate map dowsing, etc. The big witness chamber allows for the use of a larger witness to help support your work. $49.75

Lecher antenna: $390.00. Lecher antenna, also called resonance rod, is an extremely precise instrument which opens new horizons for dowsers on any level of expertise. It gives a lot more possibilities then any of the rods known before making any kind of bio-location faster, simpler and more effective. Lecher antenna is the only dowsing rod known today which is able to not only detect all possible frequencies but also create and broadcast these frequencies.

Creating frequencies has always been the main problem for science with broadcasting them being considerably more straightforward.

Lecher antenna due to its precision and strength is often called the universal pendulum of the rods. Although this is an advanced instrument even inexperienced people can use it due to its effectiveness and ease of use. It is used by placing the slider over a certain frequency marked on the scale and keep mental orientation on the problem/object/health challenge/etc.

This is the type of advanced instrument which scientists should be interested in. Precision and strength don't often go hand in hand so here is one avenue of investigation.

Mental orientation is crucial here since different problems may have the same resonance waves i.e. remains of the sacred objects of ancient Greece and some kinds of petroleum have the frequency of 8.2 units. Made of metal with chrome finish. Book about Lecher antenna is also available. It is advisable to take the workshop on extended work with Lecher antenna. To get more information about training please contact us or check event schedule.

Scientists should take this workshop. It wouldn't make much of a dent in their research budgets and would have real benefits in their study of the antenna. They could start by finding out if diesel has the same frequency as the sacred objects of ancient Greece.

Cosmos: $29.75. Cosmos rod is reminiscent of the old radio or TV antenna's, hence it is the best rod for locating and identifying vibrations from long distances. Cosmos rod is also helpful for Feng-Shui practitioners since it helps to detect geopathic stress sources inside and outside the house. But cosmos is versatile enough to also be used like a regular rod to find answers to life questions, checking compatibility with objects, etc. Made of brass with wooden handle.

This would be useful in scientific experiments which involve locating and identifying vibrations from long distances. Radio astronomers, for example, should be aware of these possibilities since radio waves are vibrations and the ones they are interested in certainly come from sources which are far away.

On the other hand science hasn't had much success recently in detecting geopathic stress so here is an opportunity for a post-doc to make a name for himself.

Universal rod: $132.00. Universal rod has been developed to help pass certain frequencies and vibrations from a source to the receiver through the use of a witness. As an example, since animals are very sensitive to smell, the use of essential oils may be more comfortable for animals if the vibration of the oil is transferred to a witness then placed on the animal instead of the actual oil.

This rod is especially useful in the case of smell sensitivity or in cases of allergies. The universal rod is heavier than other rods and because of this it is not meant to rotate by itself but is made to rotate by the dowser. Made of brass with plastic sheath.

This is thoughtful and impressive. It would be good if science in general had such consideration for animals instead of subjecting them to harmful cosmetic testing.

DNA spiral: Copper spiral is plated with 24k gold to increase conduction and emission of energy, about 16m/50 ft. diameter. The number of the coils and proper proportions make the spiral: The inner left-turning coil collects bad energy from the environment, the outer energy right-turning coil conducts and delivers universal energy and energizes the environment. With Kirlian photography we can see the emission of white light.

It is interesting that it should be the left-turning coil which collects bad energy while the right turning coil conducts and delivers universal energy. In science this is known as chirality and is related to the property of molecules to behave differently according to their handedness. Chemical properties are thereby different, so it is no surprise that the physical properties of the DNA spiral should show this asymmetry.

Taking the above into consideration, the DNA spiral promotes on-going exchange and supply of energy in one's environment. That's why from a Feng Shui point of view the DNA spiral should be placed in energetically bad spots in one's home especially where energy is stagnant: Basements, closets, garages, empty rooms, bathrooms etc.

Stagnant energy can be defined as energy which refuses to move. Closets, especially, seem very likely places for such lethargic energy to accumulate.

The DNA spiral can be placed in a central spot of the house to force circulation of energy even through the walls and to harmonize its flow in empty big halls, living rooms, stairways, basements, guests rooms etc. Generally speaking everywhere where energy seems to be stagnant dead energy, in order to create a healthy flow of energy.

Science at present has no reliable way of forcing the circulation of energy, particularly through guest rooms.

From a Feng Shui point of view one may decide to place it in the central point of one's home to influence the whole area or place it in a Feng Shui corner to increase the flow of a certain vibration only i.e. abundance, relationships, health etc.

The full spectrum of waves is a work in progress for science and though some progress has been made gaps remain. The vibrations for abundance, relationships and health are poorly understood and research has stalled in this area.

By creating a kind of energetic vortex the DNA spiral minimizes or, in many cases, cuts off negative influence of black streams, lay lines, emf or any other noxious energies. It also strengthens the human energetic system so we are less vulnerable to these negative vibrations.

And this is exactly the type of inspiration science needs. Black streams are beyond mainstream science at present though emf is known (electromotive force). Its place in the family of noxious energies however is deeply mysterious. Not so well known is the spiral's effect on strengthening the human energetic system though preliminary tests are rumoured to have been conducted on hamsters. Although the ethics of such research can and will be debated initial results confirm that small mammal species do seem less vulnerable to negative vibrations as a result.

Due to is ability to energize, the DNA spiral has a positive influence on all living organisms, especially humans: It activates one's own DNA strand, considerably increases the flow of one's vital forces; helps speed up the healing process; enhances the process of clearing energy in a room i.e. after conflicts, arguments etc. Its clearing ability prevents the environment from getting over-energized. $70.00

Activating one's own DNA strand has been the goal of decades of genetic engineering. Such research has not yet been able to reliably energise the spiral let alone increase the flow of the vital forces. This is all entirely sciences fault for not taking research to the next level. If science did, perhaps it would be able to prevent the over-energisation of the environment which is surely something we would all applaud especially in these days of global warming and climate change.

One of sciences next steps must surely be to conduct research as a matter of urgency on dowsing and the effect on the environment of the instruments available to dowsers. Lecher antenna is obviously important but I feel that the powers of the DNA spiral should take priority. With sciences over inflated research budgets and its almost incomprehensible attempts to find subatomic particles which are of no practable use to anybody $70.00 seems a small price to pay. Scientists would also have to go on a course to learn how to dowse, but again, that does not seem such a hardship compared to the endless costly conferences and lectures they already attend in the name of 'progress'.

Chapter 9: Energy Healing

Metaphysical Institute has developed the unique Human Energy Assessment Release Treatments, the HEART Energy Healing System, a simple, more universal energy healing paradigm. By communicating directly to the super-consciousness using the Delta state, this complete holistic healing of body and mind is invoked. It includes advanced Theta and Ho'oponopono techniques together with many automatic healing programs.

Good news on the medical front.

With this breakthrough no training or special information or knowledge is required, you simply run the program to heal yourself by saying 'HEART Healing, Go'. The Metaphysical Institute trained practitioner remotely transfers to your mind, while in a Delta state, the HEART energy healing system that consists of more than 500 automated diagnosis and positive change instructions. Your mind uses this like other learned programs, like when you learned to walk or talk, you really don't understand the full process; you decide and you just do it.

Better and better.

You will now heal yourself of diseases and prevent others from developing, while feeling fantastic all the time. This is only the beginning; within a short time you will see new possibilities. Much of this will happen automatically or you can achieve this by directing the automatic HEART energy healing program to remove limiting beliefs, emotional blocks and unhelpful programs that are holding you back.

Unhelpful programs are notorious for holding people back. It is a universal experience that when such programs are removed you always feel fantastic.

If you do not feel well, your energy fields are disturbed. Once you have a basic understanding of the mechanisms of these disturbances you are given simple, practical ways in which you can restore your own human energy fields.

It has long been suspected that not feeling well is related to disturbances in human energy fields. Now we have proof.

Aura or external energy source field disturbances: The first is collective negative energy. This is basically caused by a large number

of people experiencing the same emotion from a single incident. A tsunami causes massive fear and if it affects a large population the collective fear creates a negative energy field that spreads around the world and disturbs approximately 80% of people, generally disturbing their aura which affects the crown chakra, very often many of your other chakras as well.

An increase in aura disturbance as a direct result of a spreading negative energy field was noted as far back as the San Francisco Earthquake of 1906. Doubtless the effect predates that by hundreds if not thousands of years.

This is often the cause of hereditary defects that are passed on through generations. If you think of the aura as the software and the body as the hardware and if the software is corrupted or disturbed it may in time affect the hardware, and we suspect the DNA, and although this is only a relatively small change over many lifetimes it becomes significant.

The gift of 'HEART' energy healing system and being able to read life energy fields: To be able to find out the universal truths using proven kinesiology methods, of asking questions and receiving yes or no answers from the source is another humbling responsibility. The consequences of this privilege allows me to share with you the insights to many of life's mysteries. Who are we? Where do we come from? What is our purpose? What happens when we die? What causes ill health? Simple methods to heal our disturbed human and life energy fields which is more often called energy medicine.

As more is being revealed we are revealing it to you, the searchers. To keep up to date please subscribe and receive a free gift. We will inform you of updates and new information. Your privacy will be respected and you can unsubscribe at any time.

Until science has caught up, it may be prudent for research labs to take out a subscription. The free gift will obviously be declared and made available for public use if public funds are involved.

All healing takes place from within you, as you were created to heal yourself naturally. All a competent energy healer is doing, when invoking a healing, is removing any impediments from preventing this natural process.

And this is the promise for the future. Modern medicine concentrates too heavily on outside cures, failing to appreciate the natural processes in the body which promote self-healing. This is a wrong turn by science which is in urgent need of correction.

'HEART' energy healing uses the life energy field healing of your aura, chakra, hara and your core star. We assess and remove any external negative energy, as well as assessing the integrity of your life energy fields and restoring any locks and disturbances by finding the original causes, then healing them. This can be done as distance healing, as a standalone treatment or as a complement to other more conventional medical treatments.

This is good news for hospital waiting lists, which will largely disappear. It also spells the end of overcrowded doctors surgeries.

New do it yourself 'HEART' energy healing: An automatic system to live life easily and effortlessly for an investment that is less than the average price of one cup of coffee per week.

Such is the low cost involved that Governments should be able to support all of its citizens and still cut overall spending on health.

Rapid cure for dyslexia and phobias: This uses 'HEART' energy healing system to give very rapid cure to these specific problems that affect many people cured generally with one healing session.

Science has been searching for a cure for dyslexia for some time. It is gratifying that the search can now be called off but science must be careful to give credit where it is due and not, as so often, try to claim all of the glory for itself.

School of Metaphysical Healing is in the advanced planning stage. Designed to train medical professionals and other suitable alternative health providers and practitioners in the unique 'HEART' system to assess and heal life energy fields. Eventually one thousand schools will be set up around the world.

Such will be the likely demand that this may just be a starting figure.

Accredited medical research is currently being evaluated and planned in mainstream accredited research facilities. Metaphysical Institute would like to hear from others involved in this or like research and contribute to or become involved to further this aim of more and effective scientific research.

And here we see yet another example of reasonable-minded people reaching out to science in a gesture of comradeship and alliance. It is remarkable how often this happens and regrettable how often the gesture is not reciprocated by the scientific establishment. It is this kind of behaviour which gets peoples backs up about science and it really is about time that it stopped.

Many scientists aren't arrogant but they allow those who are a disproportionate voice when it comes to public discussion. One of science's next steps should be to behave more gracefully when others make important discoveries.

With new instruments and equipment such as the gas discharge visualization GDV camera and new software from people like Dr. Korotkov we can begin to see real time changes in auras as they are healed. The research project could include these studies together with less conventional kinesiology and aura readers, to prove the enormous benefit of human energy field healing.

Here we see the Metaphysical Institute behaving as a modern research establishment would: Setting up multi-disciplinary teams with clear goals and expert leadership together with the most modern instruments. This is something science should be comfortable with in engaging fully.

This research will be the first step to introduce human energy healing to more mainstream medical treatments enabling a reduction in costs and more effective and rapid healing.

And with science's cooperation this admirable aim would be achieved sooner.

With one of the major benefits being prevention as virtually all medical problems and diseases show up in the energy fields and can be cured before the physical problem is manifested.

This may threaten to put many doctors out of work but we must not let self-interest stand in the way of progress. Billions of people stand to benefit from having their diseases cured before they happen.

Ralph Waldo Emerson: Is it so bad, then, to be misunderstood? Pythagoras was misunderstood and Socrates and Jesus and Luther and Copernicus and Galileo and Newton and every pure and wise spirit that ever took flesh. To be great is to be misunderstood.

This heartfelt plea is the position at present. Although the language is somewhat flowery the sentiment is not. To be great is almost invariably to be misunderstood. Science must take the lead here, after all it does claim to be *the* system of knowledge.

The Metaphysical Institute offers some unique services based on 'HEART' energy healing system that were developed and are taught by the Metaphysical Institute. All services come with a money back satisfaction guarantee.

There could be no better evidence that this system is a serious and proven technology. One wishes that all medicine came with a money back satisfaction guarantee.

All Metaphysical Institute services are non-intrusive and performed by analysing, checking and restoring the clients life energy fields. In most cases there is no physical contact with the client and many services are done as distance healing.

I should think most doctors would see the advantage of having no physical contact with their patients.

Distance healing is possible because we are all connected to each other through our life energy fields. This allows the Metaphysical Institute trained practitioner to be able to test and assess the integrity of your life energy fields; namely your aura, chakras, core star, tan and your tien hara group fields from anywhere in the world. We heal and restore any blocks and disturbances in your life energy fields by finding the original causes then healing or removing them.

Group field theory is a theory of quantum gravity. It is closely related to background independent quantum gravity approaches such as loop quantum gravity, spin foam and causal dynamical triangulation. A group field theory is, technically speaking, a quantum field theory living on a Lie group, whose Feynman diagrams correspond to spin foams and simplicial manifolds (depending on the representation of the fields). Thus, its partition function defines a non-perturbative sum over all simplicial topologies and geometries giving a path integral formulation of quantum spacetime.

Life energy fields, also called Chi in Chinese medicine or Hado by Dr. Masaru Emoto in Japan, are present in all life in the Universe. Dr. Lo, a quantum physicist, is working to prove the existence of Chi through quantitative, scientific means.

Again, we note the level of expertise which is being deployed by the Metaphysical Institute.

Why would I need 'HEART' energy field healing? Energy field healing will cure, or help cure most diseases, whether physical or mental ailments, any pains, injuries, viruses, cancers, dyslexia, allergies, food allergies, bacteria problems, yeast related problems, chronic fatigue, recovery from surgery, problems or discomfort with all organs, heart, lungs, kidneys, liver, pancreas to name but a few. A 'HEART' energy healing will also compliment conventional treatments speeding up healing and recovery.

Conventional medicine has long been stumped by yeast related problems so here is another breakthrough.

You will feel energised and fantastic afterwards and results happen in most cases in hours and days, generally much quicker than conventional treatments.

Which only begs the question, for how much longer must conventional treatments continue to be offered?

To effectively assess your condition it is important that we talk to you by telephone. We want to understand what you are concerned about; to inform you and help you understand the process so we can effectively find and remove the original cause of your problems. This first consultation is free of charge and with no obligation on your part.

If you choose to proceed with a 'HEART' energy field healing session we will set a mutually convenient time, you will be asked to agree to a simple terms and conditions contract and complete a simple questionnaire where all your information will be kept confidential. Payment is requested in advance; a secure PayPal invoice will be e-mailed to you; you simply pay by PayPal or credit card.

All of this is standard medical practice.

Pre-treatment - 10 minute information call is free to ascertain whether this treatment would address your specific needs.

If this treatment had been thought in any way to be a money-making exercise then here is the rebuttal. Unlike modern medicine, which charges for everything, the pre-treatment 10 minute information call is completely free.

'HEART' energy healing – single: $110 for your initial consultation.

'HEART' energy healing – single: $80 for follow-up consultations.

'HEART' energy healing – family: from $260 per family of 4 or by arrangement. Ask about couples rates.

It is difficult to imagine a more reasonable price structure.

D.I.Y. Automated 'HEART' Healing Go, self-care: $110 for 6 months for an individual - $160 for six month for families - instant ongoing 'HEART' energy healing, unlimited DIY healing. Use it every morning when you wake up, then as often as necessary. Available only after an initial 'HEART' energy healing.

For an unlimited service this is extremely competitive. How many doctors can claim similar reasonable remuneration for their services?

'HEART' energy healing - monthly care $160 per month includes email support and phone support. Disturbances can be in many layers and occasionally a person may need extended sessions to peel off all these layers.

Such support is invaluable yet here it is, offered at an economical price. Email and phone support is almost unheard of from today's doctors so it should come as no surprise if this service is heavily oversubscribed. The occasional person who requires extended sessions to peel off all of their layers might well consider their money well spent when one considers the alternatives.

As this newsletter goes out to you who had the insight to subscribe, you are witness to an event that is unique in human existence. The creator has chosen Ian Stone and Diana Gilbert, two ordinary people with no formal scientific qualifications, to bring to all humanity a system of healing and a way of instilling values into all humanity.

The last time this action was taken was when Jesus Christ died upon the cross and rose again 3 days later to give all humanity a path to salvation.

At 11 pm Melbourne time on 29 June 2009, 'HEART' healing and other important changes were bestowed to all humanity that ever was, is, or will be. This was only possible from work started by Ian Stone some two and half years ago.

Within three days of the above date all souls on Earth will have been bestowed the golden 'HEART' aura and golden soul energy fields which are added to those you already possess. All new souls and physical beings throughout the Universe and beyond will also have these two new energy fields added to their existing structure.

Science will hopefully be able to confirm that the two new energy fields have been added to the existing structure throughout the Universe, though confirmation beyond the Universe is outside observational parameters at present. The successor to the Hubble Space Telescope, the James Webb Telescope, could have this task as a priority once initial calibration protocols are completed.

Each internal chakra controls specific organs and mind and body functions. The crown chakra is our connection to the creator, the creator of the Universe, and every other living thing in the

Universe and beyond. We also have many other smaller energy fields, probably an almost infinite number as we are basically all energy and only a small amount of solid matter; this, scientists have demonstrated by using the most powerful microscopes.

Scientists have indeed been using the most powerful microscopes available.

A patient may seek an energy healer hoping to find relief from pain or painful symptoms, wanting a cure for a specific disease, virus, tumour or cancer. The patient will receive much more than that. The focus is not only to eliminate the back pain, cancer or the tumour but to also work with the patient to find and heal the root cause of the original symptom or disease.

The 'HEART' healer will achieve even much more than this; they will restore the body, mind and soul to perfect health and balance combining the ultimate in technology from all known healing methods by using the information as used in creation of the human species.

The human aura is now also a golden aura, an energy field of 8 layers that surrounds the physical body in an egg shape or bubble of spiritual light, colour, sound, feelings and life energy. Each person's aura vibrates within their own frequency at different levels of strength and intensity. Because the aura constantly gives off and absorbs energy, it can also be affected and reacts to other energies and frequencies such as another person's aura, where exchanges of energy take place all day, with everyone you meet.

As you can see above you can heal yourself very easily. However most of us have many hundreds of blocks and disturbances that affect both our mental and physical health and behaviour. The best way to clear all the past and present life disturbances is to have an initial 'HEART' energy healing where a Metaphysical Institute trained healer will clear and cleanse you of all these hundreds of external, internal, past and present, energy disturbances that show up within your human life energy fields.

Energy disturbances are becoming more and more recognised by leading surgeons as the main obstacle to a person's health. Historically, the task has been to remove these but tools such as scalpels have only treated organic problems and in a not too subtle way. Because scalpels are made of metal they have, as a by-product, sometimes conducted energy fields away from the affected site but the effect has inevitably been hit and miss.

A Metaphysical Institute trained healer, on the other hand, is well placed to clear and cleanse you of the energy disturbances that show up within your human life energy fields. This is a breakthrough and one which science should immediately seek to confirm. At best, today's surgeons can only remove one or two blocks or disturbances per operation. It is a quantum leap in technique to remove hundreds at a time.

They will also remove viruses, bacteria and yeast problems and more, as well as deal with any other specific conditions you bring to their attention. During this consultation they will explain the process, instruct you and answer your questions to help you in the process of healing yourself in the future.

Medical science over the previous century has made some progress in removing virus and bacteria problems though not as much as had been hoped or claimed. They have unfortunately made no progress at all in removing yeast problems.

To change some habits, behaviours and addictions and more complex problems it may also be necessary to perform other actions within the physical body including altering neuron paths. This is best done by a trained 'HEART' energy practitioner.

Brain surgery has made considerable advances in its short history but even its most skilled experts will not claim to have successfully altered neuron paths. In the light of this it seems advisable to allow a trained HEART energy practitioner priority in clinical cases where neuron path alteration is indicated.

These advanced healing techniques also involve the life energy fields, namely the core star, tan tein and the hara group. These are essential to human life and not found in other species on Earth. When these function correctly this can have a profound effect upon our life. Dysfunction of these essential life energy fields can have major detrimental consequences.

No serious scientist has ever debated this.

The tan tien is our connection with the Earth as the tan tien actually resonates to the exact frequency of the molten centre of the Earth and the energy field that is responsible for holding our body in the shape of a body, instead of a gelatinous mass.

Energy scientists are working closely with body specialists and their preliminary results point overwhelmingly to evidence that there is an energy field that is responsible for holding our body in its shape. Unconfirmed reports suggest that when this field is removed the body reverts to a jelly-like mass though these results are classified.

There is also what some have referred to as fourth dimensional healing or more correctly, creative healing. Where the healer, in combination with the creator, miraculously creates new organs and tissues in days and weeks to restore that chosen person to good health. Another book 'The Fourth Dimension' by Dr. Paul Yonggi Cho listed in our links section is good reading on this subject.

Dr. Paul Yonggi Cho offers a sober, comprehensive and meticulous analysis of creative healing as one would expect. Science today can create crude copies of organs and tissues but the process is complicated and protracted. Dr. Cho is showing science the way forward in his ground-breaking book and scientists should order themselves a copy and set aside suitable reading time without delay.

The value of our own healing by Amber Agha: How much value do you place on your own wellbeing and healing? In these financially precarious times it is tempting to skimp on what can be seen to be a luxury. But skimping on healing or resenting paying for it defeats the purpose.

Amber makes an important point here.

Healing of any sort involves an exchange of energy. It means that when we receive healing and we pay for it in whatever form/to whatever value we can afford, then we place that value on our own healing. We say me, my body, my well-being is worth this amount and I am prepared to invest that in myself. Strong in the belief that money is energy and when we approach that energy with fear or need all we end up doing is pushing that very energy we want away from us. If we focus on lack we attract lack.

If we focus on how we cannot afford that massage, that treatment, we focus on not having that. But what if we said yes this feels right, and I am confident that my needs will be met in funding this, I trust that the Universe will help me provide for this because my intention is to heal, to do myself some good.

The belief that money is energy has recently united the fields of Economics and Physics. This new synthesis promises much. The Gaia hypothesis extends the paradigm and the Copenhagen interpretation of Quantum Mechanics is fully in tune with the idea that the Universe is indeed working towards positive intentionality. These ideas are at the cutting edge of modern Science.

Now I'm not saying re-mortgage your house or go into debt for weekly treatments. It's about priorities. Sometimes in life we have to

put others first and sometimes we get the overwhelming feeling that we are running on empty and need to tend to ourselves. When that time comes the most disrespectful thing you can do to yourself is start to then look for the cheapest option.

What does that say about what worth you put on yourself? Save up - put aside £5 a week until you have enough. Sell those books you no longer read, consider doing a swap. What can you offer a practitioner that they may need? Can you repair something, offer them a discount on a service you provide and so on. Be truthful, speak from the heart and keep your intention to heal and doors will open. They may not be the ones you had in mind, but if you trust they are part of the healing journey it's a good place to start.

Even practitioners have to make a living and it is disrespectful not to take this into account. Until Science puts its own house in order and investigates ways of making treatment less expensive then this is the way it has to be.

It is Science's fault that it has not explored these avenues of treatment sooner and you might like to make this point to your elected representatives when they next vote to allocate your taxpayers money to yet another billion pound experiment to find yet another subatomic particle that you have never heard of and which can have no possible impact on your life. Such money could be redirected to practitioners whose only aim is to heal you.

The most important thing you can do is take care of yourself. We are spirit having a physical experience. So make sure you keep nurturing yourself along the way.

Or the journey will seem gridlocked in rush hour and not an open freeway with the top down and some cool tunes playing, destination unknown.

Venergyer distance/remote healing £60. Indian head massage treatment/gift £50.

These are very modest sums when one considers that the Large Hadron Collider cost 2.6 billion pounds just to construct. And that doesn't include the enormous cost of running it. Or the sums of money paid to the scientists who work there and the money needed to refit and replace worn out parts and so on. In fact, the LHC is a very large money drain with results that can only be described as esoteric as best.

Of what purpose is the discovery of the Higgs Boson? How can it help people? Of what use is the LHC in ordinary peoples everyday lives?

These are questions which science needs to answer and it should begin to do so in the near future as a matter of urgency. It is not that the science is worthless it is just that the money involved in 'big science' has become too inflated. There is too big a call on the taxpayers purse when the benefits perceived are so small. There is nothing that the LHC has provided for its huge investment apart from detecting a particle which has no practical use. There are no spin-offs, unlike the Apollo program which developed velcro and non-stick frying pans and food in a bag.

Taking everything into account it seems that one of sciences next steps should be the redirection of funding away from the LHC and into Energy Practitioners. In line with their underlying principle, many philosophers agree that we are indeed spirit having a physical experience.

Aristotle may not have put it exactly this way but in a modern context there can be little doubt he would have any problem agreeing that if nurturing ourselves failed to happen then our journey would seem gridlocked in rush hour and not an open freeway with the top down and some cool tunes playing, destination unknown.

If more proof were needed as to relative worth one may like to divide the cost of the Large Hadron Collider by the price of an Indian head massage. The answer is how many people could be treated and it is an answer which science does not like to publicise.

Theta healing: Also known as theta DNA healing or the Orion technique, is the simplest, most elegant and most effective healing technique that I have ever come across. It combines both science and spirituality and yet is extremely practical and pragmatic in its application.

Perhaps the best thing about it is that everyone can do it. It was developed by naturopath and medical intuitive and healer Vianna Stibal, who created a model of what she instinctively did when healing her clients and then refined and added to the model over a period of years to give us the technique we have today. It is a technique that is at the same time very ancient but also very modern.

Medical Intuitives are undervalued in today's theraputical paradigm. Professionals in general would do well to listen to the advice of those who have no formal training. It is surprising how often intuition is ignored by specialists who take it for granted that because they went to this prestigious medical school or studied that particular field of medicine they are automatically best placed to provide the answers to complex medical questions. As usual, ancient people were often way ahead in their knowledge. When allied to modern techniques, as in the case of medical intuitive Vianna, the results can be remarkable.

The process involves entering a theta brainwave state in which it is possible to connect to the source energy and by the application of quantum physics to change in an instant thought patterns that might have been held for years or even generations and by changing these thought patterns or beliefs to manifest a different reality.

Quantum Physicists will be quick to confirm that change in an instant is one of the defining characteristics of their theory.

It is now becoming accepted that our thought patterns change our body chemistry and even our DNA and that negative thoughts and beliefs can create disease and emotional problems so, once negative and unhelpful beliefs or thoughts are changed, the disease or problem they have created can no longer persist – things have to change. It is also possible to use the theta technique to scan inside the body and to promote healing once the underpinning beliefs have been changed. It is truly amazing.

Acceptance that our thought patterns can change our DNA has been slow in coming but now that science is on the right track final proof may be only a matter of months away. Until the announcement is made it is only right that people should be informed that they only have to think in order to change their DNA. For some this will be easy but for others the process may be more difficult and they may find themselves having to think hard, and in some recalcitrant cases, really hard.

Kinesiology assessment: During this energy healing a Metaphysical Institute trained practitioner will ask specific questions of your spiritual side connecting using the Delta state, while using a kinesiology technique to determine whether these answers are yes or no. They will not read your mind. These questions will be like: Does this person have any disturbances affecting their aura?

Science should make it a priority to enable all doctors to apply the same methodology by dissemination of best practice as soon as is feasible.

Kinesiology, although not widely understood, has been researched and tested to be repeatable with large numbers of people. Even when the people are not aware of the subject, large groups of people all come up with the same answers.

This is the very basis of the statistical method.

'HEART' Healing report 10/7/2008 8:24 am, Melbourne Australia. External energy disturbances have not come from within you and you have little control over them. They affect our aura and often our crown chakra.

This is not good news.

Negative energy – Most comes from our Earth but recently some has come from the Universe and most people are still disturbed by this event.

Negative energy's recent arrival from the Universe was missed by Astronomers. This in itself is disturbing.

Internal energy disturbances: You will see major, minor and micro. These are like three levels and signify the importance or severity of disturbance. Each is a negatively associated emotional incident that is changed to a positive from the moment of the original incident to the present and into the future. All are causing disturbances to our life energy fields.

Modern analysis has confirmed that major, minor and micro are not just like three levels, they actually are three levels.

During the clearing and removing of these disturbances the modern version of Ho'oponopono is also used. This has the effect of not only cleaning yourself but also invokes clearing others with the same problem.

It is an advance that the modern version is being used. The old version was nearly right but the modern version has obvious advantages.

Problems detected and removed – virus 1, bacterial 0, yeast 2: These consist of viruses, bacterial and yeast infections that are present in your body at the time of the healing. You acquired these when your immune system was compromised when your aura was disturbed.

This person was lucky that no bacteria noticed. The outcome could have been significantly worse.

Other conditions that may require treatment - pathogens, parasites, worms, toxins, heavy metals, radiation: These can be present or have built up within our bodies due to exposure or the environment where we live.

Worm build up would be particularly disappointing. One may consider moving house to change one's environment if this occurred.

Please note: The numbers of disturbance are those showing up at this time of this healing. For your ongoing good health you may consider a standard 'HEART' energy healing about every three to

six months to re-energize and clear any problems that become apparent during this period. The number of events will vary from person to person and depend on many factors. If you have severe physical or mental problems you generally have more disturbances.

The lower estimate would obviously be indicated here.

The latest breakthrough: The amazing, do it yourself 'HEART' energy healing, automatic program. Live life easily and effortlessly by having access to your own 'HEART' energy healing program to heal yourself, wherever and whenever you choose, to feel fantastic all the time. You are now in control, clearing your own blocks and healing yourself.

It is natural for doctors to feel side-lined but they must accept progress. It is no use trying to deny people access to this method of healing just because one feels one's career may be jeopardised. There is no immutable law which says doctors have to be involved to cure patients.

To believe and behave otherwise is a Luddite position and I am hopeful that doctors, well-educated as they are, will not attempt to smash the mills of energy healing in a futile attempt to hold on to their status and standing in society. Older doctors could retire. Younger ones may be able to retrain.

What will you feel during an energy healing and after a 'HEART' energy healing? This will depend on how sensitive you are to changes in your body. After a 'HEART' energy healing treatment many people will feel some or all of the following particularly after a sleep, within a couple of days: A sense of peace, a feeling of lightness as if a weight has been lifted from you, more relaxed, feeling more energized, ready to get on with life, more enthusiasm, more confident in a quiet reassuring way, fear and doubts seem to have disappeared, happier in life and relationships, less anger and not as easily disturbed, cleared of blocks that make them hesitate, procrastinate or give up, much readier to make decisions and take action, life just seem easier and effortless, easier and more confident to make peace with people, greater awareness and respect for all life and our connection to all others, more awareness of the balance of nature and the connectedness and importance of everything created.

Who can honestly claim any of this after a visit to the doctor?

The Metaphysical Institute is not receiving and is not looking to profit from any recommended materials or internet links. No affiliate programs will be promoted.

This is reassuring.

Much 'HEART' energy healing treatments and teaching has been done, free of charge, as teaching, information and awareness, not profit, is our main purpose and motivation.

Providing a sharp contrast with drug companies, for example.

What is distance healing? Distance healing is possible because we are all connected to each other through our life energy fields. Distance healing has been practised for centuries, however Western scientists are only now seriously looking at it.

Distance healing could have been a firmly established part of modern medicine had not so much money been spent on the effort to find a smaller particle than the last one, a search which threatens to degenerate into a 'Russian Doll' expedition where the discovery of a small particle merely sets the stage for the attempted discovery of an even smaller one nestling inside. Ordinary people, rightly, ask themselves what is the point and particle physicists continue to come up with no convincing answer.

Other services - please ask: Soul mate advisory and soul mate 'HEART' healing - are you with your soul mate? Special 'HEART' healing for soul mates.

Modern medicine will admit defeat when asked to positively identify soul mates.

Relationship alignment - align yourself and others that are special to you.

Alignment with others that are special to you is the holy grail of alignment research and if it is true that it has finally been achieved science should immediately make the discovery public.

Removing past relationship ties and bonds - past relationships can be difficult to break because of energy and emotion ties and bonds.

Molecular scientists will confirm that bonds are difficult to break. How much more difficult they are to break on the human level can therefore be imagined.

Phobia cure - suffering from a phobia or panic attacks: We offer cure as part of our distance healing.

Science has been searching, unsuccessfully, for a generic phobia cure or at least a cure for panic attacks. Finally, one has been demonstrated and it is no use science hiding its head and pretending that such a thing does not exist.

We don't profess to be able to cure everything as many conditions are in an advanced state of disease, for instance organs may have deteriorated to a point where surgery is necessary, however energy field healing will greatly aid recovery and remove the original cause of the problem often preventing a reoccurrence. Most transplant surgery has a limited life; this is because the original cause is still unresolved.

This is the main problem with transplant surgery. If the original cause is still unresolved then a transplanted organ will inevitably have a limited life. In the past this has been misunderstood as rejection of the transplanted organ due to the recipients immune system attacking the donor organ. A more modern view is that rejection actually occurs because the original cause of an organ's deterioration has not been detected nor resolved. This promises a revolution in transplant surgery.

To understand: The aura energy field is the software and if it is disturbed (corrupted) or malfunctioning in some way, fixing the body, the hardware, will generally only be temporary as the aura program will continue to create a corrupted body part.

Computer scientists should be seconded to medical schools to confirm this analysis.

The healer will not read your mind, however they will ask specific questions of your spiritual being and receive either a yes or no answer to these questions. They will use a kinesiology technique, themselves, to determine whether the answer is a yes or a no. The answers will not come from your conscious mind. To do this they only need to think of you and the connection will be made with your spiritual being so you can be on the other side of the world it really makes no difference.

It is an obvious advantage that the healer can do this even if you are on the other side of the world. Contrast this with less advanced medical practitioners who continue to insist you appear personally in their surgery before they will make a diagnosis. This can be most inconvenient.

If you have a consultation in person, physical contact in the form of muscle testing using a kinesiology technique will most likely be used to demonstrate that a disturbed condition was removed. This is only for your benefit to help you understand that the original condition causing the problem has been removed or healed.

This is fast becoming the principal way of determining if and when a disturbed condition has been removed so it is no surprise that the technique is referenced here. The practise of muscle testing is expected to become widespread in future as *the* way of confirming cures.

Will I need ongoing healing sessions? This a good question, the answer is most likely for other energy field disturbances. The original incident that was removed in a healing session will be permanently healed. However you may have very similar symptoms from a lesser disturbance that happened at a different time. This has happened to me early on when Dr. B. was treating me. I would go and get healed, feel great and sometime later I could feel things weren't ok. I would book another appointment only to discover something else had caused this problem.

The message is simple: Keep going back to your energy practitioner. He will advise you if things are not right with your energy field. He will be happy to help. This contrasts with doctors who dismiss patients making repeated appointments saying they are hypochondriacs. An energy practitioner would never do this.

Let me explain. If you are going about your daily routine and someone or something triggers a thought or memory of an incident that affected you sometime in the past, your aura is going to apply the same learning to this thought or memory that you taught it to associate to the original incident. If that was negative and never replaced with a positive, your energy fields will become disturbed.

Trying to avoid situations which may trigger such thoughts will be of benefit here. But if it is not possible then your energy practitioner will not turn you away.

From Robin Brister in Fort Worth Texas, USA: This note is to tell all how wonderful Mr. Stone is. Mr. Stone you have been a blessing to me and to Lolo my dog. I will never forget how wonderful you were to me when you were on vacation with your family, but never forgot about me even while you were on vacation. You know I have told you over and over how sorry I felt at that time but today is another day, you still are one of the most caring people any one could ever know in their life. Words can't say enough. Thank you for all you have done for me and continue to do for me up till today. Robin Brister, Ft. Worth, Tx.

How many doctors have received a testimonial such as this?

From Steph in Northern NSW, Australia. (Hand written letter, with this was a cheque for $900 that was not asked for).

The accident: I fell and broke a hip, which is the upper part of the femur. At the hospital the medical staff said it would take approximately 1 year to heal fully and 2 years before I had full muscle strength.

What happened: I felt the response of my body to the 'HEART' energy healing was not only rapid - every day, moment by moment, my body was busy repairing. It i.e. my body, demanded my full awareness of its needs and what I had to do in order for it to be A, Comfortable while I was lying flat on my back, B, Helped to move more easily, C, Taken care of by Physio's, doctors, Nurses, even the cleaners.

My body felt so alive, present, articulate that I learned a whole new appreciation of how it functions, what it needs, how it wants care delivered, when and why that care needs to be given - everything was/is so precise. In other words I learned at the ripe age of 63 how to listen and respond with love to my beautiful body - a first as this had never happened to me before. Every day I feel my body getting stronger and more resilient again. The doctors, nurses and hospital staff were all amazed by my progress and lack of pain.

It's only just gone 3 weeks since the operation (plate and pins into upper right femur) and I'm walking around with ease on my crutches - many times I've stopped using them in the last week as my body is getting so normal, I forget and just walk normally (gently and a bit lopsided), especially around the house. I've even found myself at a neighbour without aid. I can hardly believe it myself. Thanks again, your help and love are a beautiful and cherished gift.

Evidence such as this is exactly what science has been seeking.

From Tara in France - I have been running my program and my health is well and my HEART is, well let's just say in growth phase! Also from Tara July 09 about 2 months later - Dear Ian - when I initially came I wanted to heal physical symptoms, I had allergies and infections but what I received was so much more! By running the 'HEART' healing program I feel I am aligning my energy with source and hence I find that I can control and heal negativity and any sickness that may be lurking. I am still growing but on the right path. :) Thank you so much!

Evidence such as this is difficult for science to discount.

From Chastity USA - Hi Ian, I feel so great today. Today was the best day of my life, quiet. Everyone like my boyfriend, mom, and sister all took a walk in my shoes today. It was the most peaceful day, no one bothered me, my hand does not burn. I felt so great I forgot to take my meds today. On my way can't wait.

Forgetting to take meds is a recognised precursor to a cure.

From Wanderley, Brazil - Original email request 27/10/08: Dear Ian, greetings. I don't have money. If you can, please, send distant healing for me - condition - depression, arthritis, and anemia, mental and emotional confusion.

Reply 27/11/08: Hi Ian, excuse my delay. Gratitude to you. I'm very well, better and better day by day - no pain, emotional tranquillity, joy, little fatigue. Thank you for your generosity. Wanderley.

This is almost certainly proof that distance healing works.

From Connie USA: Ian, I wanted to thank you again for your wonderful gift of healing to my guys at the prison. I wanted you to know that the men you directed your healing to have shown great progress in their lives. They are much more calm and peaceful and productive. They are better able to focus on whatever task lies before them. I think of you often and your kindness to men you have not met yet. I so appreciate what you have done. Consider yourself hugged!

Again, it will be possible to confirm that the guys in the prison have been helped by distance healing. A double blind study should be arranged whereby some guys do not receive such healing while other guys do. It should then be a formality to confirm that the guys receiving distance healing show great progress in their lives and are much more calm and peaceful and productive. This could be demonstrated, for example, by statistics showing the deployment of riot police amongst the two groups.

After ear operation from H, Southern NSW Australia: Thank you for the healing you gave me after my operation. You were very available and generous and helped me very much during this time when I was feeling awful. I am much improved and the effects of the anaesthetic seemed to disappear.

Doctors will know that the effects of the anaesthetic do not always seem to disappear so this is good evidence.

Spiritual awareness from Mark, Northern Suburbs, Melbourne Australia: I found your website very interesting. Isn't it strange that

some things you've instinctively known all along? I felt this as I read through your text. The site looks great!! Very professional! Formatted good and easy to read. Menus are well set out and you've also given direction and suggested topics to be next read. I've many questions and would like run through them over the phone if possible. Here's a thought! Isn't it weird how material things blind us from the truth?

Mark appears to have hit the nail on the head here. Knowing things instinctively is something which science should investigate because it often appears that science simply ignores peoples instincts and instead goes off on its own path, one which is not at all instinctive. Science (and scientists) should consider a radical rethink of their methodology in the light of this.

One of science's next steps should surely be to encourage scientists in all fields to stop concentrating on material things. There are things not material which deserve investigation.

Chapter 10: Feng Shui

If you are like most people you probably have a wide variety of electronic appliances and devices in your home. Unfortunately, electronics emit the kind of energy that can be detrimental to the positive energy of your home. It is important to take electronics into consideration when arranging furniture and decorating in your home. With a little care and thought, you will be able to keep your essential electronic devices and yet still practice good Feng Shui practices within your home.

Science is well on the way to confirming that electronics emit unhelpful energy.

Electronics and the bedroom: In most cases, you would do well to keep electronics out of the bedroom as much as possible. The energy emitted from electronic devices can be quite detrimental to sleep, relaxation and rest. Instead of an electronic alarm clock on your nightstand try using a wind-up clock. You should also avoid having a computer in the bedroom if at all possible. Unfortunately, many people do not have an extra room in their home to use as their office or study area and as a result they place their desk and computer in the bedroom. If this is the case, you should at least cover the computer monitor with a cloth at night so that you are shielded from the yang energy.

Anyone who has tried to sleep in a room with a computer will be well aware of this.

The newer flat-screen monitors are also preferable since they have a matte finish and are less mirror-like as compared to the older cathode ray tube style of monitors.

This is interesting. Science should determine, by experiment, exactly why a matte finish confers such an advantage.

Place your desk and computer in the bedroom so that it does not directly face the bed.

Be aware that science has determined emissions can travel through walls so it would be appropriate to warn your next-door neighbour that you are potentially irradiating him with yang energy.

Electronics in the living room and family room: It is almost impossible to avoid having at least some electronic elements in the living room or family room. Instead of trying to eliminate them completely, focus instead on arranging them so that they will provide the least amount of disruption. If you frequently utilize the living room for relaxing activities such as reading or conversing with family and friends you might want to consider removing the television, stereo and any game systems from this room.

Instead, place these devices in the family room or game room. That way you will still be able to enjoy them whenever you want to but will be able to leave the room when you are finished. This will allow the other rooms of your home to remain free of electronics. If you must have a television in the living room it can be helpful to enclose it in a cabinet that has doors. This can reduce its disruptive influences when the television is not in use.

Science should investigate with urgency the level of disruption caused by a television which is not in use.

Whether you are simply sick with a common cold or someone in your home suffers from a chronic illness Feng Shui can often help. Because Feng Shui is based on energy you can harness the power of this energy to improve a person's health. Energy is constantly moving throughout your home, whether it is good energy, bad energy, or both. When you are trying to improve your own health or the health of another family member you will want to focus on ways to increase the good energy while decreasing the bad energy in your home.

Reducing the effects of negative energy: If there is illness in your home you'll first want to identify where you feel the negative illness energy is residing in your home. If more than one person is already sick or if there has been chronic illness in your home chances are the negative illness energy is actually surrounding the front entrance to your home. You can help protect others in the home by temporarily using another entrance so that those entering the home will not walk through the path of the negative energy.

This seems like good common sense.

Then reduce the level of the negative energy around your home's entryway by hanging some bright metal objects around the front door. Shining a bright light in the room towards the front door can also help to dissipate the negative energy.

Being careful not to dazzle anyone who is still using it, of course.

If only one person has become ill in your home you might be able to prevent the negative energy from spreading throughout the rest of the house. For example, if they have been confined to bed you may be able to attack the negative energy directly in their bedroom. Place bright metal objects or a bright light in the hallway outside the door to prevent the negative energy from permeating the rest of your home. If the weather permits open a window in the bedroom to allow the negative energy to escape.

Explaining to the ill person that this is only for their own good.

Get good Chi energy flowing by making sure everything works. Fix broken things.

Most people will already be aware that when things break they almost invariably generate bad energy in the person who was trying to use them.

Crystals are fruit of the earth. They are gifts from Mother Nature to both delight us and help us feel more secure. Any man who wants to make his woman very happy knows that crystals are always a good choice. They are also a good choice for other challenging situations: To slow down energy that is moving too fast. If you have stairs that pour out your front door or long hallways, crystal light covers will slow down that racing energy.

It is high time science made the discovery that crystals can slow down energy more widely known.

To speed up energy that is moving too slow: If you have a room that is full of so much clutter the energy can't move enough for you to even begin clearing junk out hang a 50-60 millimetre crystal from the ceiling and feel the weight lift off your shoulders so you can get started.

Would 70mm improve things still further? These are the areas science should involve itself in.

To deflect energy that is charging at you: If you sit with your side to the doorway you can place a chunk of crystal on your desk between you and the entrance and you'll be able to concentrate without feeling blind-sided by distractions so often.

Energy charges at us, of that much science is certain. If it is electromagnetic energy the charging is done at the speed of light which is 300,000,000 metres per second. There is no chance that anyone could move out of the way quickly enough so a chunk of crystal is a sensible solution.

There are many types of crystals. Choose your crystal by whether or not it feels like it is yours. You'll know. If you don't feel

like you need a certain crystal it is not the one for you. **Different crystals resonate at different frequencies.**

This is good science. Many people will have had similar feelings when handling crystals. You just know.

I carry a small pouch of crystals and gemstones with me at all times. Here is a list of some of which I carry and why: Quartz crystal: Amplifies my energy, doubles my auric field, dispels static electricity and negative vibrational frequencies, dissolves karmic sees, enhances psychic awareness and even soothes burns.

Quartz has been dismissed by science as merely sand but it is obviously much more than that. Doctors should consider rubbing sand into burns.

Turquoise: Promotes spiritual attunement, dispels negative energy, heals spirit, promotes problem solving, alleviates pollution on all levels.

Pollution is a pressing problem in all countries yet they continue to try to tackle it without the use of turquoise. This must change.

One last note about crystals. They'll stay with you for a time and then they'll leave. Sometimes you'll be inspired to give one to someone who needs it. Sometimes they'll disappear entirely. I've had crystals disappear in Texas and reappear in Colorado over a year later. I don't know why they do this but when one of yours exits your dimension know it is headed where it is needed.

Theoretical physics already deals with different dimensions. Modern mathematics is largely the study of different dimensions and this study has led to breakthroughs such as string theory and parallel Universes. So it is not surprising that crystals can enter and exit different dimensions.

Author's bio: Trisha Keel is a Feng Shui Master and Environmental Energy Expert.

Science can boast many of the latter but few of the former. Does science have the balance right?

It was so great to meet you - I had a blast. The day after fixing the money corner my husband got another share option grant. We were able to refinance our house at half our current mortgage rate and approval for our car port came through. Liz B. Katy, Tx.

Fixing the money corner is a sophisticated feng shui technique and should not be attempted by anyone unfamiliar with advanced methods.

Energy: what is this stuff? Everything in the Universe is made of energy: Every person, place, event, and thought. Everything. There is nothing that is not energy.

Physics will confirm that there is nothing which is not energy.

Energy is always moving. It is either moving into a new form through expansion or it is returning to its original form through decomposition. Simply put, everything is either growing or it's dying. Nothing rests, nothing stays the same.

The principle of conservation of energy corroborates this. 'Energy cannot be created nor destroyed but it can be changed into different forms.'

Everything has its opposite and every opposite is the same, differing only in degree. The opposite of light is dark; darkness is only the absence of light. Without light there is no dark and vice versa. The extremes meet in the middle, somewhere in between shadows and highlights.

Science has confirmed that darkness is the absence of light and several papers have been published on the subject.

The dark side of the symbol is called yin. Yin represents the pure potential of an energy but not its manifestation. Cool, dark, receptive, horizontal, multiple, and feminine are all yin characteristics.

Horizontal and feminine have been linked in many cultures.

The light side of the symbol is called yang. Yang represents the pure manifestation of an energy but not its potential. Hot, light, directive, vertical, single, and masculine are all yang characteristics. The most pure yang energy is seen in the Sun. All energy is manifesting and emitting and spraying out in full force every single minute without ceasing.

Vertical, masculine and spraying out in full force will hopefully be found to be the case for the light energy released by the Sun when the SOHO solar observatory is in position.

The Tomorrow's Key Feng Shui Practitioner Certification Program is the only training that combines the ancient Eastern metaphysics of Tibet, China and the Orient with the most current discoveries of Western Scientists' Quantum Physics. This dual basis for your learning ensures your education is not based on superstition alone. We will explore the realm of the quantum in order to understand why Feng Shui works.

The realm of the quantum is one of the most currently investigated aspects of modern science. Its relation to feng shui has not yet been explored by science. To make up for this, Trisha offers the following course for those interested:

We will work through actual rituals in class and in the field. You will be specifically trained in how to prepare for a consultation, how to evaluate energy at a distance and how to protect yourself energetically.

Combine the ancient principles of Feng Shui with the modern concepts of quantum physics to learn how and why to manifest your intentions. Tomorrow's Key Certification Program:

Day 1: Energy is everything. Western study of energy: quantum physics. Eastern study of energy: metaphysics. Personal energies and intuition. The energy of numbers. Video presentation: Transmitting energy.

Day 2: Bagua energy maps for life and for death. Mantras and verbal energy focus.

Day 3: Dowsing with wands and pendulums.

Day 4: Protecting yourself energetically.

Day 5: Energy clearing: Preparation and performance. Electromagnetic fields: Identifying, shielding, and physical influence on humans. Writing your business plan. Establishing your business. Marketing strategies and resources. Additional professional resources. Certification ceremony.

A full and comprehensive program.

Dear Trisha, that was the most exhilarating week of my life. I loved the class and it was so great to get to know you. You are awesome. Thanks again, Debbie Edwards.

People don't have to write these sorts of things so it is very telling that Debbie should.

Course prerequisites: All students are required to have read and to be familiar with the following books:

'The modern book of Feng Shui': by Steven Post.

'Move your stuff, change your life': by Karen Rauch Carter.

Course benefits: Accelerated learning: As a certified master teacher, I have designed both a curriculum that delivers thousands of years of information in only one week and a delivery method that engages all your senses during the instructional periods. By pairing abstract learning with actual hands-on consultations your training

places you at the professional level from which you will go forth into your new life path. You will actually be a Feng Shui practitioner from the moment you enter the first day of class.

Science could learn a lot from this. Instead of it taking years for scientists to acquire their degrees wouldn't it be much simpler if professors followed Trisha's example and taught courses which delivered thousands of years of information in only one week?

Professional listing: You will receive a free listing in the directory of professional Feng Shui practitioners on the Tomorrow's Key website, including links to your email and website. There are hundreds of hits each day from people seeking Feng Shui information and advice. Prospective clients can click directly through to your website, if you have one, or be able to telephone or email you directly.

Scientists, when trained using Trisha's methods, could quickly become established and have their own website to which people could click directly through. Also, a listing in the directory of professional scientists would be an obvious advantage to those at the start of their career.

Mind Master Support: After certification you may email for help on any project you may have. I am happy to provide on-going support and information to assist you in any way possible. In addition, your classmates and others that I have trained are also accessible to you, as you are to them. We all work together to be more than we are apart.

This is also something which science could learn from. Although professors do support students once they branch out on their own that usually spells the end of professorial assistance. How much better would things be if they could make themselves available for contact via email, as and when required.

Materials: Each student receives their own dowsing pendulum, copy of the I Ching and Chinese coins for divination; cinnabar, joss papers, red envelopes, sage sticks for smudging and more.

The use of Chinese coins for divination is something which science should take a closer interest in. It never ceases to amaze me how much science turns its head away from real people and their experiences.

Resources: Each student receives a professional Feng Shui practitioner resource guide that I have written and compiled. These guides are available only to my own trainees. They include: Overview of the nature of energy from the Western point of view

including quantum physics, the law of attraction, the ancient universal laws of the Kabbalah, and the energies of emotions. Overview of the nature of energy from the Eastern point of view including the energies of the celestial animals, the five elements and the Bagua energy maps for the living and the dead.

Of these energies only quantum physics is at all understood by modern science and that only imperfectly. One of sciences next steps must surely be to ally the methodology of Western science with the ancient universal laws of the Kabbalah as there is probably much to be gained. Science, disappointingly, has yet to get to grips with the energies of the celestial animals.

Registration and tuition: Course tuition is $1,499. A $300 non-refundable deposit reserves your place on the course. The balance of the tuition is due 14 days before the first day of class.

This is very reasonable and scientists should start their research into Feng Shui by enrolling on Trisha's course. Every space scientist in the world could be allocated a place for a fraction of the money currently spent on the International Space Station.

Dear Trisha: My husband and I are buying a new house. He was born in 1975 and I was born in 1978. Which way should our front door face?

This may appear at first sight to be a straightforward question but appearances can deceive. Even such a seemingly 'easy' opening query has many ramifications but Trisha is well equipped through her advanced study to answer it. Her remedy begins:

According to quantum physics, there are no such things as time or space. You might get a physicist to use the term space-time if you pushed the point. Considering that time is a mental construct of man and that all space is one space we don't even want to get into string or m-theory here we turn to the understanding that we create our reality by our own thoughts and the meanings we attach to our encounters or the material things in our lives.

Trisha is surely correct that there is no need to get into string theory at this point. Perhaps later. She continues:

There is only now. And now. And now. You can only be here. Your point of power is in the here and now. Place your attention upon how you feel right now right here. If you feel uncomfortable then you are uncomfortable, regardless of your door or head or what year you were born.

It is by our intention that we define our world, through the power of choice we create our lives as each day evolves. Therefore, I suggest to you not to worry about when you were born or which way your front door faces. Don't be concerned with the direction your head should point while you sleep. Focus on creating an environment that is delicious. Focus on filling your life with joy. Thanks so much for your question, Trisha.

A full answer backed by the latest science. It would be pleasant if modern day professors could answer questions from the public so clearly.

Dear Trisha: I'm designing my new house. I love the floor plan but the master bath is over the kitchen. Can I keep it this way or is it really bad, as I have read, to have the bowels over the kitchen? That makes sense but the plan looks so good the other way. Eb.

Eb makes an important and often overlooked point here.

Dear Eb, there are always ways to counterbalance the energies but not everyone can manage the cost sometimes. One of the worst parts about the bathroom over the kitchen is that if you're downstairs cooking, you will hear the toilet, shower, or bath drain while you are cooking. I've been in a kitchen standing talking to a client while her husband ran upstairs for a second. We could hear every sound he made through the pipes. Very disconcerting for her, embarrassing for me. It'd be nicer for you not to have to experience it.

Having said that, I will tell you that you can choose to keep your bathroom over the kitchen. From a safety point of view, try to align the toilet so that it's not directly over the stove or any food preparation area. If there were ever a leak you wouldn't want anything dripping down into your kitchen.

Sound, practical advice.

From an energy point of view, you really don't want to have water energy from the bathroom over fire energy of your stove. If you have to place the toilet over the stove can you use slate, limestone, marble or some other natural earth material for your bathroom floor? Fire energy rises and fire energy generates earth energy. Water energy falls, and earth energy dams water energy. So the earthen material will help separate the two energies from conflict. It'll probably look really sharp too. Trisha

Energy management *and* style!

Trisha. Feng Shui? This is unscientific hokum. I'm amazed and disturbed that AIA (American Institute of Architects) would touch it just because one or two percent of the population might believe it. Next they'll offer a course in creation science. Scary. John, Architect, Houston.

And Trisha replies:

I know that there are con artists and predators who prey on the weak, ignorant and superstitious. I mean no disrespect to any of my colleagues or to any of you architects, designers and builders. But what I'm teaching has nothing to do with superstition or religion. In fact, the study of how all matter is made of energy is all about quantum physics and theory of relativity.

This is undeniably true.

Here is my reply to the unnamed architect above: Hello John, thank you so much for your reply. I completely understand your point about unscientific hokum. The entire superstitious craze about bamboo, Buddha's and three-legged frogs with coins in their mouths makes me absolutely want to throw things. This is not what my course is about. There is nothing on my website or in my teaching anywhere trying to sell you superstitious trinkets. The course has nothing to do with superstition or religion. In fact, it's much more about quantum physics. The Heisenberg uncertainty principle is taught in every course and seminar that I give.

The Heisenberg uncertainty principle is another cornerstone of modern-day physics. Its application to Feng Shui is clearly something which science needs to study further.

Energy makes up everything in the Universe. Humans not only feel it, they can harness it. Even if they can't see it. Just because you can't see electricity or radio waves doesn't mean they aren't there. Nor does it mean you're subscribing to superstitious nonsense if you have GPS or satellite radio in your car.

I encourage you to keep questioning those in Feng Shui who are all about selling trinkets and good luck charms. I thank you for making the point for me. I invite you to simply consider that there might be something of value to learn in a five thousand year old study of what makes humans comfortable. That knowledge might make a difference in your next home design. Sincerely, Trisha.

Five thousand years is a long time and it should be no surprise that methods have been refined and techniques developed. To put things into perspective, relativity theory is only one hundred years old and quantum theory less than that.

Trisha, thanks for your thoughtful clarification. My own experience has been with the Feng-charlatans, one of whom recently killed the sale of a $525,000 loft condo in an early modern restoration 'good brick award' building downtown because of bad Chi or some such nonsense. I think she was trying to solicit a bribe, which if paid, no doubt would have revealed good Chi whatever that might be. Regards, John.

My final reply to him: Oh my goodness, John. That's a horrible story. If whomever this person was had been the real deal, she'd have known how to manipulate that energy to make the Chi beneficial. There is no such thing as good or bad energy. Energy is energy.

Tell you what, the best way to outwit a snake is to have more knowledge than they do. If you take my course you'll know how to counter the next Feng-charlatan that tries that trick on you. I'd venture to guess that you'd know an amazing percentage more than anyone playing this game of I see it, but you can't understand. If you'd rather not take the course, at least call me so I can give you enough information to save your next half million dollar sale. Best of luck. I hope I get a chance to meet you one day. Trisha Keel.

This is an appropriate place to leave the debate. Notice though that Trisha is happy to discuss the problem of Feng charlatans over the phone for no fee whatsoever. This is an example of actively wanting to pass on knowledge and it is something that scientists should take on board.

Far too many stay locked up in their ivory towers and refuse to engage with curious members of the public. Scientists should establish a phone protocol where they disseminate their home and workplace numbers, perhaps in the local press or on television via short advertisements. Then they could accept phone calls from members of the public. In this way, science could become part of people's lives. As one of science's next steps, this is surely one of the most important.

Chapter 11: German New Medicine

Dr. Ryke Geerd Hamer started his cancer research when he developed testicle cancer after his son Dirk was shot dead. He wondered if his son's death was the cause of his cancer. Subsequently he investigated and documented over 15,000 cases of cancer and always found the following characteristics to be present which he termed the iron rules of cancer.

Every cancer and related disease starts as a DHS that is a Dirk Hamer syndrome which is a serious acute-dramatic and isolating conflict-shock-experience. It manifests simultaneously on three levels, psyche, brain and organ. The theme of the psychic conflict determines the location of the focus or Hamer Herd in the brain and the location of the cancer in the organ.

The idea that a shock to the system can lead to cancer sounds intuitively correct as many people will affirm. The location of the shock in the brain being responsible for the exact site of the cancer in an organ of the body is a vivid insight.

At the moment of the conflict-shock a short circuit occurs in a pre-determined place of the brain. This can be photographed with computed-tomography CT and looks like concentric rings on a shooting target or like the surface of water after a stone has been dropped into it. Later on, if the conflict becomes resolved, the CT image changes, an oedema develop and finally scar tissue.

Doctors who regularly use CT scans should be in a good position to confirm the presence of these concentric rings.

How specific and precisely located these brain lesions are may be seen from the following: After a professional lecture a doctor handed him the brain CT of a patient and asked him to explain it. From this Dr. Hamer diagnosed the patient to have a fresh bleeding bladder carcinoma in the healing phase, an old prostate carcinoma, diabetes, an old lung carcinoma and sensory paralysis in a specific area, in addition to the corresponding emotional conflicts.

This is good evidence for Dr. Hamer's theory.

Amazingly, Dr. Hamer was able to show that at the same time as the concentric brain lesion appears, the target organ CT may show such a concentric lesion. According to Dr. Hamer this happens instantly when the psychic shock hits the subconscious level and this same second is the start of cancer. However, other diseases can be caused by the same mechanism. How severe a disease becomes may depend on other psychological, energetic and nutritional factors but its nature and location are determined by the content of the conflict shock.

Given the success of Dr. Hamer's diagnoses (see above) there can be little doubt that this is indeed the case.

Interestingly the German New Medicine (GNM) can predict when alternative medicines and chemotherapy will work and when they will not. For example when the conflict that causes a tumour is already resolved the tumour will heal naturally and the alternative therapist or chemotherapist gets the credit. If the conflict is not resolved no diet or no amount of chemotherapy will cure the tumour.

And with the several thousand case notes produced by Dr. Hamer he appears to be justified in making these statements.

According to Dr. Hamer it starts with the biological conflict shock, which has to be unexpected and dramatic. This can be experienced for example when you lose your job or when a loved one has to go to the hospital. On a brain CT-scan the impact of the conflict shock is visible as a series of concentric rings. This starts the disease. The biological shock will usually cause a great deal of emotional distress also. But you can have emotional distress without getting a disease.

It is the experience of many people that emotional shock can herald the onset of disease but it has taken the research of Dr. Hamer to codify this.

What we call disease, including cancer, Dr. Hamer calls Meaningful Special Biological Programs. These programs are controlled by the body and the brain. The purpose of the special biological program is to help you resolve the conflict. Only when the conflict is resolved does the second part of the program start, called the healing phase. De-stressing or thinking away the emotional stress does not get you in the healing phase.

In other words, there is no shortcut.

There are people who are miserable with their lives and they don't have cancer. What about them? Explain then a little kid getting leukaemia. What kind of emotional distress causes that? A lot of people get stressed and have dramatic things happen to them and they do not have cancer.

Dr. Hamer is well aware of these objections and has an answer to all of them.

According to German New Medicine leukaemia is caused by a self-devaluation conflict. A child could experience this for example after falling off a swing and breaking a shoulder. However, leukaemia is already the healing phase. The ultimate purpose of the program is to strengthen the bone.

Doctors have missed this almost entirely because they are incorrectly concentrating on curing the leukaemia.

Dr. Hamer has found for every cancer case he examined a corresponding biological conflict. So he has no need to suppose that there are substances that cause cancer. However there are toxic substances that can make you sick or kill you so it is wise to eat a healthy diet and avoid environmental toxins. There is absolutely no proof that carcinogenic substances have a direct effect on an organ without first going through the psyche and the brain.

It does seem that empirical proof is lacking.

The practice of animal experimentation and vivisection in the name of medical research is one of the darkest chapters in the history of mankind and in the history of medicine in particular. The number of animals dying of torture through the practice of vivisection is estimated at around 500,000 a day world-wide.

Through ruthless animal testing medical science has assembled over 1,000 alleged carcinogens proving only that researchers have found a thousand ways to distress animals with conflict shocks. During the experiment the helpless animals can suffer a number of biological conflicts: Attack conflicts, death-fright conflicts, existence conflicts, territorial conflicts, abandonment conflicts, self-devaluation conflicts and so forth - all causing cancer.

This is a terrible indictment of modern medical research.

Modern medicine is largely based on theories, statistics and countless new studies that are often contradicted by the latest research. Even though medical doctrines such as the concept of an immune system or of malignant and metastasizing cancers have never been scientifically verified the guardians of the existing medical paradigm have found effective ways to turn pure assumptions into truths, leading the public to believe that the unproven results are good science.

In this paragraph Dr. Hamer firmly puts modern medicine in its place. It is true that it is based on theories and statistics.

German New Medicine is founded on natural laws rather than on theories, on inductive reasoning rather than on postulation. Dr. Hamer's discoveries offer a complete scientific system that serves as a basis for an entirely new understanding of diseases.

Natural laws are obviously superior to theories and inductive reasoning has long been shown to be superior to postulation.

With the knowledge of GNM we are able to identify the causes of diseases, accurately predict their development and recognize symptoms that indicate healing. Empowered with the understanding of the five biological laws of the new medicine we no longer have to fear disease and thus can free ourselves from the controlling grip of established medicine.

This is very good news. Science should take steps to verify the five laws as soon as possible.

Interactive workshops on Dr. Hamer's scientific discoveries conducted by Daniel Miron and Julie Chrétien: Interactive training within the reach of everyone: A training where the trainer interacts with love, intuition and passion. A dynamic training where the grand themes of new medicine are addressed in a practical manner using case histories and the participants' real life experiences allowing them to progress quickly. A practical training which can then be used in day-to-day life.

This is a far step from the dry, academic nature of much of today's medical training. Dynamic training and quick progress are the keywords here. It takes far too long for today's moribund medical schools to turn out medical practitioners.

Some of the themes which will be addressed: The iron rule of cancer and the five biological laws of Dr. Hamer's new medicine. The phases of disease. The human body's anatomy and physiology according to new medicine. All the conflicts contained in Dr. Hamer's chart.

Disease: The brain's perfect solution. Conflicts: Low self-esteem, abandonment, territory, separation, fear of dying, dishonour, repugnance, resistance, etc. Conflict diagnosis: Disease, allergies, eczema, psoriasis, fibromyalgia, chronic fatigue, multiple sclerosis, paralysis, Parkinson's, diverticulitis, arthritis, rheumatism,

infections, migraines, ulcers, kidney stones, hepatitis, etc. Cancers and several other themes.

This seems comprehensive and stands comparison with best practice in modern teaching hospitals.

Duration and cost of the training: Two five-day modules at a cost of $600.00 per module. This cost includes 1,100 pages of pedagogical material, the Medicina Sagrada course manual and the précis of biological decoding of ailments and diseases. A $50.00 non-refundable deposit is required and is credited toward the cost of the first module. Payable in cash or by postal money order only.

And when we consider the enormous cost of medical training in the modern day these fees are extremely reasonable.

Daniel Miron and Julie Chrétien. Daniel: A great lover of nature, Daniel Miron has been observing human, animal and vegetable behaviour, as they relate to natural laws. A health professional for over ten years, Daniel is a Naturotherapist and consultant of New Medicine. He successfully integrates his knowledge of New Medicine with his facilitation activities. He keeps abreast of all of Dr. Hamer's recent discoveries and applies them in the therapeutic process with his clients.

Julie: Julie Chrétien holds a diploma in physical education and has studied psychology at the University of Ottawa. She is a practitioner of New Medicine, kinesiology, energy medicine, massage therapy and she is a reiki master. She now devotes all her time to her New Medicine consultant activities.

For over eight years she has been seeking her own truth, in her heart, in her relationships and in her work. She naturally blends her strong intuition with her deep knowledge and life experience to awaken human consciousness and encourage beneficial awareness in others. Participants greatly appreciate her constant support and availability.

Science should recognise the importance of Daniel's observations of vegetable behaviour and Julie seeking her own truth. Such talents have been overlooked so far in the battle against cancer.

In 1981, Dr. Hamer announced that cancer, contrary to what was believed, came from an unexpected psychic shock, a shock that catches the human or the animal on the wrong foot, what he calls the biological conflictual shock; that cancer develops, as long as the

biological conflict exists and could disappear spontaneously as soon as the conflict ceases if we let nature run its course.

This is a good restatement of the basics of Dr. Hamer's ideas.

Letting nature run its course means, for example, accepting the presence of mycobacteria, not intervening on the periosteum, not doing any chemotherapy, radiation, morphine. All of these treatments are non-biological and disturb the cycle of nature that has been operating for millions of years. These treatments are responsible for the statistics from the German centre for research on cancer in Heidelberg, namely: 98% of cancer patients who are treated die within 7 years and 95% die within 5 years.

These are telling statistics. Science should spread the knowledge that chemotherapy, radiation and morphine disturb the cycle of nature.

With German New Medicine, 95% of patients who are not treated survive.

Perhaps Dr. Hamer might consider renaming his insight 'German No Medicine'.

Before 1981 we had numerous theories regarding the origins of cancer but no one conceived it was possible that it could arise from a conflictual biological shock that was extremely brutal and dramatic.

This is undeniably true. But they were less enlightened times.

However, this hypothesis was already postulated several centuries ago and before that, in antiquity, but it has since been forgotten and considered non-scientific.

Yet again, we see here that the wisdom of ancient peoples far surpasses modern so called knowledge.

But another important point must be noted: It is the subject of tracks, which we discuss in more detail later. Everything that the individual feels at the moment of the DHS, whether it affects sight, hearing, smell or touch, as well as all the different aspects of the conflict, stays anchored within the person and can be seen more or less clearly on a cerebral tomography.

Scientists have simply not been looking closely enough at brain scans.

For example, a right-handed, 40 year old woman catches her husband in a sexual act with an attractive 18 year old woman in the marital bed. If she loves her husband she will likely have a sexual conflict but she will also have a conflict of self-esteem with respect to her partner resulting in a decalcification of the right shoulder.

If it is not an attractive young woman but a prostitute the sexual conflict will still be there but a conflict with the partner will be added which brings about a right breast cancer and a nausea conflict since a prostitute was lying in the marital bed and a disgust conflict hypoglycaemia because of the nausea experienced.

This is a real step forward.

All these different aspects that function almost like the biological conflict itself with different tracks are now present. What is to be understood by track: Each time we again experience one of the elements of the conflict we relive the initial conflict and return on that track.

For example: The colour of the hair, the shape of the face, the silhouette of the chest of the husband's mistress, her odour, her perfume, her voice etc, are reminiscent of the initial conflict. In the case of a later meeting with a woman reminiscent of this mistress the wife instantly finds herself on the track once again. This awakens the entire conflictual complex.

Which would neatly explain relapses.

It is impossible that the patient not see the DHS on his brain scan if he understands what to look for and the same applies to the medical doctor. One cannot avoid seeing a target configuration on a cerebral tomography. At the precise second of the DHS the organ is also affected by a cell multiplication or, in other cancer cases, by cell loss that leads to an ulcer or a necrosis tumour of the skin or of a mucous membrane.

This is a good guide as to what to look for. If a doctor still can't see it then the patient should point it out to him.

This criterion makes New Medicine since its beginning a science in the sense of the physical, natural and biological sciences. It allows us to reconstruct and reproduce each medical case which has never been possible with conventional medicine.

This is entirely the fault of conventional medical practice.

In order to pass their exams, all doctors had to memorize the 1,000 diseases of medicine with their symptoms and their therapies. Doctors knew approximately 500 cold diseases such as cancer, angina pectoris, multiple sclerosis, depression and mental diseases, and approximately 500 warm disease infections, rheumatism, leukaemia, osteosarcoma, Hodgkin's disease and so on.

What doctors were told were diseases were only one phase of the SBS for the cold diseases of the sympathicotonia. They had not seen

the warm phase of the vagotonia flu or other infectious diseases and for warm diseases they had missed the cold phase altogether or had mistakenly identified it as a disease in its own right. Therefore they could not understand disease and were unable to treat the patient truly scientifically.

This is extremely important and science should embark on a research programme to sort diseases into the cold and warm categories as soon as possible.

How ignorant we were, having presumed ten diseases in cases of bone cancer, even though it was in fact a single SBS: Bone cancer = osteolysis = osteoclast metastases: Plasmacytoma, osteosarcoma, acute articular rheumatism, primary chronic articular rheumatism; osteoblast metastases: Anemia, polycythemia, leukopenia and leukaemia.

It is true that modern day medicine makes things much more complicated than they need be. Science should take a leaf from Dr. Hamer's methodology and simplify things.

We can realize why we could not understand cancer, as long as we had not understood all the whys and wherefores, and first of all the mechanism of creation of our evolution in relation to our conflictual biological programs. That is why, in our ignorance, we had always considered that cancer was elusive, malignant, that it was a totally uncontrollable phenomena that evolved savagely and one that no-one could understand. All this is false.

Patients should certainly challenge any doctor who claims their cancer is malignant.

Cancer, and all the other so-called diseases that we now understand as SBS or Significant Biological Special programs, could not be more meaningful, logical and understandable. All is governed by the five biological laws of nature, which is sound from a scientific point of view, especially compared to conventional medicine's 5,000 unproven and unprovable hypotheses.

It is a little known fact that conventional medicine has 5,000 unproven and unprovable hypotheses and science should really acknowledge this.

We consider microbes as harmful agents that we must destroy at all costs. It's ridiculous. We have an urgent need for microbes, all the microbes that are present at our latitude. If, for hygienic reasons, we no longer have mycobacteria tuberculosis, we can no longer evacuate

our tumours in the healing phase. This has disastrous consequences for a great number of tumours.

It is indeed ridiculous that we seek to destroy microbes. It is important that science recognises this and ensures that tuberculosis microbes are not destroyed. To those doctors who question this, a patient might like to ask which is more dangerous – tuberculosis or tumour?

Mycobacteria have been in existence for nearly as long as monocellulars, therefore much longer than animals or humans. They have a well-defined role to play: They must caseify and destroy the tumours governed by the brain stem at the very beginning of the healing phase conflictolysis.

Doctors who insist that mycobacteria can cause diseases such as leprosy should ensure they quickly attend a course on German New Medicine.

Poor us, having believed that we needed to destroy tuberculosis. Indeed.

Nature's regulatory circuits can no longer function if we play sorcerer's apprentice and destroy certain elements. Almost everything we have done as modern-day doctors has been pure nonsense.

And it is the tragedy of modern day medicine that we continue to fail to realise this.

We then see that the microbes integrate completely in the biological process of the SBS. They grew like us and for us. They are also a link in the chain, something we ignored. That is why we blindly attempted to destroy these allies by the use of antibiotics and sulpha drugs. Microbes are not what kill us but rather the enormous edema that is formed in the brain if the conflicts last for too long.

This paragraph should be compulsory reading for medical students everywhere.

Viruses: These are not living organisms per se like bacteria but rather protein molecules of a complex nature that multiply exclusively during the healing phase, after the resolution of the conflict, and that help reconstruction of the skin ulcers or mucous membranes. They seem to be the friendly catalysts, as in chemistry, substances that, by their presence, have an effect but do not transform the chemical process. Viruses will be repressed when their work is done.

Science should institute research programmes with the aim of discovering just how friendly viruses are.

If we believed at one time that we had to push away all viruses this no longer stands: We must even make sure that viruses corresponding to a certain healing phase are really present.

Certain hospitals are already working on this as we speak.

We don't yet know if viruses are transmissible or if they can be produced by our own organism. They double in a protein environment, that's obvious.

Science should also make it a research goal to discover if viruses are transmissible.

It is astounding to realize that disease is a special program that is biological. This challenges not only the symptomatic therapies but also shows them to be absurd. Who would still want to intervene after this discovery of the marvellous cycles of nature, in this special program, in the biological sense of nature?

Doctors of all specialties would do well to ponder this message. It is a mistake to intervene. Instead, the proper function of a doctor should be to stand back and applaud nature's progress.

Diseases do not exist and have never existed in the sense that we understood them. They are simply different phases of a meaningful special biological program conceived by nature.

We now understand why 80 to 90% of animals heal spontaneously on their own, even in the case of cancer. Humans also healed spontaneously before, in the same proportion, before the existence of chemotherapy, radiotherapy, morphine, all this medicine where cynicism and potassium cyanide rule, and where the patient panics and thus becomes ill. And we now also understand why 98% of these terrified patients die, whereas 95% of patients treated with New Medicine survive.

Doctors looking to improve their statistics should stay calm and speak in a reasonable tone of voice. Avoiding sudden movements would also help.

For example, a fear will often create a nervous tic and a separation will affect the skin. Resistance will have an effect on the pancreas. Marking of one's territory will manifest in the bladder. Therefore, each disease carries its own conflict.

It is surprising science hasn't already noticed these correlations.

Disease is there to save our lives temporarily and therefore has its reason for being: To resolve the conflict. The greater the conflict the more serious the disease. There is a perfect correlation between each conflict and each disease on the one hand, and the intensity of the conflict and the seriousness of the disease on the other hand. Biology is perfectly logical and precise, to the specific cell. Disease does not come from the outside like an enemy but from inside, in reaction to a situation that will become deadly unless it is resolved. The origin of everything that happens is within us, so it is within ourselves that we must look for conflicts.

Microbiologists will shortly be able to confirm that disease does not come from the outside.

If we become aware of the conflict and work to resolve it in concrete ways, we reprogram our brain to a new reality. Thus, the brain will drop its former program and adapt to the new one. There will then be no stress when a similar event happens in the future, since the reaction will not be as it was before.

Computer scientists have long suspected that the brain is a sophisticated computer.

New Medicine, one of Dr. Ryke Geerd Hamer's scientific discoveries, was made in the early eighties. Basing his work on the study of brain scans, he treated over 40,000 patients and confirmed that so-called diseases were recorded in the brain, thus creating a pathology affecting an organ.

It is through brief psychological talk therapy with each patient that he verified that all those who had the same pathology had a lesion on the brain at exactly the same place and that the conflict at the origin of the pathology was the same for each of the patients. Dr. Hamer discovered, demonstrated and proved the five biological laws, the foundation of New Medicine. To this day, not a single case has demonstrated an exception to these biological laws. This is 100% verifiable, it is not a statistic, it is a law.

This differentiates New Medicine from the medicine practiced today in our hospitals. There, everything is statistics, from surgical techniques to patient response to drugs. New Medicine does away with all of this by introducing laws into medicine for the first time.

One single DHS can impact more than one brain area resulting in multiple 'diseases' such as multiple cancers, erroneously called metastasis. For example: If a man unexpectedly lost his business and the bank takes all his assets he can develop colon cancer as a result of an 'indigestible morsel conflict' ('I can't 'digest' this!'), liver cancer as a result of a 'starvation conflict' ('I don't know how to provide for myself!') and bone cancer as a result of a 'self-devaluation conflict' (a loss of self-esteem). With the resolution of the conflict all three cancers will go into healing at the same time.

'Metastasis' has long been suspected to be shorthand for 'the cancer has spread but we don't know why'. But now there is no need for such confusion. What could be more satisfying as an explanation than the above?

Imagine a world in which three cancers can be cured by a visit to the jobcentre. This is the promise of German New Medicine. Imagine the money it will save and the wasted lives it will renew. Verifying GNM must become one of science's priorities in the very near future.

Chapter 12: Golden Ratio

Evolution uses golden ratio neural patterns by Randolph Fabian Directo: Leonardo Fibonacci of Pisa revealed a numeric sequence in which each subsequent number is the sum of the previous two: 1, 1, 2, 3, 5, 8, 13, 21, 34, 55, 89 and so on. The sequence in turn gives rise to several unique ratios including .618, .382 and 1.618 aka the phi or Fibonacci ratio.

These ratios are well known to mathematicians but their extension into neural patterns has not so far been the study of science. This may be about to change.

These ratios exist throughout nature in the form of fractals or golden sections. Although they have been employed in man-made structures like the pyramids, in sacred geometry they have always existed throughout all of nature, living and non-living. We can reveal systems of fractals based in the golden ratio.

Fractals are a relatively new invention of mathematics but again it seems as though ancient people have beaten us to it. Sacred geometry is but the earliest example and science should have pricked up its ears way before today.

Robust fractals permeate all life forms; they are a combination of self-identical and indefinite fractals. Trees are branching robust fractals as are living circulatory, bronchial and nervous systems.

Although late on the scene this is all now part of mainstream science.

Inventor of the Neurophone, Patrick Flanagan, conducted experiments to determine the effect of various phi-based geometries on muscle strength. Pyramid shapes based on the golden ratio made people very strong. The Neurophone is superior learning technology based on the golden ratio as well as energy healing devices.

Here we see Flanagan pushing the boundaries. Science should be able to confirm his results concerning phi-based geometries and muscle strength though the effort will undoubtedly have to take the form of a multidisciplinary approach since it is unlikely that any single scientist by himself will have the necessary knowledge from the fields of mathematics, physics and biology which confirmation seems to demand.

This is part of the problem with science in the modern era. Individual scientists have become too specialised and have therefore lost sight of the bigger picture which researchers like Flanagan possess. A team of specialists should be assembled to quantify the strength of people exposed to pyramid shapes based on the golden ratio. A second team should be formed to confirm the superior learning technology of the Neurophone and this team should be expanded to include leading educationalists, sociologists and experts with experience in the field of energy healing devices.

Penrose and Hameroff provocatively suggest that consciousness emerges through the quantum mechanics of microtubules. Microtubules are composed of thirteen tubulin and exhibit 8:5 phyllotaxis. Clathrins, located at the tips of microtubules, are truncated icosahedra, self-identical fractals or golden sections.

For those interested Penrose has authored a book specifically about microtubules and their possible link to consciousness titled 'Shadows of the Mind'.

DNA exhibits a phi resonance in its 34:21 angstrom Fibonacci ratio and the cross-section through a molecule is decagonal, a double-pentagon made of golden sections.

Here we see the basis of life in terms of the golden ratio.

The human interface: Our brainwaves or neural patterns can be entrained by external stimuli through the frequency following response. For example, a short time after listening to a steady rhythm like a drum beat our dominant brainwaves follow that beat. The evoked cortical response also causes our autonomic functions to follow that rhythm like our heart beats.

This sounds feasible.

The human body's neural system contains a synchronization pulse or synch pulse like a metronome. The synch pulse represents our normal state of equilibrium. Music enters through our ears and creates a pulse position like a shadow, shape, or colour contrast with regard to the synch pulse. Our nervous systems utilize pulse positions so we can sense depth in the physical world.

Depth sensation was once crudely thought to be a function of stereoscopic vision but clearly there had to be more to it than that.

One neuron processes thousands of signals or pulses from different sources of stimuli simultaneously. While each neuron transmits our resonant life beat or synch pulse, each one also transmits all sorts of semi-random pulses as our organism interfaces with the environment.

That synch pulse is a set of floating harmonics based on the golden ratio, the integral of our brainwaves combined with our other life processes.

This is very new and it should be an urgent goal of science to confirm these discoveries.

Knowing the golden ratio: Unlike the fuzzy logic we use to identify people, places, and things, we are far more subjective when we sense the golden ratio. We tend to see beauty and feel pleasure when we encounter forms and functions based on this ratio which is integral to creation and life; this is why we're attracted to the intense sensory experience provided by nature.

Amongst the interdisciplinary teams we now need to include psychologists.

People have recognized this and built structures based on sacred geometry like Stonehenge and other megalithic structures, not only because of their strength and beauty; they also resonate with the strength and beauty within our minds. We want to achieve perfection because we realize that nature is not perfect.

And also archaeologists. They will provide accurate measurements taken at the sites of megalith construction which will confirm the ground plans and elevations of ancient structures to be based on sacred geometry.

Physical nature dances around the golden ratio; like a spinning gyroscope interpolating it's central axis, physical nature is not exactly cantered on perfection. We humans are the ones who measured, interpolated, and extrapolated this fundamental nature of perfection. We are the ones who focus on perfection provided by the golden ratio. This begs the question: What happens when we resonate with a function that focuses on perfection?

Resonance is a well understood concept in physics but only when it relates to physical objects such as bridges oscillating under the influence of wind (Tacoma Narrows) or people walking (Millennium Bridge, London). But science has been found wanting when it comes to describing how people themselves resonate with a function that focuses on perfection.

According to Scott Olsen, instructor in philosophy and comparative religion at Central Florida Community College, in a state of phi-induced quantum coherence one may experience Samadhi, cosmic conscious identification with the awareness of the Universe itself.

Quantum coherence is at the cutting edge of current research in physics though up until now it being phi-induced has been overlooked by all the research teams I am aware of. Yet it seems that Scott has hit the nail on the head here and his discovery should form the basis of one of sciences next steps forward. It is difficult to think of a more important link than that between the ratio of two lengths and cosmic conscious identification with the awareness of the Universe itself.

Therefore, when we are entrained by stimuli based on the golden ratio we transcend our lower level consciousness to be one with our bodies and our true nature; this has been our goal for ages.

And now it should become the goal of mainstream science.

Author's bio: Randolph Fabian Directo, known as HealingMindN on the Web, researches neuroscience and psychoenergetics. He counsels people with various chronic ailments caused by emotional barriers. Most people have mental barriers that prevent their psychic awareness.

OceanScape GhostShip: Altered States. A neural entrainment program based on the golden ratio helps to remove those barriers and induce your full potential. Also see his power nap and sleep induction programs based on the golden ratio available through HealingMindN.

As this program is based on the golden ratio it would seem that the scientific foundations for OceanScape GhostShip are firmly secure.

OceanScape, GhostShip: Altered States helps you achieve vivid extra sensory perceptions using the looming power of the ocean riding upon mathematically precise brainwave entrainment.

A major goal of neuroscience has been a reliable method of training brainwaves in a mathematically precise way. The looming power of the ocean has, of course, been a research project of oceanographers for many decades now.

An ocean ambience is gently modulated from alpha rhythms into a theta meditation session with gamma bursts. Gamma bursts during meditation is often found in mediators who have lucid visualizations or out of body experiences. Key frequencies are also calculated from the Schumann resonance using the golden ratio; this induces a very other-worldly feeling.

Key frequency calculations will be familiar to undergraduate students. Post-doctorate research concentrates on calculations incorporating the Schumann Resonance though even advanced researchers have not yet formally announced the use of the golden ratio to peer reviewed journals, though their utilization has been rumoured for many years. It is now time for science to come out into the open and admit that Randolph Fabian Directo simply beat them to it. There is no shame in following prior precedent.

Spiritual manifestations occur in many forms. Upon listening to OceanScape GhostShip: Altered States, you will notice unusual synchronicities coming into your life - which will never be the same.

Imagine yourself walking along your favourite seashore in the early morning. The Sun denotes an eerie glow to the cloud cover and the water. You walk further and deeper along the surf than you ever have. In the distance, you hear a steamship.

The ship's horn seems to get closer the further you walk along the shore. When you finally see the ship it looks abandoned and adrift. You wonder to yourself if anyone is on it, then you suddenly find yourself sailing deep into the ocean adrift on this ship.

As you sail deeper and deeper into the sea you feel the ship rocking and creaking with the waves; it looks very old and you wonder if it will hold together as there seems to be a storm approaching. You search the ship and discover that you're not alone. The storm has many messages, many stories to tell you.

You wonder if you're dreaming or awake but the feeling is unmistakable. The ship rocks you back and forth like a cradle at sea that somehow makes you feel safe. The passengers on this ship want to tell you stories as well. In that time, you come to understand. As the storm passes you feel yourself returning to the shore. Your friends return with you, but only you can sense them...

Modern Science should stop its obsession with the overtly physical phenomena it is currently concerned with. There is much more to be learned from studying the Golden Ratio and the unusual synchronicities which follow from it. Now that the Higgs Boson has finally been discovered the Large Hadron Collider (the most powerful synchrotron in the world) could be put to even better use investigating these synchronicities.

Caution: Do not operate heavy machinery or dangerous equipment while listening to this recording.

This is undoubtedly wise advice.

Chapter 13: Homeopathy

NaturalNews: In this article, I would like to dispel a plethora of myths surrounding homeopathy which have been used to discredit this highly efficacious healing art and science. Homeopaths are given few opportunities in the media to defend their profession so a lot of misconceptions abound. The medical profession in general presents a fierce and blinkered opposition, yet as Big Pharma is learning of all sorts of amazing cured cases they are determined to stamp out competition via EU regulation.

Big Pharma is undoubtedly making lots of money by denying homeopathy its rightful place in medicine and science should stop being at the beck and call of these large drug companies and stand up for itself.

Myth No. 1 – Homeopathic medicines cure nothing: Homeopathy works by stimulating the body's own healing mechanisms through like for like. A substance that would cause symptoms in a healthy person can be used to cure the same symptoms in a sick person by giving a minute, highly potentised dose of that substance acting as a catalyst to jump start their own healing mechanisms. Every one of us has our own natural innate healing powers. All that is needed is the correct stimulus to kick start it.

This seems completely rational.

In healthy people this may just be rest and good food but many people become 'stuck' in their physical, emotional or mental illness and cannot recover.

We all know people like this.

Of course there are different levels of health and the choice of potency given should reflect that. Low potencies are given for very physically ill people and higher doses for those whose problems are emotional or of the mind. Homeopathy is very successful in treating emotional problems such as stress, anxiety and fears.

Again, this seems uncontentious.

Unlike orthodox medicine, outcomes of homeopathic treatment are measured by the long term curative effects and the eradication of the disease state culminating in complete restoration of health. If we could have two year trials of outcomes for conditions such as asthma, arthritis and other chronic diseases, this could be proven.

It is of course shocking that science refuses to complete long term studies of homeopathic cures. Big Pharma flexing its muscles again, one can only assume.

Myth No 2 – Homeopathic medicines are just water: Homeopathic medicines are not made using only dilution. Dilution alone would do nothing whatsoever. Many homeopaths are getting tired of reading this highly inaccurate reporting in the media. All homeopathic medicines are made by a process of dilution and Succussion (potentization through vigorous shaking, 100 shakes between each potency i.e. between a 1c and a 2c, between a 2c and a 3c potency, between a 3c and a 4c, etc., etc.).

Of course mere dilution would do nothing. But the point is that it is *succussion* which is of paramount importance. Shaking the diluted medicine holds the key and it is high time science devoted its considerable resources to examining shaking diluted medicine more fully.

Most homeopathic medicines can be bought in either 6c or 30c from Boots or from health shops. Higher potencies of 200c and 1m (1000c) can be obtained only from homeopathic pharmacies. Succussion is nowadays done by machines, originally by hand. Succussion brings out the formative intelligence of the substance and imprints it upon the 60% distilled water + 40% alcohol medium used to make homeopathic medicines, alcohol acting as a preservative.

The formative intelligence of substances has, sadly, not been studied by science as much as it could have been.

Myth No. 3 – Homeopathic medicines are unscientific: Homeopathic medicines undergo a scientific 'Proving' where a control group of 50+ healthy volunteers ('Provers') are instructed to keep taking a remedy under trial until they develop symptoms which they must record in detail. Substances that have been rigorously tested include nearly everything on the Periodic Table - metals, minerals and gases, as well as plants and even things like snake venom.

This is the scientific method in practice.

The Provers are given a bottle of a new remedy being tested in the 30c potency and must keep taking it until they develop symptoms, which must be carefully recorded and then submitted to a database. The Provers must be healthy and symptom-free to start with so that the symptoms they experience are new ones caused by

the remedy. They must keep a careful daily note of what happens and not discuss it with any of the other Provers.

Whatever symptoms the Provers all experienced in common become the black type symptoms of the remedy which are then added to the Materia Medica of homeopathic medicines and Homeopathic Repertory (encyclopaedia of symptoms). Thus the curative indications of a remedy are obtained for clinical use.

All of which is compliant with today's best practice.

Symptoms have also been obtained through historical records of accidental poisonings, such as Arsenic and Belladonna. For example, poisoning by Arsenic causes vomiting, diarrhoea, restlessness, anxiety and extreme chill. Therefore you might get a patient in this state (possibly after food poisoning) and Arsenicum in a homeopathic tablet will quickly alleviate them. All of the remedies tested have been diluted and succussed (potentized), so they are not toxic like modern drugs.

This is a telling point against modern drugs and in favour of diluted shaken medicine.

The Homeopathic Materia Medica and Repertory are extremely large books divided into volumes. The Repertory is divided into sections in this order: Mind, Vertigo, Head, Eye, Vision, Ear, Hearing, Nose, Face, Mouth, Teeth, Throat, External Throat, Stomach, Abdomen, Rectum, Stool, Bladder, Kidney, Prostate Gland, Urethra, Urine, Male, Female, Larynx, Respiration, Cough, Expectoration, Chest, Back, Extremities, Sleep, Dreams, Chill, Fever, Perspiration, Skin, Generals. Obviously some sections are bigger than others.

A large volume of supporting literature is a feature of a mature, well-developed science and homeopathy clearly passes this test.

In the various Repertories, remedies are listed alongside the full range of symptoms (rubrics) in abbreviated form, all information being systematically taken from Provings and clinical practice. Every human state of mind, emotion and body is listed.

Symptoms that would mean nothing to a medical doctor can be looked up and the curative remedy found in these huge books. Homeopathy is a study of human nature and how negative states of mind and emotions affect the physical body culminating in illness. Nowadays many homeopaths use computer software programmes which contain all this information.

Bringing things right up to date via the use of software is a natural progression.

Myth No. 4 – Homeopathic practitioners receive inadequate training: In fact, all qualified homeopathic practitioners undergo a four year training course at accredited Colleges, which includes Anatomy and Physiology, as well as Pathology and Disease, Materia Medica, Homeopathic Philosophy and study of the Homeopathic Repertory. Yet medical doctors and nurses treat after much shorter courses. To be really good, you need to study intensively for about 10 years. Homeopathy is a lifetime's work and you never stop learning.

Accredited Colleges teaching a four year course yet medical doctors allowed to practice after much shorter courses. The evidence is adding up.

Myth No. 5 - There are no studies that prove homeopathy works: In the past 24 years there have been more than 180 controlled, and 118 randomized, trials into homeopathy, which were analysed by four separate meta-analyses. In each case, the researchers concluded that the benefits of homeopathy went far beyond that which could be explained purely by the placebo effect. Another meta-analysis found that 65 of the 89 trials analysed had produced an effect way beyond placebo, (source: What Doctors Don't Tell You).

And that is the clincher.

A study of 6500 patients at the Bristol Homeopathic hospital was conducted showing that over 70% of patients reported complete cure or significant improvement of their symptoms.

6,500 patients and 70% cure or improvement is statistically very significant.

Historical records show that epidemics such as cholera and typhoid were treated successfully using homeopathy in the 19th century with very high success rates compared to orthodox medicine.

These successes have, of course, been buried in the mountain of unfavourable comments generated over the years by opponents.

Homeopathy can never be tested properly through conventional trials because each prescription is individualized as every person is unique. Therefore 10 people with arthritis, for example, may all need a different homeopathic medicine. So it is far from ideal to follow the allopathic trial paradigm to test homeopathy. In orthodox medicine trials, all are given the same medicine to be tested. In homeopathy, all may be given different medicines.

Even modern medicine is moving towards individual, tailored treatments, having belatedly recognised that people are not the same.

Anybody who has an understanding of the principles of homeopathy can be left in no doubt that we are dealing with a scientific therapeutic method in the best possible sense: It is based on observation, facts and phenomena and follows the rules of inductive logic that can be tested in daily practice. It is a comprehensive and comprehensible mode of therapy which in some countries is first line treatment for the whole range of acute and chronic conditions. It has been proven abundantly that it is superior in the treatment of epidemic diseases to allopathy.

Allopathy is the name given to modern medicine which is drug based.

'It is amazing how people, who like to see themselves on the side of unprejudiced evaluation, can be so blinkered. People pass judgment on homeopathy who have never bothered to study it. Like any science it takes time to learn (especially to learn it correctly) and years of practice to master but the rewards for patients, practitioners and the NHS (the British National Health Service) purse are great.

Before those who preach pure science come down on therapies like homeopathy too heavily, they should ask themselves how many of the accepted treatments within the NHS have a scientific evidence base?' (Peter Morrell. Hon. Research Associate, History of Medicine, Dept. of Sociology, Staffordshire University, UK.)

Here we see a distinguished member of a department of Sociology finally breaking through the wall of silence built up by mainstream science. Where Sociologists lead many doctors may soon follow.

With every homeopathic medicine we know exactly the substance it was made from, unlike most modern drugs where we have no clue of the ingredients. This is ironic too as all natural health products, whether vitamin, mineral or food supplement, must clearly state on the label every single ingredient.

When we go to the supermarket or health food shop, we hold up the packet or bottle and read what is in the product, yet people happily swallow prescription drugs with no idea whatsoever what they have taken! They could contain cyanide or any poison and the patient would be none the wiser.

How can ordinary people be expected to know what is in their allopathic medicine? To give an example here is a list: Acetaminophen, brompheniramine, chlorpheniramine, dextromethorphan, guaifenesin, pseudoephedrine. What are these? They are ingredients in a common cold remedy.

Those who claim to be scientists should have the ability to at least try to understand different paradigms. If not, they look more like people who have settled on a comfortable view of the world which might soon look very outmoded indeed. As the great musician and conductor Sir Yehudi Menuhin once said: 'Homeopathy is one of the few specialised areas in medicine which carries no disadvantages but only advantages'.

Science should listen more to Sir Yehudi.

Myth No. 6 - Homeopathic hospitals are a waste of money: There are 5 homeopathic hospitals in the UK -- in London, Liverpool, Tunbridge Wells, Bristol and Glasgow. They cost the NHS around £6 million a year. Compare that to the £100 billion for the total 2008 annual NHS budget! These homeopathic hospitals save money for the NHS as the Smallwood report commissioned by Prince Charles has demonstrated

Prince Charles is well known as a staunch proponent of homeopathic medicine. He is beyond reproach, cannot be influenced by big drug companies and has enough wealth to be able to afford any treatment in the world. Yet he speaks up for homeopathy.

Myth No. 7 – Cure with homeopathy is simply the placebo effect: When Prince Charles treats his farm animals at Highgrove with homeopathic medicines do they know that a remedy has been put in the water they drink? Farmers successfully use homeopathic medicines for their cows suffering from mastitis.

Does a tiny baby know when their fever drops dramatically using Belladonna or Aconite, that they have been given a homeopathic medicine?! As anyone who has treated animals and babies with homeopathic medicines will tell you, homeopathy works even better on them than it does on adults! If proof were needed, this is it. Not placebo.

No-one could possibly doubt this proof. It is now up to science to release the findings to the public with their full endorsement. They must come out of their shelters and take on the big drug companies.

Myth No. 8 - Homeopathic medicines contain no molecules: Any remedy under a 12c or a 24x potency still contains the original molecules of the substance and this is known as Avogadro's number. These low potencies are most suitable for physical illness of long

duration as well as to heal specific organs that are not functioning properly.

Avogadro's Number is a well-known constant in science. Its value is 602,214,150,000,000,000,000,000. This is a very large number of molecules.

Myth No. 9 – 'Anecdotal Evidence' does not constitute scientific evidence: Most medical, surgical procedures and drug usage are not backed by studies, only by anecdotal evidence. According to the U.S. Government's Office of Technology Assessment (Congress of the United States, Office of Technology Assessment: Assessing the efficacy and safety of medical technologies. Washington, DC: US Government Printing Office, 1978), only 10-20% of all medical procedures and off-label drug usage are backed by clinical studies. Strong anecdotal evidence among informed professionals is actually quite reliable, at least as reliable as clinical testing.

In fact, it makes one wonder why science bothers with clinical testing in the first place. Perhaps they would be better off just asking around.

The problem isn't with the use of anecdotal evidence. It's with the double standard applied by the establishment (medical and regulatory) that holds complementary medicine to an absurdly higher standard, allowing medical doctors to do pretty much whatever they want. If informed anecdotal evidence is allowable for 85% of all medical procedure and drug usage, why is alternative health held to an impossible 0% standard?

This could be because big drug companies want things to stay the way they are.

Millions of people worldwide testify that homeopathy cures their illnesses yet apparently that cannot be construed as 'evidence'. If a person were to walk out of their house to the town centre and witness someone having their bag snatched or witness a car accident then when they relay this information to the police or to their friends and family, it is anecdotal evidence. If someone goes on holiday, stays at a nice hotel, eats delicious food, swims in the sea, comes back home and relates the holiday to their friends, that is anecdotal evidence. Does that mean that the above never happened? According to the detractors of complementary or alternative medicine, yes it does!

Which demonstrates how silly the detractors of complementary and alternative medicine can be.

Millions of people have been cured of their diseases and afflictions using homeopathy, herbs, healing, vitamin supplements, special diets and on and on. Yet according to orthodox medicine all of these cures are anecdotal evidence and as such do not merit any further investigation, study, or validity. As far as orthodox medicine is concerned, these cures never happened.

Yet what if someone witnessed a car accident and the police wanted them to make a statement? Would the statement in court be dismissed as anecdotal evidence? Would the police, even if they arrived at the scene of the accident to find the person still there comforting the passengers or trying to help, say they had not been there and their evidence is non-existent? I don't think so.

And neither would any reasonable person.

So how for so long have we put up with the top dogs in the medical establishment dismissing our cures as total nonsense, figments of our imagination, placebo cures, or outright lies? How, when millions are cured around the world using homeopathic medicines, can these cures be dismissed as unworthy of attention, simply 'anecdotal evidence'.

To repeat what is becoming increasingly obvious, there is a good chance this is happening because of the malign influence of big drug companies aided and abetted by compliant medical doctors.

Orthodox medicine implies through this that all cures with alternative medicine are untrue or simply imagined. When all the evidence is put before them they become angry and even aggressive, simply refusing to see or to listen.

Everyone will recognise the way orthodox medicine behaves when challenged.

All the case notes in the surgery show that Mr. A had arthritis for 5 years, had been on anti-inflammatory medicines, yet after homeopathic treatment the arthritis is cured. The reaction of the doctor is either disbelief or an attitude where they will not talk about it and do not want to know.

And even when presented with case notes such as those from Mr. A doctors continue to behave in the same way. Yet the case notes show that Homeopathy cured Mr. A. so something is very wrong at the heart of modern medicine and it is high time it was exposed.

Of course there are some orthodox doctors who practice acupuncture, homeopathy or herbs themselves and who do believe

that these therapies cure the patient but they are in a small minority. The opposition is always the top cancer specialists and professors whose lives and vested interests are the most challenged by the idea that anything other than pharmaceutical drugs or surgical interventions can cure the patient.

Naturally. They should be made to see reason.

Very often the doctor's prognosis can create enormous fear in a patient making them much worse, striking terror in their hearts and creating a mental block to healing when they are told by 'experts' they will never get better.

Everybody knows someone for whom this is the case. Doctors should not strike terror into patients hearts, instead they should explain things gently and in a reasonable way and then prescribe Homeopathic medicine.

Pharmaceutical drugs cure nothing. They merely suppress the symptoms driving them deeper into the body of the patient. Believe it or not, the disappearance of symptoms does not equal cure! Very often a new and deeper set of symptoms are created which are even more serious. Pharma drugs work through the Law of Opposites, e.g. Antibiotics, anti-inflammatories, anti-convulsants, anti-hypertensives, anti-depressants, anti-psychotics, etc, etc.

And this brings us to the crux of the matter.

Hence the eczema patient whose skin symptoms have been suppressed, goes on to develop asthma. The arthritic patient whose joint pains are suppressed, eventually will go on to develop heart disease.

There can be little doubting the careful demonstration of cause and effect delineated here.

The doctor makes no connection whatsoever that their drugs have created these deeper illnesses but just goes on to give the patient more and more powerful drugs, making the patient sicker still. Then when they die, they say, 'We did everything we could'. Yes and you killed the patient!

When put like this it seems reasonable for everyone to question their doctor and his prescribing habits very carefully indeed.

So there is no question that dismissing cures as anecdotal evidence through the use of natural medicine is nothing more than a whitewash and a desperate means of concealing the knowledge of those cures from the public as a whole.

The drug companies, in cahoots with the medical profession as a whole, are very desperate indeed. This is clearly demonstrated by the whitewash they so freely apply to cures obtained through the use of natural medicine.

Samuel Hahnemann: 'Hahnemann was a doctor but gave up his practice because he was appalled at the poisonous side effects of most available medicine. He started experimenting and did something rather novel -- he took some quinine, while perfectly healthy. He observed that the effect on him was identical to a malarial attack: alternating fever with heat and chills. This is where homeopathy started: A substance, given to a healthy individual, causes symptoms. If given to someone who suffers those symptoms, it will thus neutralize the sickness.'

This magnificent leap of the imagination is, of course, totally ignored by modern doctors.

Hippocrates stated that there were two laws of healing, the law of Opposites (allopathy) and the law of Similars (homeopathy). A Greek physician called Galen had laid these rules down in about 150 AD. Homeopathic theories are based on fixed principles of the Laws of Nature which do not change unlike medical theories which are constantly changing! Homeopathy is both a science and an art.

That these Laws of Nature have been around for so long gives us confidence in their truth. Modern medical theories, on the other hand, can be fairly criticised for changing far too often. This is a fault with modern medicine which science would do well to correct.

Treating the Whole Person or Holistic Healing: We are not just a collection of parts to be fixed as doctors treat us but are always operating as a whole person all of the time. In other words medicines are chosen that treat the whole person and not just the part. This may seem strange to grasp and yet doesn't it in fact make total sense?

Do we leave our sore throat on the desk of the physician as we leave the surgery? Or our arthritic knee behind? No, every single tiny function of our body operates as a whole, all of the time. You cannot treat one thing and not affect the rest. That is why pharma drugs are so dangerous, as for example, in treating a cancerous tumour, the medicine will affect and disturb the other systems of the body.

This makes sense.

We are all energy beings. The electricity in our bodies transmits messages to all parts/systems of the body. Illness is caused when these messages are not getting through.

Modern doctors have been slow to pick up on this important insight. It seems intuitive that illness will be caused when messages fail to get through and it is suspected that electricity in our bodies is responsible for message transmission.

All systems of the body are communicating with each other at all times. Water is a great conductor of electricity and it transmits the electrical current. This is how homeopathic mediums work, by communicating a current/pattern/frequency of energy via the whole human body to jump start the body's own inherent healing mechanisms.

The ability of water to conduct electricity is well known. As such, it is sensible to conclude that it is the water in our bodies which transmits the electrical current.

Homeopathy treats different sorts of people with distinct characters and personalities as well as different physical looks and natures. It individualizes each person treating their diverse pattern of symptoms looking at them as a whole.

It seems reasonable that good-looking people should be treated differently.

Is it not true that no two people are alike? That every person is unique? This is why you could line up 20 people with asthma and they might all need a different homeopathic medicine. There are in fact about 250 homeopathic medicines for asthma but the correct one for each person must be selected taking into consideration such things as what makes the condition better or worse, what time of day it comes on, whether the person is hot or cold, worse for damp, in need of fresh air or prefers the windows closed and so on. You would be amazed at how each person's symptoms are so different and yet all have been diagnosed with asthma.

Today's doctors simply refuse to take account of whether their patients prefer the window closed. This is the tragedy of modern medicine.

After homeopathic treatment, careful analysis is taken of the Direction of Cure of the patient's symptoms. Constance Hering was a converted sceptic of homeopathy. He was the first to talk about the Law of Cure which says that symptoms are cured from above down, from the inside out and in the reverse order of their appearance.

This has stood the test of time in clinical experience. A simple example would be after a curative remedy is given for eczema all over the body, we would see the eczema start to move down and when it is only on the ankles, we know it is nearly cured.

Brain surgeons have followed suit by starting at the front and working backwards (provided the patient is lying down).

The other important point about homeopathic treatment is that very often the appearance of an illness stems back to an important event in the life of the patient such as a shock, fright, loss or grief suffered by them. The homeopathic practitioner will always enquire whether there was a life changing event that severely affected the patient.

It is extremely common to find that the onset of a condition coincided with a major event. The homeopath will select a remedy that corresponds to the way the patient reacted to that event, mentally, emotionally and physically, in order to clear the state which caused the illness. In other words you treat the cause to remove the effect.

Cause and effect is one of the mainstays of science so here again we see the scientific method clearly infusing the practice of Homeopathy.

Examples of remedies for people who have undergone severe shock would be the following: If a person becomes very tired after a shock it would be Phos. Ac. When they become indifferent to their surroundings and loved ones, if they are just sitting there, not moving, staring in front of them, not speaking, it would be the homeopathic remedy Opium. If they become terribly restless and anxious it would be Aconite. Arnica is for shock when people say they are fine but they are obviously not.

Ignatia would be for patients in floods of tears, hysterical, slamming doors and telling people to go away. Platina would be for very proud, angry and indignant patients. These are all possible ways people can react from a shock and homeopathy must treat the individual. The trick is to work out how people are behaving and which one they need!

Modern doctors ought to be ashamed of themselves that they do not go into this level of detail with their own patients. Given its obvious superiority I would not be too surprised if these very doctors seek out homeopathic remedies when they themselves become ill. That they then keep quiet about it is nothing short of a scandal.

Homeopathy works on fixed principles that correspond with the Laws of Nature. The body has its own intelligence. That is why the human race has survived. When a baby is conceived, Nature chooses the best genes from both parents in order to create a stronger, healthier human.

Modern-day geneticists will hopefully soon confirm that on conception, nature steps in to select the best genes from both parents to create a stronger, healthier human. This is known as Natural Selection.

The human race has survived because we all have an innate healing power in our bodies. In homeopathy this is called the Vital Force. Homeopathy stimulates the Vital Force to heal the body, through like for like (using a potentized substance that would cause the symptoms but in a tiny dose acts as a catalyst for healing).

Biologists are investigating the Vital Force in increasing numbers, partly as a response to the success of Homeopathy and partly because they have been trying to locate the innate healing power in our bodies.

If people want to improve their looks, homeopathy does just that. When you are healthy and well, you obviously look better! Homeopathic practitioners believe in prevention, having treatment can prevent illness rather than leaving it to the surgeon's knife. There are thousands of homeopathic medicines which treat every ailment known to man, truly the most wonderful science on this planet.

And such a well-kept secret. Not for the first time one wonders about the power and reach of large drug companies who are able to ensure the secret remains secret.

Many people buy self-help books or think they can treat themselves with over the counter remedies. This is a short term solution. The reason is as stated above. You cannot treat individual symptoms without taking into consideration the rest of you! Only a qualified and experienced homeopath who will spend 1-2 hours taking your full medical history and all of your symptoms can prescribe the remedy that fits best.

In other words if you have hay fever, the homeopath will take into consideration all other physical symptoms as well as your personality to come to the correct prescription. Itchy, watery, red eyes, worse morning and evening would be Sulphur but only if all the other things about you fit the Sulphur picture.

This is where people will make the mistake of thinking they can prescribe Homeopathic medicine for themselves and thereby take a shortcut in the hope of saving money. But this is a false economy. Only an expert Homeopathic practitioner will have the necessary expertise and of course that expertise must be paid for, as in any other walk of life.

You cannot prescribe for yourself as you cannot take all of it into consideration at once. So, if for example you buy Natrum Mur. for your hay fever, it may work for a bit if you are healthy but the hay fever will come back, will not be cured for good, because it was not the remedy that fitted best.

Only an expert can take everything into consideration all at once and a layman would be foolish to try.

Homeopaths believe that illnesses manifest for three reasons: Firstly, they are genetically inherited from our parents, grandparents and forefathers. Secondly, they can be caused by a traumatic event such as death of a loved one, divorce, job loss or any event that has a serious impact upon the person. Thirdly, they can be caused from drugs taken by our parents (passed on to the foetus) or by ourselves. There are also of course accidents and injury.

In recent years science has been moving away from the outdated germ theory of disease towards this new paradigm.

Inherited disease can be traced back to one or more of what homeopaths call Miasms, these are syphilis, gonorrhoea, psora (scabies), tuberculosis and cancer. We are all a mix of all of these. However one or more of the miasms is uppermost in a person and is an important aspect of the case-taking to determine the appropriate medicines.

It may startle otherwise respectable people to learn that they are a mix of syphilis, scabies and cancer but disease is no respecter of persons so they will just have to get used to it.

So many people are in ignorance of the vast amount of study needed to become an expert in this field. Also there are hundreds of homeopathic books only available at specialist bookshops, many printed in India, where homeopathy is more popular than orthodox medicine

Scientists should organise book buying expeditions, especially to India, so that the shelves in their laboratories contain definitive homeopathic texts.

The truth is that homeopathy is getting ever more popular and the drugs companies are putting their spin in overdrive through their science and media PR operation outlets to counteract this in any way they can.

Scientists should stop allowing themselves to be taken in by the drug companies. Perhaps if they took time off from trying to look for the successor to the Higgs Boson and read Homeopathy books they would be better able to counter the incessant spinning.

The reason there is this assault in the press against homeopathy is because Pharma wields enormous power over the media and because the popularity of homeopathy has been increasing due to the side effects of conventional medicine. Also, unlike other natural therapies, it is dispensed in pills and so is in direct competition.

And it is precisely this direct competition which is so worrying Pharma.

Homeopathy attracted support from many of the most respected members of society in the U.S. such as William James, Henry Wadsworth Longfellow, Louisa M. Alcott, Mark Twain and former American Presidents James Garfield and William McKinley. In Britain, among its supporters were George Bernard Shaw, Charles Dickens, W.B. Yeats, William Thackeray, Benjamin Disraeli and Yehudi Menuhin. Other famous supporters were Dostoevsky, Johann Wolfgang von Goethe and Mahatma Ghandi.

Their support is good evidence for Homeopathy.

Nowadays, celebrities using and supporting homeopathy are many and include among others: Catherine Zeta-Jones, Tina Turner, Whoopi Goldberg, Pamela Anderson, Jane Fonda, Cher, Rosie O'Donnell, Martin Sheen, the Red Hot Chilli Peppers, Jane Seymour, Lesley Anne Warren, Mariel Hemingway, Lindsay Wagner, Paul McCartney, Axl Rose, Linda Gray, Susan Blakely, Michael Franks, Cybil Sheppard, Dizzy Gillespie, Vidal Sassoon, Angelica Houston, Boris Becker, Martina Navratilova, David Beckham, Priscilla and Lisa Marie Presley, Cliff Robertson, Jerry Hall, Diane von Furstenberg, Ashley Judd, Naomi Judd, Olivia Newton-John, Julianna Margulies, JD Salinger, Blythe Danner, Pat Riley (coach of the Miami Heat) - The list of famous people who supported homeopathy is endless...

It is particularly telling that Pat Riley, coach of the Miami Heat, adds his endorsement.

Homeopathy is practiced nowadays in countries all over the world and is especially popular in France, South America and India where there are around 250,000 homeopathic doctors! In a recent Global TGI survey where people were asked whether they trust

homeopathy the following percentages of people living in urban areas said YES: 62% in India, 58% Brazil, 53% Saudi Arabia, Chile 49%, United Arab Emirates 49%, France 40%, South Africa 35%, Russia 28%, Germany 27%, Argentina 25%, Hungary 25%, USA 18%, UK 15%.

As ever, the UK lags behind the rest of the world. But these figures do not show the full extent of support for Homeopathy. India is a huge country with a massive population and 62% of it completely dwarfs 15% of the population of the UK. The same can be said for Brazil, Chile and the United Arab Emirates. When the numbers are added up there is an overwhelming level of support for Homeopathy.

A lot of people today are confused about what homeopathy is (and isn't). This situation is not helped by the sceptics of homeopathy who go to incredible extents to exaggerate and misconstrue what homeopathic medicine is. It is more than a tad ironic that these 'sceptics' hold themselves out as 'defenders of medical science' and yet they commonly exhibit an embarrassingly poor scientific attitude when evaluating homeopathy and when determining what scientific evidence does and doesn't say about it.

It would make a pleasant change if scientists took the trouble to find out what they were talking about before opening their mouths.

Because many sceptics of homeopathy today indulge in spreading misinformation , this article is addressed at setting the record straight and is packed with references to confirm the veracity of what is being asserted here.

This approach gives us confidence that the author is presenting a researched case, as opposed to scientists who mostly say the first thing that comes into their head.

First, to clarify, advocating for or using homeopathic medicines does not preclude appreciation for or use of selective conventional medical treatment. Advocates of homeopathy simply honour the Hippocratic tradition of 'First, do no Harm' and therefore seek to explore and utilize safer methods before resorting to more risky treatments.

Hippocrates is known as the Father of Medicine and his values and writings form the basis of all medicine practiced today. Homeopathy seems closer to his principles than mainstream medicine, which is a point in its favour.

This perspective has historical and international roots, and it is thus no surprise that American health care, which has been so resistant to homeopathic and natural therapies in its mainstream institutions, is presently ranked 37th in the world in the performance of its health care system. In comparison, the number one ranked country in the world is France, a country in which around 40 percent of the population uses homeopathic medicines and around 30 percent of its family physicians prescribe them.

All the spin in the world cannot disguise these numbers.

The evidence is there: The fact that homeopathy became extremely popular during the 19th century primarily because of its impressive successes in treating the infectious disease epidemics that raged during that time is a fact that is totally ignored by sceptics. It is highly unlikely that a placebo response is the explanation for homeopathy's notable successes in treating epidemics of cholera, yellow fever, scarlet fever, typhoid, pneumonia, or influenza.

These are facts often conveniently forgotten.

Sceptics are wonderfully clever in trying to make up stories and excuses for the good and often amazing results that people get from homeopathic medicines. Most often, however, they simply say that 'old news is no news,' as they brag about not learning from the past, as though this is a good thing.

It is rather arrogant of scientists to say these and they should refrain from doing so.

It is readily acknowledged that the pharmacological process of making homeopathic medicines is often misunderstood or inadequately understood. Homeopathic medicines are made with a specific process, called potentization, that is unique to homeopathy. Each medicine is made in double-distilled water in a glass test-tube, diluted in a 1:10 or 1:100 solution that is vigorously shaken 40 or more times.

Then, this process of dilution and Succussion (vigorous shaking) is repeated 3, 6, 12, 30, 200, 1,000, or more times. Although one would think that one is diluting out whatever was in the original solution, the immense worldwide experience using homeopathic medicines over the past 200 years proves otherwise.

This detailed methodology shows the meticulous preparation of Homeopathic medicine. It is surely intuitively correct that immense worldwide experience automatically trumps the naïve assumption that the original solution is simply being 'diluted out'.

There is a body of intriguing but not yet fully verified theories about how homeopathic medicines work. These theories are too technical for this article, though I sincerely hope that the 'good sceptics' out there will work to explore and help figure out the many mysteries that may explain homeopathy, rather than repeat the old reactionary mantra that 'it cannot work.'

Scientists really should stop repeating old mantras. It makes them seem unprofessional.

For instance, the 'silica hypothesis' is particularly intriguing, especially in light of the fact that approximately 6 parts per million of 'silica fragments' or 'chips' are known to fall off the walls of the glass vial during the shaking process. In addition, the shaking process generates nanobubbles and transient localized regions of high pressure topping 10,000 atmospheres that have been hypothesized to alter the water in a significant and persistent way.

There is lively debate surrounding a full theory of Homeopathy. The 'silica hypothesis' is vying with Nanobubble Theory and the HPT (High Pressure Topping) Model to see which will win out in the competition to explain its success. The smart money is on the HPT Model, especially if string theorists become involved and the model is extended from Classical to Quantum HPT.

Just as a 'C' note of a piano is hypersensitive to other 'C' notes, living organisms are hypersensitive to extremely small doses of medicines that are made from substances that cause the similar symptoms that the sick person is experiencing.

Keyboard scientists will readily confirm this.

This ancient principle, 'like cures like,' was heralded by the Oracle at Delphi, the Bible, and various Eastern cultures, and the fact that modern-day immunology and allergy treatments derive from the primary principle of homeopathy, 'the Law of Similars,' provides additional substantiation to this system of medicine.

As scientists working in the field of archaeology are continually discovering, ancient principles are usually superior to modern ones.

Actually, a better description of this principle of Similars is the 'principle of resonance,' which any student of music knows has both power and hypersensitivity. The additional wisdom of this homeopathic principle is that its use leads to the prescription of medicines that mimic, rather than suppress, the symptoms and the innate intelligence of the human body. In this light, homeopathy can and should be considered a type of 'medical biomimicry' and a 'resonance medicine.'

It is high time modern medicine stopped prescribing drugs which suppress the innate intelligence of the human body. There is no clinical need for this, the practice just seems to have started and no-one has thought to end it.

Homeopaths may not yet adequately understand precisely how their medicines work but the body of historical and present-day evidence and experience is simply too significant to ignore.

To take another relevant example, Physicists working at the Large Hadron Collider aren't completely sure how it works but they still discovered the Higgs Boson.

The fact that so many highly respected people and cultural heroes over the past 200 years have used and advocated homeopathy provides additional evidence for this medical system. Some of these cultural heroes include eleven U.S. Presidents, six popes, JD Rockefeller, Charles Darwin, Mother Teresa, Mahatma Gandhi, and scores of literary greats, corporate leaders, sports superstars, world-class musicians and monarchs from virtually every European country.

Particularly impressive is the endorsement of six popes.

It is also important to acknowledge that hundreds of thousands, even millions, of medical doctors who learned conventional medicine have used homeopathic medicines in conjunction with or (commonly) as replacement for conventional medicines. In comparison, the number of medical professionals who have trained in homeopathy and then stopped using these medicines is extremely small.

This is impressive, but given the facts, hardly surprising.

The fact that homeopathic medicine represents the leading medical alternative in Europe and in significant portions of Asia (especially India and Pakistan) provides additional support for this often misunderstood medical science and art. In fact, over 100 million people in India depend solely on this form of medical care. Further, according to an A.C. Nielsen survey, 62 percent of current homeopathy users in India have never tried conventional medicines and 82 percent of homeopathy users would not switch to conventional treatments.

Such statistical analyses provide good additional evidence.

The Homeopathic Revolution: Famous People and Cultural Heroes Who Chose Homeopathy — a NEW book by Dana Ullman, MPH. Dear Friends of Homeopathy, many of you know the various books that I have written on homeopathic medicine since Everybody's Guide to Homeopathic Medicines was first published in 1984.

I have written many other books since then but my newest work is by far the most important work of my life. This is NOT any 'ordinary' book on homeopathy…as you shall soon see. I am honoured that the foreword to this book was written by Dr. Peter Fisher, the Physician to Her Majesty Queen Elizabeth II, but what is inside the book is even more intriguing and compelling.

It is telling that the Physician to Her Majesty Queen Elizabeth II has taken time to contribute the foreword. Her Majesty has the choice of treatment from every system of medicine in the world yet chooses Homeopathy. What more proof do scientists need?

This book documents many famous people of the past 200 years who have been known to use and/or advocate homeopathic medicine. I am not simply talking about 'celebrities.' In fact, I have already uncovered the following amazing facts:

Darwin might not have been healthy enough to have written 'Origin of Species' without the homeopathic and water-cure treatments that he received from Dr. James Manby Gully (according to Darwin's own letters).

This is from one of the world's greatest scientists so provides notable evidence.

Numerous leading conventional physicians and scientists who have had extremely positive things to say about homeopathy include Sir William Osler ('the Father of Modern Medicine'), Emil Adolph von Behring, MD ('the Father of Immunology'), Charles Frederick Menninger, MD (founder of the Menninger Clinic), August Bier, MD ('the Father of Spinal Anaesthesia'), C. Everett Koop, M.D. (former Surgeon General), Brian Josephson, PhD. (Nobel Laureate & Cambridge Professor).

This is an impressive list.

At least eleven American Presidents used homeopathic medicines or sponsored legislation to allow homeopathic practice (Lincoln, Tyler, Hayes, Garfield, Arthur, Harrison, McKinley, Coolidge, Harding, Hoover, & Clinton)…and two British Prime Ministers (Disraeli and Tony Blair).

The list of famous endorsers goes on…

Many of America's literary greats advocated for and often wrote about homeopathy including Washington Irving, Louisa May Alcott, Nathaniel Hawthorne and Mark Twain... And European greats such as Goethe, Sir Arthur Conan Doyle, Lord Alfred Tennyson, and George Bernard Shaw...and many modern greats including J.D. Salinger and Gabriel Garcia Marquez.

…and on…

Numerous sports greats have told of their use of homeopathic medicines including David Beckham, Martina Navratilova, Boris Becker and many more.

…and on…

Many world-class musicians have greatly appreciated homeopathy including Ludwig van Beethoven, Robert Schumann, Frederic Chopin, Sir Yehudi Menuhin, Cher, Tina Turner, Paul McCartney, George Harrison, Pete Townshend, Annie Lennox, Bob Weir, Paul Rodgers, Axl Rose, Moby, Jon Faddis, and Dizzy Gillespie.

For those not sure, Bob Weir is in the Grateful Dead.

Numerous movie and TV celebrities have benefited from homeopathy including early stars such as Sarah Bernhardt, Douglas Fairbanks, Jr., Marlene Dietrich and John Wayne... and some of the modern-day stars, including Catherine Zeta-Jones, Lesley Ann Warren, Pamela Anderson, Jane Seymour, Suzanne Somers, Lindsay Wagner, Michael York, Dr. Phil and Robin McGraw, Priscilla & Lisa Marie Presley, Jennifer Aniston, Jade Jagger, Tobey Maguire and Orlando Bloom.

Hollywood stars are acknowledged as some of the best judges of medical practice today.

Order The Homeopathic Revolution. Order a copy of this book or order a personalized autographed copy of the book and support a major media campaign. Please help us tell the world about this book and about homeopathy!

This request is difficult to resist!

'Drawing upon the extensive use of homeopathy by historical figures, founders of modern medicine and current celebrities, The Homeopathic Revolution documents the long standing efficacy of homeopathy. Given the research breakthroughs in the biological and clinical effects of nanopharmacology it is a certainty that homeopathy has an evolving scientific foundation in the Integrative Medicine of the future.' Kenneth R. Pelletier, PhD, MD.

Modern scientists have gone a little quiet about nanopharmacology. It would be uncharitable to suggest that drugs companies had anything to do with it but one could be forgiven for thinking so. It has also been noted how wealthy the doctors who oppose Homeopathy are. Is this coincidence?

'How do we reconcile the fact that the American Medical Association has long held it to be worthless when so many people have reported they found help for their ills in homeopathy? The AMA responds that patients are unqualified to evaluate the treatment they receive.

I think that attitude is phenomenally patronizing and I think Dana Ullman has done us a major service in showing that many of the smartest and most successful people in recent history have publicly expressed appreciation for the benefits they have received from homeopathy.' John Robbins, Author of 'Diet For A New America' and 'Healthy at 100'.

The AMA should stop patronizing people. It is unprofessional.

'Dana Ullman takes the reader from the origins of homeopathy as a 19th century system of medicine into a 21st century science. Through this fascinating ride of medical history he shows us that many of the world's most famous and respected people of the past 200 years have advocated for and appreciated this other approach to medicine. I am personally amazed that so many of my cultural heroes have benefited from this misunderstood science and healing art. Homeopathy deserves a definitive place in health care today.' Leonard A. Wisneski, MD, FACP.

Here, a doctor acknowledges the importance of a system of medicine having the support of famous people in order to become validated. Science should stress this approach more.

'As the interest in alternative approaches to medicine continues to multiply, Ullman's The Homeopathic Revolution provides an absolutely compelling case for the past, present and future of this potent approach to health and healing.' Ken Dychtwald, Ph.D.

Author, Bodymind, Healthy Aging, Age Wave, Age Power.
Consultant to many Fortune 500 Companies.

And finally, if any further proof were needed, we see a Consultant to many Fortune 500 Companies adding his voice to the chorus of approval for Homeopathy.

How science can continue to fly in the face of so much hard evidence is the really astounding thing about the Homeopathy debate. It seems obvious that, in this case, there really is 'something in it' and science should prioritise research into the methodology and practice of Homeopathy with all possible haste.

There is more than enough proof presented here to make that urgent task one of Sciences Next Steps.

Chapter 14: Medical Products

Everyone is interested in their own health. Over the years there have been many 'quack' cures which science has investigated and rightfully disregarded. But science has not investigated all of the treatments available and has therefore potentially missed some important contributions to medicine. Science should take up the challenge and begin investigating all of the possible remedies available today.

Hulda Clark Orgone energy zapper standard model. The current does not pass uniformly through the body. With regular zapping the current passes mainly through our liquids, such as our lymphatic and vascular system, a small fraction reaches every organ and tissue of our body. Blood and lymph are the most important locations to zap.

Zapping has been largely ignored in the medical literature up until now.

It takes three treatments to kill everything. Why? The first zapping kills viruses, bacteria and parasites. But a few minutes later bacteria and viruses, different ones, often recur. I conclude they had been infecting the parasites and killing the parasites released them.

It is intuitive that killing an organism would release other organisms within which remained unaffected as they were protected from the original zapping.

The second zapping kills the released viruses and bacteria but soon a few viruses appear again. They must have been infecting some of the last bacteria. After a third zapping I never find any viruses, bacteria or parasites, even hours later.

Those organisms were missed by the second zapping and again this is fully conversant with modern biological theory. The new dimension here, and one which science should undertake to study fully, is that there is a third release. This is something which was never suspected and can be viewed as one of the multitude of tricks which bacteria and viruses have up their sleeve in order to keep propagating.

A first seven-minute zapping is followed by an intermission lasting 20 to 30 minutes. During this time bacteria and viruses are released from the dying parasites and start to invade you instead. Each parasite has its own bacterial and viral escapees. The second seven-minute session is intended to kill these newly released viruses

and bacteria. Again, viruses are released, this time from the dying bacteria. The third session kills the last viruses released. A fourth and fifth session may be very beneficial too, to eliminate prion protein streaming from killed salmonella bacteria.

This lays out the protocol required. Note also the additional requirement which should also be observed in acute cases. Prion protein streaming prevention promises to be one of the prime disease inhibitors of the 21st century.

What changes can I expect to see from zapping? Cuts, scrapes, wounds, infections, etc. heal rapidly. Eliminates sinus infections quickly. Stops toothaches. Quickly removes even chronic pain. Flu and colds gone. Improved mental clarity, improved sleep and better dreams.

The advance here is clearly better dreams. Science has not been able to approach this despite decades of research.

Stops ulcers. Dead tapeworms in the toilet after a week or so with some people. Reduced flatulence. Warts shrink and disappear, especially fast with children. Tumours disappear. Carpel tunnel syndrome gone in a day or so. Yeast and Candida-type fungi gone quickly. Eyes less bloodshot. Skin tone immediately improved.

Likewise the disappearance of tumours. That alone should place scientific research on a war-footing to fully investigate and validate the zapper.

Helps remove heavy metals and organic toxins. Establishes normal pH within a few days in most cases. Acne disappears in short time. Psoriasis disappears in a fairly short time. Earache disappears. Improves organ efficiency when placed over an organ. No more PMS. No more migraine. No more depression. Stops itching, even from bug bites. Ringworm gone the same day. Scabies eliminated in one day. Stops dandruff. Breathing improves. Better joint mobility. Less body, breath and foot odour.

It is unusual for one device to have this amount of medical applications but the evidence could not be clearer.

Zapper with 33 kHz of frequency. User friendly with easy to follow instructions. Comes with Velcro wrist bands and special silicon- graphite-rubber electrodes, paper towels no longer needed.

Special features include LED timer 7-20-7-20-7 this is 7 minutes on, 20 minutes off for three continuous sessions. Includes a choice of continuous mode, LED indicator, plus a reminder to renew the 9 volt battery. **Has warning beep sound when contact is broken or insufficient. Belt clip & instruction booklet included: price $225.00**

This is a remarkable price for such an instrument.

Hulda Clark Orgone Food Zappicator: Magnetic polarization of food: Turn your food back to life force North Pole energy. Zappicator with 1 kHz of frequency. User friendly and easy to operate. Comes with built in timer of 20 minutes. Automatically beeps when program ends. Operated by a 9 volt battery, like the zapper.

What a food Zappicator can do: Not only can PCB's (Polychlorinated Biphenyls) asbestos and benzene be destroyed but a whole host of toxic molecules. This kind of 'zappication' chemistry is too new to answer many questions like how each phenomenon works. Most important is that it does not destroy most vitamins or change food in noticeable ways. It does oxidize some minerals.

One cannot expect the originators and inventors of a revolutionary device to have time to confirm its exact modus operandi. But most scientists appear to have a lot of time on their hands so they should begin to use it more profitably for the benefit of all of us. For example, scientists have built huge machines in an attempt to find dark matter. But dark matter is predicted to take many years to detect. While they are waiting, scientists should put their time to good use by investigating zappication chemistry.

Some things it destroys are: Parasite eggs and stages, Bacteria and viruses, Prions found in dairy products and meats, Phloridzin, the pituitary-destroying chemical, Chlorogenic acid, the hypothalamus-destroying chemical, Phenol, the thymus-destroying chemical, Apiol, another thymus-destroying chemical, Gallic acid, the trigger for SV 40 virus, D-phenylalanine, the malignant melanoma related chemical amino acid, D-mannitol, the abnormal sugar molecule that clogs lungs and ducts.

And what a boon to mankind the zapping of these disease vectors would be.

Magnetic polarization of food: Fruit and vegetables, leaves and flowers, even nuts and grains are north polarized when they are freshly picked or purchased. Inside, where the seeds are, the polarization is southerly.

Scientists should insist, via their relationship with Government, that polarization information is clearly labelled by food manufacturers on their products and that the 'sell-by' date is accompanied by a magnetic polarization arrow pointing clearly to the direction in which the food is polarized.

When the soft parts begin to age and wilt and show deterioration within a week of being stored in the refrigerator, the northerly polarization is changing to southerly. It happens gradually. For example, a large bunch of grapes will have some turned completely south in a few days, the wrinkled ones, while others are still completely north, the freshest looking ones. The seed does not change its polarization.

And science should insist on including a 'polarization change' date in order to inform consumers exactly when the product is likely to swap its polarization from north to south. This is particularly important for produce which, unlike grapes for which the polarization change is obvious as they become wrinkled when turned completely south, the change is not so obvious. An example is peas. It is extremely difficult to ascertain if a pea has altered its polarization just by looking at it.

The conclusion is that we were meant to eat northerly polarized food, with just a little bit of southerly food in the form of seeds. Yet most of the food we eat, even refrigerated food, has turned at least partly south. We are getting an overdose of south polarized food as well as water. That is why I recommend zappicating our food, especially when we are sick.

Too much southern fried chicken, for example, can hardly be good for you.

Zappication: Water that is simply zapped gets electrical energy, just a voltage impressed on it. We know, from bottle-copying, that water can hold very many frequencies of electrical energy. Food and our bodies are mainly water. Is it the same in food? Such research is badly needed.

Bottle-copying is a completely new paradigm and one which has not yet been fully investigated by any labs I am aware of.

Food that is simply put in a magnetic field has magnetic forces impressed on it. We can see that from making north and south polarized water. Electrical energy even generates magnetic energy and vice versa so we always receive a dose of both even when only one kind is applied. This, too, needs much more research.

And if science would stop spending so much money on more and more powerful telescopes intended to 'see' black holes then maybe we would have an answer. North and south polarized water is familiar to polarization scientists but electrical energy generating magnetic energy and vice versa leading to us receiving a dose of both even when only one is applied will come as news to everyone.

What is a Food Zappicator? A breadboard containing an electro-magnet attached to a zapper. In this way the electro-magnet exudes its own magnetic field which pushes a diaphragm back and forth to create a physical effect at the same frequency. The Zappicator combines all 3 kinds of energy mentioned above and delivers them at the same frequency. Fortunately I did not find evidence that 'good molecules' like vitamins and organic minerals were affected.

Science is aware that molecules can be affected by zapping but has so far been unable to confine the effect to particular molecules. It is therefore a breakthrough that the Zappicator can distinguish between bad molecules and good molecules.

Functions: It changes the angle of light that is passing through each molecule of food further to the left if an amino acid is zappicated. The d-amino acids are changed to l-amino acids this way. Remember, the body considers d-amino acids as allergens; it only uses l-forms itself. The food has been improved, not so allergenic, before you eat it. Of course, changes can come after you eat it. It could change back to a d-form in a southerly zone.

There is little of more importance than the angle at which light passes through each molecule of food. Obviously, the further to the left this angle can be made, the better.

It changes the polarization of the food to north, if the north side of the magnet faces the food. Food has been made 'fresher'.

Again, we see the importance of the north/south divide. Why anyone would choose to eat food polarized to the south after fully understanding this research is an important consideration for all consumers, producers and even Governments. International food agencies should make it one of their priorities to ensure that food undergoes a process of polarization to the north before it is distributed to starving people.

Supermarkets could install north polarizing devices at their food reception centres and pass all edible products through them. The same should apply at check-out, just to be sure. It is likely that the same groups who successfully introduced horse meat into the food chain will also attempt the same thing with south polarized food.

Benzene gets oxidized to phenol, at least at trace levels. PCBs disappear, no doubt slightly changed, an important step nevertheless. Phenolic food antigens disappear. Finally, food seems to taste better. Maybe changing d- to l-amino acids or alpha to beta forms or L- to D-sugars can be tasted.

Chemists will understand this passage and I have yet to find one who seriously disputes the evidence. However, the Zappicator is just one medical device which has yet to be fully understood by science. There are others which scientists have overlooked. Not only have they disregarded the scientific principles on which the devices are based, they also seem to ignore the testimonials of the people who use them. This is a common failing: why do scientists discount the experiences of ordinary people when they report their findings? Do scientists think that they are somehow above ordinary people? This is a disturbing yet familiar complaint.

Geoclense Harmonizer Orgone Energy testimonial: Thank you so much for arranging the pickup for me. I met Gerard and had a chat with him, which was very interesting. I feel so much better with the Geoclense Harmonizer in place. Within 10 minutes I felt the energy change in the house, it 'lifted', cleared, got lighter. My partner said he felt better, he had been feeling down and suddenly wasn't. Even the cat came out of hiding and played on the couch with us. She has been much more herself since then, too.

Everyone has had experience of the energy in their house 'lifting' clearing and getting lighter, but the clinching point here is that the cat was also affected. Cats being much more themselves is a sure sign that they are content.

How amazing that something so simple has had such a profound effect. I can still feel some 'hard' energy coming from one side of the house, I suspect it will clear with the purchase of a Quan Yin and a Buddha Schumann generator, which my partner didn't want to do but I will anyway. I can feel it affecting my ears, pressure in my head and discomfort in my spine. Interesting.

Science has differentiated between hard and soft energy for some years now but it is interesting to have these partitions confirmed. Despite intensive research, energy scientists are able to clear softer forms of energy but have hit a barrier when it comes to hard energy. One of their next steps should be to test the Quan Yin and Buddha Schumann generator.

Thank you, Karen, for your help, sensitivity and support. Thank you for doing the service you do for all of us by having these products available. If only more people knew the effects of all this technology and took steps to protect themselves, how many lives would be so different?

How many indeed?

Geoclense Electromagnetic Radiation Protector: Remove all EMR, geopathic stress & noxious energies immediately & protect your family & your health. Harmonize all your geopathic stress all at once. Now available overseas.

This shows the limits placed on protective devices by inadequate funding. Science, with its almost unlimited funding, would have enabled the Geoclense electromagnetic radiation protector to have established itself worldwide if only it had taken an interest in the technology.

The Geoclense harmonizes all geopathic stress and electromagnetic radiation in the whole building it is plugged into including your entire office block. To activate the Geoclense simply plug it into any power point and turn the switch on the power point on, and leave it switched on.

And the use of the technology is so simple. One can only assume, yet again, that science is having a sulk and individual scientists are not promoting the device because, as usual, they didn't think of it first.

The Geoclense works by harmonizing all the electrical wiring circuits with harmonious negative ions to balance the noxious positive ions as it increases the presence of photons and oxygen in the atmosphere.

There are few people now who would want to be surrounded in their home or their workplace by positive ions. Science has been aware of the problem but the breakthrough here is to achieve balance by increasing the presence of photons *and* oxygen in the atmosphere. Research has been sporadic but has concentrated only on one or the other, not both together.

Photon presence coupled with negative ions is believed to increase the amount of oxygen in the air, which is vital for our well-being.

Since these results were presented there has been next to no disagreement.

Positive ions decrease photon and oxygen levels. That is very evident by the heavy sickly feeling in the chest or abdomen area.

Likewise, there has been no research which disproves the theory that positive ions decrease photon and oxygen levels. This is very persuasive. Anecdotal evidence concerning the heavy, sickly feeling in the chest or abdomen area merely adds to its veracity.

Positive ions are known as a noxious unhealthy energy and are naturally created when there is current passing through the electrical circuit or there is geopathic stress created by man-made structures.

Despite years of research electrical engineers have not been able to stop electrical circuits creating positive ions. It seems they may never be able to stop it, though research continues. Medical professionals will readily confirm that positive ions are under suspicion as a noxious unhealthy energy and any responsible geologist will be happy to comment on the link between positive ion creation and the geopathic stress created by man-made structures. This is known as 'joined-up science' and it is in the public's interest that there is more of it.

Our Geoclense harmonizes this noxious energy emitted from electrical appliances as it connects to the circuit it is plugged into. Because electromagnetic radiation is the major contributor to the deterioration of the natural Earth magnetic grids such as ley lines, Benker grids and water veins, all geopathic stress is harmonized within the building which the Geoclense is operating.

That the natural Earth magnetic grids have been deteriorating is not disputed. Electromagnetic radiation is heavily implicated and Benker grids in particular have been affected.

Even though the power point switch is turned on, the Geoclense is not using any electricity but is neutralizing it simply by accessing the electrical circuit, making the Geoclense completely safe to use.

Science, as usual, has missed the boat by proposing preventative measures which do use electricity but that simply exacerbates the problem. The whole point is that current flowing in a circuit produces the unwanted effects. So the obvious solution is to plug in a device which uses no electricity.

The Geoclense has a very effective range harmonizing not only the noxious energies in your home or workplace, but is capable of harmonizing overhead electricity lines and power poles up to 200

metres away. Our Geoclense harmonizers will harmonize a building and area up to 12 acres around the building it is plugged into. We have found from actual in the field testing that a Geoclense harmonizer will harmonize a building of up to 40 floors. Powerful, effective and works immediately.

This is beyond science at present.

What is Geopathic stress? Geopathic stress is when the vibrational rate of the Earth's magnetic energies and underground water causes a lower vibration, which in turn creates stress to the human body.

Science has lagged behind in dealing with the effects of geopathic stress though the vibrational rate of the Earth's magnetic energies has at least been considered for measurement. Unfortunately, science has deemed other measurements to be more important though when these are disclosed the public may be forgiven for asking what is so important about them and how can they be more imperative than accurately measuring the vibrational rate of the Earth's underground water?

Typical of sciences 'more important' measurements are an effort to determine the exact value of the so called Fine Structure Constant. What relevance this could possibly have to taxpayers even scientists struggle to explain.

What are Earth's magnetic energies anyway? The Earth's magnetic energies consist of gridlines known as the Hartman grids and Curry grids which are generally life giving in natural surroundings but become harmful when a building is constructed on or over them, or if the land that has been poisoned by toxins, chemicals, fertilisers, land fill or other contaminants. Water veins represent where water runs in underground streams.

Those taxpayers could be well advised to ask, perhaps through their elected representatives, why research on the Hartman and Curry grids is not being actively pursued, and point out to those same representatives the life giving properties of the gridlines should their politicians be unaware of them.

Where three of the grids or water veins cross a natural vortex is formed and these are known as acupressure points of the Earth. These acupressure points, again in natural surrounds, are life giving but become harmful when contaminated by the construction of buildings or other man-made structures. These then become a danger to human health, especially if you are sleeping or spending a

lot of time over them, as this can lead to physical or mental health problems and other serious health disorders.

Contamination of acupressure points is the scourge of our age. Town planners claim to be unaware of it but they should be challenged at every available opportunity.

Geopathic stress can also be caused by fault lines and fissures, known as fault zones. These occur with the rising or falling of the Earth's crust. This can cause a lower than normal vibrational rate in the Earth's frequencies which can then create severe physical and mental states in the human body.

The Earth's vibrational frequencies, when lowered, are a largely unrecognized danger and the severe physical and mental states created in the human body appear to be a subject of avoidance in medical circles.

Many metropolitan cities around the world are affected by major fault lines they have been built on. Geopathic stress caused by fault zones lowers melatonin levels causing sleep disorders, mental instability and depression. These then lead to related problems such as alcohol abuse, pharmaceutical drug use or recreational drug addiction.

Unfortunately most of our lower socio-economic housing areas are more often placed in areas of severe geopathic stress, although it doesn't discriminate, as even some of our better socio-economic areas are built over these fault zone areas. How can Orgone and Schumann energy help?

It is a breakthrough that social problems can now be directly related to geology. Nobody with knowledge of the settlements built close to the San Andreas fault can doubt that San Francisco has a well-documented issue with recreational drug use. Those of us of a certain age have only to recall albums by Moby Grape and The New Riders of the Purple Sage to fully understand this.

Orgone energy offers solutions to cure or remove geopathic stress and fault zones in your home or business. The geopathic stress can be neutralized or harmonised with tools such as Orgone energy/ Schumann generators. A Geoclense geopathic stress harmoniser raises the vibrational rate of these magnetic grids and fault zones to a life giving energy, improving the physical vibration of your home and work environment.

Science has been slow to realise the importance of raising the vibrational rate of magnetic grids and fault zones. Sporadic research has occurred but the vibrational rate was never raised by a sufficient amount to attain life giving energy and research therefore stalled.

Other household dangers such as electromagnetic radiation from electrical appliances can also harmonized with the Schumann generators, Orgone energy products and other tools providing an energetic state rich with healthy life-giving negative ions which are naturally found in nature, returning the vibration of your space back to how it should be.

For too long thought had been that positive and negative ions would be found in equal amounts in nature but the research to prove this was never done. This is an example of the 'laziness of science' which is becoming more exposed in recent times. Why scientists, of all people, think that equal amounts of anything will be present anywhere without checking first is a mystery.

EMR mobile phone protectors and Wi-Fi protectors on your electronic equipment and computers will protect you by neutralizing the harm that they cause. Wearing an EMR Orgone energy protection pendant can help to keep you protected from these no matter where you go.

Someone recently stopped having anxiety and panic attacks in his office soon after he started to wear an Orgone pendant. A young boy stopped crying continuously and gained lots of confidence in himself from wearing an Orgone pendant and is a completely different boy now.

It is all very well for an investigating scientist to claim that he or she found no effect at all when trialling an Orgone pendant. That may indeed be so, but if the same scientist was made aware of the youngster who stopped crying continuously, gained lots of confidence in himself and is a completely different boy now that that picture would change.

The same scientist, if still unconvinced, could be challenged to say how he explains the fact that someone recently stopped having panic attacks in the office soon after he started wearing his Orgone pendant.

Estimated at an average of 7.83 Hz, the Schumann Resonance is the basic frequency of the Earth.

Many scientists have doubted that the Earth has a frequency at all. But like all scientific theories, the theory that the Earth does not have a basic frequency is open to question.

Many proponents have called it the tuning fork of the planet, claiming that it generates natural healing properties when living

things are entrained to its rhythm. **Entrainment occurs when two objects are synchronized by a common vibration. When two or more objects resonate together, they are in tune. Being in tune with the Schumann Resonance is to be in touch with the fundamental flow of one's being.**

Three ways to access this frequency are alpha brain waves, meditation and dolphin contact.

Alpha brain waves: It was Dr. Herbert König, Schumann's successor at Munich University, who demonstrated a connection between Schumann Resonance and brain rhythms. König compared human EEG recordings with natural electromagnetic fields in the environment and found that the average frequency produced by Schumann vibrations coincided with the frequency of alpha rhythms, the brain frequency of the relaxed and creative mind.

Moreover, König discovered that the dominant brain wave rhythm of all mammals in alpha state is 7.83 Hz.

This is a highly persuasive result.

In one experiment, student volunteers lived for four weeks in a hermetically sealed environment that screened out magnetic fields, specifically the 7.83 Hz frequency. These students started suffering emotional distress and migraine headaches which were immediately cleared after brief exposure to 7.83 Hz frequency. Alpha brain waves can be produced in the brain simply by relaxing and cultivating an openness in attitude and thought.

The problem with many scientists, who continue to oppose this theory even though it has been proved, is that they find it difficult to cultivate their own openness in attitude and thought. Scientists are in fact well known for maintaining a closed attitude in thought. Scientists should allow themselves to simply relax. Then their brain waves would vibrate at 7.83 Hz frequency and they would become more open to creative thoughts.

Meditation: A tuned system occurs when a transmitter is accepted by a receiver. When one oscillator resonates with another energy information is transferred and the participants become part of a field. During deep meditation, when alpha and theta waves are stimulated, resonance is created between the human mind and the planet Earth. This synchrony not only generates a rise in consciousness energy, it consolidates the benefits of improved health.

This is exactly the type of discovery which scientists refuse to acknowledge. So this is exactly the type of paragraph which they need to read again having relaxed, placed their brain waves into the alpha state and become more open in their thought.

Does my Orgone energy pendant need to be cleansed on a regular basis?

You do not need to clear your Orgone energy pendant ever as it does not accumulate noxious energies but works more by providing a protective energy frequency around you, which is a solid state and will not change. Our Orgone energy pendants provide a 25 metre dome of protection around you when wearing them.

Science has reportedly experimented with energy pendants but has only been able to produce ones which unfortunately *do* accumulate noxious energies. It is real advance to hear that this obstacle has finally been overcome. The production of a protective energy frequency which is a solid state has long been a goal of science. That 25 metres is the extent of the dome of protection is something which had previously been thought to be decades away.

What is the life expectancy of my Orgone energy pendant? Does it become ineffective after a certain time frame?

The life expectancy of the Orgone energy pendants will pretty much last a lifetime so you do not need to be concerned about them becoming ineffective.

This is reassuring but will also come as a blow to elderly people who may have thought that they could pass on a recently purchased energy pendant to their children. Science should investigate how it is that the pendant is aware of the length of its owners lifetime.

How to use a Schumann generator: A Schumann generator is a solid state negative ion generator which produces a negative ion resonance throughout a building or predetermined space. The Schumann generator for indoor applications should be placed on a hard surface preferably to enable it to perform at its best resonance over a large as possible area.

Science has experimented with negative ion generators and is aware of the concept of resonance but has failed to amalgamate the two ideas.

Uses for a Schumann generator: Place it on top of your refrigerator or a flat metal surface to harmonise and neutralise geopathic stress, noxious energies and all water veins from your home. Place it under your Brita water filter or a jug of water to

energise your water prior to drinking it as this will improve hydration within the body.

Science has been seeking to improve the hydration effect of water without success.

Place it in your refrigerator to neutralise the motor of your fridge which actually causes your food to go off. Place all your food on top of it for 10 minutes prior to consumption to energize and rejuvenate your food by making vitamins and minerals more absorbable.

Motor neutralisation has long been a goal of science. Most people will have noticed that their food, stored in a fridge, can go off when the motor operates. Research has been done involving the motor being turned off with food stored in the fridge but the results have been inconclusive.

Stand a bottle of wine on it for 10 minutes prior to consumption to improve its flavour or any drinks for that matter. Place it under a vase of flowers to keep them fresh for longer.

This is a boon to people who can only afford cheaper vintages or whose flower budget is limited. On the other hand it will not find favour with vendors of more expensive wine nor florists in general. But this is an example of the dual nature of scientific discovery. Inevitably there will be winners and losers.

Place it on your desk and stand all your coffee or water on it while you are working to energise them. Place it in your handbag so that you keep a nice energy field around you everywhere you go.

Place it on your bedside table before you go to bed to get a better night's sleep, especially good if anyone is experiencing nightmares. Take it with you when you stay overnight anywhere and place it on the bedside table before you go to bed to remove any noxious energies while away from home.

The problems of noxious energies are so prevalent that there is no need to rehash them here. Any advance in the science of noxious energy removal is to be welcomed.

Place it in an area of your home or office where you know there is noxious energy to harmonise and neutralise it. Place all skincare, hair care and beauty products on it for ten minutes prior to using them. Place it under any indoor plants or place it in your vegetable garden and increase growth. Place it in your compost bin to break it down faster and improve it.

Biologists have been seeking a way to improve growth but have been side-tracked into the cul-de-sac of genetic modification.

Place it in your sewer or toilet to harmonize and neutralise the water veins and noxious energies it causes around your home.

Science is not good at producing practical solutions to the everyday problems ordinary people face so removing the noxious energies toilets can cause is a real boon.

Use it as a rainmaker, you will need specific instructions to do this.

As with all new technology one should read the user manual carefully before using it. That said, the rainmaker application promises much to people living through a drought or simply in an area where they feel they have not received as much rain as they would like. Of course, they should check that other local users of the generator are not attempting to make rain at the same time in order to avoid counterproductive monsoon conditions.

Place one anywhere else you can think of to improve the energy of an item, food, drink and the surrounding environment.

The use of the Schumann generator is limited only by the imagination of the user. How refreshing! Normally, scientific discoveries come with strict guidance which leaves little or no room for individual resourcefulness. Why should people have to follow a hidebound rulebook written by scientists who all too often lack the creative spark?

The dangers of Wi-Fi internet technology: Wi-Fi technology sends a beam from the router direct to the laptop that is around 400mm in diameter and is a very noxious energy field. The same occurs for mobile phones, hand phones, cell phones and portable phones as well as computers, notebooks & laptops connected to and using Wi-Fi technology. If a person is making a call on any of these types of phones a beam is set up between the phone and the nearest phone tower or towers. To a person with energy sensitivity or low defences this beam will seriously affect them. This may also damage DNA whilst in use.

It is entirely remiss of science to not warn people of the dangers of this beam.

If a person is using a mobile phone or wireless laptop computer in a busy city environment then the noxious energy beam created may affect others in the surrounding area. This means that people need to have a sense of environmental responsibility when using such

technology to ensure that they are not affecting the wellbeing of other people that may be affected.

It is very poor social responsibility as well as environmentally suspect to use a mobile phone or wireless computer in a busy city environment.

Cell phone radiation shields not only protect the user of the phone or laptop computer but also ensure that others are protected at the same time because the cell phone radiation shield also harmonizes the noxious energy beam.

Cell phone companies have been slow to promote the importance of radiation shields and science as a whole is to blame for not flagging up the benefits sufficiently. However, now that the importance of harmonizing noxious energy beams is becoming more widely understood science should take the lead in spreading the message.

Benefits of using the Geoclense harmonizer: Our Geoclense would certainly neutralize and harmonise all the Earth's magnetic radiation, electromagnetic radiation and geopathic stress affecting your home and protect you from your entire electrical system and electrical appliances.

This is an all-encompassing claim yet science has never denied it nor provided any proof whatsoever that the claim is unfounded.

Our Orgone energy pendants: Organite pendants with the metals, crystals and coil all produce a negative energy which is harmful to the body and interferes with our chakra system causing them to go out of balance. If you are sensitive to energy you can feel great pain and discomfort in your body if you wear one of these. Some are better than others; however, even the better pendants seem to all have one side positive and one side negative.

Science is well on the way to proving the scientific case against Organite pendants. Metals and crystals have already been implicated and it is only a matter of time before the coil is shown to have an equally pernicious effect. To date, there has never been an independently refereed study which showed conclusively that even the better pendants didn't have one side positive and one side negative. This is good evidence against Organite.

Our Orgone energy pendants do not use any such ingredients and they get their double positive qualities from the way that they are made.

And, as one might expect, there has never been any scientific study which has failed to confirm that Orgone pendants have double positive qualities.

Our Orgone pendants are what we call solid-state negative ion generators, whereas Organite products are solid-state positive ion generators of noxious energy. Our unique Orgonium technology engineers this out.

Solid-state physicists have become increasingly excited by the discovery from pendant based research that negative ion generators are possible.

Amazing benefits of wearing an Orgone energy pendant and using a Geoclense: They also prevent you from getting drained from radon from concrete and fluorescent lights in shopping centres and supermarkets.

Draining produced by fluorescent lights in shopping centres and supermarkets has been completely overlooked by science and this is very much to sciences detriment. Who can honestly say that they have never felt drained after visiting the supermarket? Yet science goes blindly on, ignoring the subjective feelings of thousands of people.

But even the most aloof scientist will admit that statistics are amongst his most powerful tools and that sheer weight of numbers often informs his opinions. The same statistics apply to people feeling drained by supermarket lights as to the evidence for the existence of the Higgs Boson. In both cases the same scientific methodology is at work. Why scientists continue to find meaning in one but not the other is something only they can answer.

Earth magnetic radiation & grid lines: An Earth magnetic grid line is a standing wave of Earth magnetic radiation which generally runs in the north-south, east-west direction at varying widths and spacing. The most common types of Earth magnetic grids are the Benker grid and the 400-meter grid. The Benker grid is 1100mm wide and is seven 7 metres apart. The 400-meter grid is 6 metres wide and is 400 metres apart. These grid lines emanate upwards to the ionosphere and are polarized to the noxious positive ion energy.

These grid lines were once called the crystal grids which resonate to negative ions. But due to our electrical systems the polarities of the grids worldwide have now been changed to positive ions. The noxious energy from the grids can be harmonized back to negative ions with the Schumann generator or the Geoclense.

Science's culpability in not warning the public of the danger is a sorry state of affairs. Stories of scientists installing equipment at the taxpayers' expense in their own homes in order to harmonize noxious energy have circulated on the Internet for some years now. If these prove to be true class-action suits could well follow.

Polar Fleece Orgone Blanket: All external electromagnetic energies and Earth magnetic radiation is blocked and reflected. You become aware of subtle shifts of energy as the blanket reflects back the body's own infrared and ultra-high frequencies. One may feel a period of warmth focused around and in the body that usually passes within minutes. You will experience a sense of inner peace and quiet which leads towards a state of peace, oneness, vitality, calmness, harmony, joy, bliss, relaxation & even enlightenment. Price $159.95

This seems most reasonable. Especially as the blanket reflects back the body's own ultra-high frequencies. Science has only managed to produce a blanket which reflects back the body's infrared so this is a real step forward. Obviously, reflecting just infrared will not lead to oneness nor harmony, let alone bliss.

Orgone Energy Pendant: Our Orgone protection pendant is energetically superior to many other energetic pendants as there is no negative polarity. We have engineered out any negative energy polarity to make them double positive with our unique Orgonium technology, which means that it does not matter which side the pendant is facing. This makes the Orgone EMR protection pendant completely safe to wear no matter what your sensitivity levels may be. They are completely safe to wear at all times. Price $41.95

Science has yet to produce a pendant with zero negative polarity though there have been hints in the literature that certain teams are getting close. It had been supposed that a Nobel Prize awaited the first group to completely engineer out negative energy polarity but the production of a double positive was thought to be impossible on theoretical grounds.

Orgone Harmony Wear Pendant & Pendulum Multi-facet Clear Heart: Absolutely unique Orgone energy glass pendants. Why are these not made of crystal? Well, crystals accumulate frequencies and need to be continually cleared or cleaned. Using glass, Orgone energy pendants prevents this need to have them cleaned so that they work magnificently all the time, every time, keeping you protected no matter where you are or what you do. We have also especially

engineered our Orgone energy pendants to have double positive sides which cannot be done with crystals. Price $59.95

As you will no doubt have suspected, given science's track record, modern research labs have continued, doggedly, to experiment with crystal even when the evidence was available that glass held by far the most promise.

There are literally hundreds of examples of science simply sticking with what it thinks it knows and refusing to open its mind to a fresh approach. If it wasn't for science getting in the way and slowing down progress we could be more advanced by decades if not hundreds of years.

Orgone Harmony Wear Pendant & Pendulum Pinkish Amber with Silver: Our harmony wear glass Orgone pendant is energetically superior to many other energetic pendants due to not having any negative polarity, as we have been able to engineer-out any negative energy polarity using our secret technology to make them double positive. This makes our harmony wear Orgone pendants completely safe to wear. Price $59.95

And here we have the final damning indictment of science. Of course the Orgone pendant technology is secret. Science is so miffed that it didn't discover the technology first that it refuses to engage with the originators of any alternative technology and simply buries its head in the sand.

If science refuses to cooperate in this way why should anyone else cooperate with science? This situation cannot be allowed to continue.

One of sciences next steps should be to ensure that it stops thinking that all of the important advances can only be made by scientists. This view is simply wrong. Scientists should acknowledge it, acknowledge it in public, and acknowledge it as soon as possible.

Chapter 15: Medicine

Alternative treatments left a lasting impression on actress Gwyneth Paltrow - literally. In 2004, Gwyneth showed up to a New York film premiere in a low-cut black dress and a line of circular welts across her back, according to BBC News.

Practitioners of 'Cupping' an ancient Chinese remedy, apply a cup, or a cone, to the skin and create a vacuum with hot air or simply through suction at the top of the cup. The suction brings blood to the surface and, according to Chinese medicine, can have various benefits on the body from stimulating blood flow, relieving pain and drawing Qi energy to the area.

That this would stimulate blood flow seems reasonable enough and no scientist has yet come forward to deny this. However some have doubted the existence of Qi energy. This is unfortunate, and such scientists should remember the history of science itself.

When radioactivity was discovered the nature of the radiations was unknown so scientists gave them the names alpha, beta and gamma. Later, as science progressed, it was realised that alpha was a helium nucleus, beta was an electron and gamma was high frequency electromagnetic radiation similar to light but more energetic.

Science is in the same position today. The ancient Chinese call their radiation 'Qi' and it undoubtedly exists, it's just that science hasn't yet identified exactly what it is. No doubt with further investigation things will become clearer. But that doesn't make the ancient Chinese wrong any more than Marie Curie was wrong when she named her three radiations alpha, beta and gamma.

Michael Flatley has not been Lording it over the dance floor for several years because of a mystery virus that forced him to cancel his Celtic Tiger show. But the Celtic dancer and choreographer seemed to find a cure in the Plexus system, a treatment that utilizes so-called bio-energy, the web that unites all reality and works by rebalancing the life force energy within and around the body, according to the Plexus bio-energy website.

Flatley told the Irish Independent newspaper that Michael O'Doherty, one of the founders of the Plexus system, treated him and that he is close to 100 percent fit again.

Michael has become world famous for inventing 'Riverdance' and later becoming 'Lord of the Dance'. He has his choice of physicians and is unlikely to have the wool pulled over his eyes, especially when his career depends on him maintaining full fitness. No doubt the Plexus system costs a lot of money but in his case it has been money well spent as he is now close to 100% fit again. This appears to be another case of waiting for science to catch up.

For those in the public eye, life balance can tip decidedly toward excess. Many celebrities such as Madonna, Goldie Hawn and Cherie Blair turn to alternative Eastern medicines such as Ayurveda to find balance. Supermodel Christy Turlington is one of the most staunch and vocal practitioners of Ayurveda, a system of traditional medicine that originated in India.

'Through my practice of yoga I was drawn to Ayurveda, a sister philosophy, and also to things that are generally better for me', Christy said in an interview with Psychology Today magazine. 'Ayurveda is a very complex science that is 5,000 years old. We all have three doshas and five elements of nature in each of us, in varying degrees, which makes each of us an individual'.

The three doshas are called Vata, Pitta, & Kapha and they are each made up of complementary combinations of the five elements of nature which are space, air, fire, water and Earth.

As Nourished Magazine explains: 'According to Maharishi Ayurveda each one of us is comprised of a unique mix of three mind/body principles which are responsible for our unique physical, mental and emotional characteristics. These doshas are the governing principles of intelligence that literally govern everything in the Universe and therefore each one of us'.

Christy has clearly researched Ayurveda and is helping to spread the word. As part of this effort to educate and inform Christy is also one of the creators of the Ayurvedic company Sundari, which markets a collection of distinctive anti-aging skincare products. Science should confirm that Christy's company is on the right lines and that her creams do indeed decelerate the aging process in humans.

With a name derived from Buddhist mythology Uma Thurman should have no trouble with serenity but the Oscar-nominated actress uses a few other tricks to keep herself centred. Uma is one of several Hollywood heavyweights to ascribe to Gem therapy.

Believers say that certain gems radiate specific energies that can affect mood and well-being. Uma wears a carnelian necklace for vitality and happiness according to an article in Redbook Magazine.

The use of therapeutic gemstones has its roots in such diverse cultures and systems as Ayurveda, Chinese Medicine, and Native American Shamanism, and is reflected in the more recent practice of crystal healing.

You are a being of light: From your innermost essence to your physical body you are made of and nourished by a continuous flow of life force. This life force is a current of light which springs from the source of life itself. When it flows freely you experience joy, vitality and vibrant health. In the natural course of living however, various blockages accumulate within this flow. These blockages create dark clouds within you. They obscure your radiance and give rise to disharmony, pain, and disease.

Gemstones embody life force: Like us, the Earth is enlivened by life force. As the Earth was forming, life energy was infused into the planet's crystalline matrices. Thus the Earth's gemstones came to embody and express the life force within the planet.

Gemstones dissolve blockages: Gemstones radiate their life force with great power. When used properly they can bring light into areas of darkness and neutralize blockages within your being. As these blockages dissolve the light of the life force can once again shine through to enliven, nourish, and heal you. Each type of gemstone expresses a different frequency; therefore it can address a different kind of blockage.

This all seems very clear. That life force exists seems obvious. That we are alive is indisputable. That gemstones are part of our planet is also indubitable and it is a small step from there to allowing that gems could also embody the life-force.

Gemstones are already known to radiate light. We have only to think of lasers which use rubies to generate light of extraordinary power. Rubies are used for red light so the earliest lasers were red in colour. Then came green lasers and finally today we have Blu-ray, each of the colours mentioned relying on a different gemstone in order to generate its own particular colour. Colour is related to frequency by the equation 'speed of light is equal to wavelength multiplied by frequency'. So science already agrees that different gemstones express different frequencies. Lasers can also remove blockages. It is therefore unsurprising that gemstones can do this.

Actor Nick Nolte is known for taking chances and when it comes to his health he is no different. Nick, who in an interview with Larry King on CNN spoke about different ways to slow the aging process and stay healthy for as long as possible, said he tried a controversial treatment involving trioxygen O_3 or Ozone, the name for three oxygen atoms bonded together.

Ozone is found in the upper atmosphere where it filters damaging ultraviolet rays from sunlight before it reaches Earth. Doctors have experimented with a variety of uses for ozone since it was discovered by a homeopathic doctor in 1856. Ozone is injected into the body in controlled amounts, not inhaled. It's a treatment that oxidizes the brain, Nick said in the interview. And on a brain scan, you can see that the brain is more metabolized.

It is well known that the brain requires oxygen. A brain deprived of oxygen can expect to live for only 4 minutes before irreparable damage occurs. So it is safe to say that oxidising your brain is a good idea. That this leads to the brain becoming more metabolized is almost certainly correct. It should be a small step for science to confirm that increased brain metabolism slows the aging process and helps a person to stay healthy for as long as possible.

Television actress Suzanne Somers, best known for her sitcom roles as Chrissy Snow in Three's Company and Carol Lambert in Step by Step, has in recent years become a proponent for healthy living and alternative therapies. Susanne stirred up controversy in 2001 when she disclosed on CNN's Larry King that she was using a homeopathic drug to help treat her breast cancer. Following conventional surgery and radiation therapy, she chose a therapy using mistletoe injections rather than pursue the recommended chemotherapy after her treatment.

While some mistletoe products are used in other countries for breast cancer treatment there is no evidence that such injections can cure breast cancer said Barrie Cassileth, the Chief of Integrative Medicine at Memorial Sloan Kettering Cancer Centre in an interview for the abcnews.com OnCall+ . Still, Cassileth said, mistletoe products might cause some breast cancer patients to feel better but there's not enough scientific data on the subject.

The proof of course is that Suzanne, happily, is healthy, well and still with us. Now one case doesn't make a proof, as science will insist on telling us, yet here we have one high-profile survivor of cancer, the most dreaded of diseases, who can unflinchingly point to mistletoe injection as the prime cause of her recovery. That Suzanne willingly chose to forego chemotherapy in favour of mistletoe is something which must weigh heavily in the balance.

It seems, given this, rather misleading of Barrie to say that there is no evidence for the cure of cancer but perhaps he puts his own comments into context when he confesses that there is not enough scientific data on the subject.

Undoubtedly this is true but yet again we must ask ourselves the question *why* isn't there enough data? If she had been asked, I am sure Suzanne would have volunteered herself to be a guinea-pig, the rather unflattering description given by scientists to those who should really be thought of as pioneers, and would have been happy to show how the programme of injections restored her health on a day by day basis. But for some reason science rejected the chance.

In Europe, mistletoe Viscum Album is the most commonly used complementary therapy in cancer care and is integrated into conventional cancer treatments. Mistletoe therapy has been developed since 1917 and forms the backbone of medical care for cancer patients in anthroposophic medicine.

To those unsure of what anthroposophic medicine (AM) might be here is a guide from its founding father, Rudolf Steiner:

AM not only thinks that the vital spirit plays a major role in health, a view it shares with homeopathy, but it also brings into play other metaphysical entities it refers to variously as the etheric body, the astral body and the ego. AM thinks the soul, the senses and consciousness are beings that exist independently of the body and that such things as herbs and essential oils can bring these things into harmony with each other and with the physical body.

AM is certainly in harmony with my basic approach to reality which is that I have special powers to see directly into occult realities without the bother of tests or replication by others. When you believe you have clairvoyant powers you don't feel the need to prove your claims the way other scientists do. I approach medicine the same way as I approach everything else from Astrology to Atlantis to education to farming to metaphysics.

Of course, scientists without special powers should continue to run tests as usual.

Sarah Ferguson, the former Duchess of York, can add health writer to her copious resume. According to an article in the March 2004 issue of W magazine, Fergie and her two daughters, Princesses Beatrice and Eugenie, received bio-energy treatments from Russian energy healer Alla Svirinskaya. Fergie was so impressed with the treatments she wrote the foreword to Alla's 2005 book 'Energy Secrets: The Ultimate Well-Being Plan.' Alla is a fifth-generation healer and practices in Notting Hill, a neighbourhood in West London.

Alla has now branched out.

There are many practitioners of alternative medicine in the world increasingly leaning towards more natural therapies so it is often difficult to know who is doing what. Alla & Katy [Appleton] are a rare breed. Alla is a 5th generation energy healer with a background in orthodox medicine and Katy is an ex ballet dancer who has now taken to yoga and is the creator of two bestselling videos and author of two books.

This is a natural step to take. Energy healing allied with yoga promises much in synergy.

Alla & Katy in the past few years have combined the expertise of their different fields to bring people an unbeatable team. The effects of their work are tangible, often immediate, and their success rate is astonishing. Both command long waiting lists, regular big media publications, as well as counting many A list celebrities as their clients. As their latest project Alla and Katy have put together their exclusive retreat 'The Power of Three' specially designed with a focus on holistic wellbeing, preparing your mind body and soul for the specific attention of Dr. Alla's energy healing and Katy's yoga instruction.

It is difficult to imagine a more promising combination as far as healing is concerned. Medical science should watch their progress carefully and learn as much as it can.

Dr. Alla has this to say about her methods: 'Physicists have proven that everything is made up solely of energy - ourselves included. With this in mind, any imbalance or blockage of energy within can create a condition or disease, psychological or physiological and in some cases both.'

Physicists have indeed proven that everything is made from energy. But they failed to take the important step of going on to prove that energy blockages can create medical conditions.

Working with this energy is the principle behind bio-energy healing. Bio-energy healing involves examination of the patient's aura to determine where the problem lies. When the causes of the disease have been identified, she begins to channel the energy to imbalanced parts of the aura.

By adjusting a person's aura she harmonises its subtle energy which in its turn improves the blood circulation in affected organs. Healing gives the patient an opportunity to have a better understanding of their own bodies and then realise their hidden self-healing resources.

Modern surgeons will agree that improving the blood circulation in organs is an often vital part of treatment. Surgery however carries the risk of infection. Science has partly remedied this through the use of keyhole surgery but the aim has always been to do away with surgery altogether. Dr. Alla's subtle energy harmonisation protocol appears to have achieved this.

Her treatment opens blockages, removes toxins and releases negative energies from the body. She treats a very wide range of aliments including depression, asthma, eczema, digestive problems etc. The success rate of the healing is high. Often patients notice the improvement after one session.

This is a very respectable success rate and one which deserves further investigation. Blockage specialists will be particularly interested.

Alla studied at medical school in Russia and was awarded a scholarship to travel to India and Sri Lanka where she was nurtured in Eastern philosophy, founded on an understanding of energy fields around the body known as the aura, and energy flows through the body, chi. 'I learned that each organ has its own energy field and frequency as does each illness'.

It is a fact that Eastern philosophy is largely ignored in Western medical colleges. As it now appears, this could be a mistake.

According to Alla a person's aura has seven layers, three of which represent your physical, emotional and mental energy. Alla's bio-energy balancing is about harmonizing people on these three important levels. Such is the power of her treatment, if clients don't see a significant improvement by the end of the second appointment then she terminates the sessions.

Only an extremely powerful treatment would be discontinued if no improvement was seen by the end of the second session. This is good evidence in favour of Dr. Alla's methods.

And she's not on a crusade to convert people to her belief system: 'I'm not a new age fascist and I don't require that my clients believe. I absolutely understand when people have a healthy scepticism. I don't think alternative medicine is the only way. It's about picking the best of both worlds - orthodox and alternative. I very much stand for a pro-complementary approach. I work on what route is the best for my clients.'

It would make a pleasant change if modern medical practitioners took the same view. Orthodox doctors should follow this example and also pick the best of alternative medicine to complement their treatments. Perhaps they could start by channelling some bio-energy to imbalanced parts of their patient's aura.

Unsurprisingly, Alla's success rate is very high; her current waiting list for new clients stands at seven months. In order to reach out to more people she has put her ideas and teachings into two internationally best-selling books: Energy Secrets and Your Secret Laws of Power. 'I have an obligation to help as many people as possible, which is where the books come in.' Her books have been translated into 11 different languages worldwide. Energy Secrets will be re-issued later this year.

As well as running her hugely successful private practice and jetting regularly to the US to visit her A-list roster of clients, Alla acts as a senior consultant to leading spas around the world. She has written as a guest columnist for UK's leading spiritual lifestyle glossy magazine Spirit and Destiny and her first CD of meditations will be launched in 2010.

Doctors should ensure they read these books. More forward looking doctors will also order the CD.

Asparagus - who knew? My mom had been taking the full-stalk canned style asparagus that she pureed and she took 4 tablespoons in the morning and 4 tablespoons later in the day. She did this for over a month. She is on chemo pills for stage 3 lung cancer in the pleural area and her cancer cell count went from 386 down to 125 as of this past week. Her oncologist said she does not need to see him for 3 months.

Who knew? Certainly science didn't and if a review of the literature is an accurate guide science still doesn't know.

Several years ago I had a man seeking asparagus for a friend who had cancer. He gave me a photocopy of an article entitled

Asparagus For Cancer printed in Cancer News journal, December 1979. I will share it here, just as it was shared with me:

I am a biochemist and have specialized in the relation of diet to health for over 50 years. Several years ago I learned of the discovery of Richard R. Vensal, DDS (Doctor of Dental Surgery) that asparagus might cure cancer. Since then, I have worked with him on his project and we have accumulated a number of favourable case histories.

Case no. 1: A man with an almost hopeless case of Hodgkin's disease (cancer of the lymph glands) who was completely incapacitated. Within 1 year of starting the asparagus therapy his doctors were unable to detect any signs of cancer and he was back on a schedule of strenuous exercise.

It is telling that other doctors were involved who independently corroborated the evidence.

Case no. 2: A successful businessman, 68 years old, who suffered from cancer of the bladder for 16 years. After years of medical treatments, including radiation without improvement, he went on asparagus. Within 3 months examinations revealed that his bladder tumour had disappeared and that his kidneys were normal.

Given case evidence such as this one would expect that scientists would at least show interest.

Case no. 3: A man who had lung cancer. On March 5th 1971 he was put on the operating table where they found lung cancer so widely spread that it was inoperable. The surgeon sewed him up and declared his case hopeless. On April 5th he heard about the asparagus therapy and immediately started taking it. By August X-ray pictures revealed that all signs of the cancer had disappeared. He is back at his regular business routine.

When the evidence mounts up surely it is remiss of science to continue to show complete indifference .

Case no. 4: A woman who was troubled for a number of years with skin cancer. She finally developed different skin cancers which were diagnosed by the acting specialist as advanced. Within 3 months after starting on asparagus her skin specialist said that her skin looked fine with no more skin lesions.

This woman reported that the asparagus therapy also cured her kidney disease which started in 1949. She had over 10 operations for kidney stones and was receiving government disability payments for an inoperable, terminal, kidney condition. She attributes the cure of this kidney trouble entirely to the asparagus.

But when the evidence is all but incontrovertible surely science must start an investigation. As case after case becomes public and scientists do nothing it begins to look like dereliction of duty.

I was not surprised at this result as 'The Elements of Materia Medica', edited in1854 by a Professor at the University of Pennsylvania stated that asparagus was used as a popular remedy for kidney stones. He even referred to experiments in 1739 on the power of asparagus in dissolving stones. Note the dates.

We would have other case histories but the medical establishment has interfered with our obtaining some of the records. I am therefore appealing to readers to spread this good news and help us to gather a large number of case histories that will overwhelm the medical sceptics about this unbelievably simple and natural remedy.

It is typical of the medical establishment that they seek to supress such knowledge. Dependent as they are on costly operations and expensive drugs they have an incentive to supress a cheap and readily available cure.

For the treatment asparagus should be cooked before using and therefore canned asparagus is just as good as fresh. I have corresponded with the two leading canners of asparagus, Giant and Stokely, and I am satisfied that these brands contain no pesticides or preservatives. Place the cooked asparagus in a blender and liquefy to make a puree and store in the refrigerator. Give the patient 4 full tablespoons twice daily, morning and evening. Patients usually show some improvement from 2-4 weeks.

It can be diluted with water and used as a cold or hot drink. This suggested dosage is based on present experience but certainly larger amounts can do no harm and may be needed in some cases.

This seems an uncomplicated regime.

As a biochemist I am convinced of the old saying that 'what cures can prevent.' Based on this theory my wife and I have been using asparagus puree as a beverage with our meals. We take 2 tablespoons diluted in water to suit our taste with breakfast and with dinner. I take mine hot and my wife prefers hers cold. For years we have made it a practice to have blood surveys taken as part of our regular check-ups.

The last blood survey, taken by a medical doctor who specializes in the nutritional approach to health, showed substantial improvements in all categories over the last one and we can attribute these improvements to nothing but the asparagus drink. As a biochemist, I have made an extensive study of all aspects of cancer and all of the proposed cures. As a result, I am convinced that asparagus fits in better with the latest theories about cancer.

It goes without saying that the opinion of a biochemist in these matters is significant. He has taken wisdom from the past (what cures can prevent) and allied it to the personal experience of his wife and himself. In addition, he has involved an independent expert to perform a blood survey and undertaken his own research studying extensively all aspects of cancer and all of the proposed cures.

This is the scientific paradigm. As a result, he is convinced that asparagus fits in better with the latest theories about cancer. Science should follow in his footsteps and fully corroborate his findings.

Asparagus contains a good supply of protein called histones which are believed to be active in controlling cell growth. For that reason, I believe asparagus can be said to contain a substance that I call cell growth normalizer. That accounts for its action on cancer. It has been reported by the US National Cancer Institute that asparagus is the highest tested food containing glutathione which is considered one of the body's most potent anticarcinogens and antioxidants.

Here we have a Theory of Asparagus. The substance now identified as cell growth normalizer should be isolated by science and be made available to all cancer sufferers. It is not good enough that doctors who are aware of the benefits of asparagus continue to keep the knowledge to themselves. Shame on science.

Please spread the news. The most unselfish act one can ever do is paying forward all the kindness one has received, even to the most undeserved person.

Anyone reading this should be happy to spread the news.

The modern day Internet has become a ferment of ideas that science is struggling to keep up with. Yet a quick trawl through its contents reveals a treasurehouse of cutting edge concepts which are leaving traditional science in their wake. Here are ideas which challenge modern scientific orthodoxy so much so that it is science which now appears to be unorthodox.

This is hardly surprising. For centuries, science was able to close down new ideas and only let them out when drug companies, for instance, were ready. However, social media means that new, important ideas can be quickly disseminated. The truth will out.

It is this simple fact which makes social media (Twitter, Facebook, YouTube) the next great step forward in scientific research. We no longer have to wait for news to leak out. It is already there. All we have to do to seek the next scientific breakthrough is go and google it. Science can no longer supress the truths that millions of ordinary people already know. Together with internet discussion groups people are beginning to slip the restraints of orthodoxy. This promises to be the scientific revolution of the future.

To end with, here is a discussion bringing together Bob Beck (BB), a retired physicist, Russ Torlage (RT), a certified technician and power engineer and Leading Edge Newspaper (LE) in a wide ranging discussion.

(BB) Electrification of the blood: After enough looking around, I want people to see that the electrification of blood can cure all parasites, virus, bacteria...that means Aids, cancer...

What Bob calls looking around refers to years of in depth research.

(LE) He uses 27 - 33 Volts ac, 3.92 Hz square wave bi-phasic to electrify the blood and also uses a magnetic pulse generator.

Narrowing down treatment to specific voltages and frequencies is how modern medical research proceeds.

(BB) The second step is externally applied magnetic resonance of lymph, spleen, kidney and liver which helps to neutralize germinating, latent alien invaders and thus blocks reinfection. This quickens disease elimination, restores the immune system and supports detoxification. Detoxification is essential because you are throwing off millions of dead and dying bugs.

Magnetic resonance is already understood by science. What is new here is the isolation of these resonances to specific organs. This is what makes this new science so exciting.

(BB) We also have a unit called the magnetic pulse generator which has a tremendous magnetic field. We had a fellow stand about 20 feet away with a tri-field meter and it actually moved the meter at 20 feet but only when pulsed.

Science has not so far used tremendous magnetic fields in medicine but those days may be over.

(BB) Permanent magnets, no matter how strong, will not nor cannot scavenge pathogens with electromotive force.

No scientist has ever come forward with evidence against this.

(BB) It is important to be able to deal with rapid detoxification without producing tremendous discomfort. One of the worst things you can go through is detoxification if you are the slightest bit sick. If you go too fast and don't detoxify, you have done more harm than good. We go by the motto, first do no harm. There are ways to prevent this discomfort and we will get into that in a few minutes.

All too often, and in many people's experience, modern medicine charges in blindly attempting to cause rapid detoxification. This outdated protocol is clearly contraindicated.

(BB) The third step, which we found worked amazingly and synergistically well, was silver colloids. Pennies-per-gallon, self-made perfected colloids greatly assist in eliminating all known pathogens and preventing opportunistic infections. This has been known for a long time.

The wisdom of the past is again evident here.

(BB) The fourth step is drinking ozonized water for rapid, safe, totally natural cell oxygenation without free radical damage. I get high from drinking ozonized water every day. It's like drinking a martini and a half.

Difficult as it is to believe, taking medicine can actually make you feel good! No wonder modern day pharmaceutical companies don't want you to know about it. There appears to be a conspiracy amongst them that medicine should taste bad and not give an immediate feeling of well-being.

(LE) I would like our readers to know about how you became involved in these technologies, which enhance the immune system and kill every organism that isn't supposed to be there.

Here Leading Edge newspaper asks the uncompromising, hard-nosed questions.

(BB) When I attempted to find a copy of this paper to see what it said, I found that they had all vanished or were cut out of the proceedings. We hired a private investigator who got a personal abstract copy from one of the conference attendees. I also did a computer search and found that the only other mention of this

technology was in Outer Limits in Longevity magazine, in the December 1992 issue.

It stated that Steven Kaali, MD, from Albert Einstein College of Medicine, had found a way of inhibiting Aids in blood, but that years of testing would be required before the virus-electrocuting device was ready for use. In other words, they discovered it and then tried to cover it up.

And you might like to consider the possibility that worldwide drug conglomerates have conspired to keep it quiet. A virus electrocuting device would have led to no billions of dollars in research money, no expensive antiretroviral drugs to sell and no dragged-out 'cure' with enormous profits made from drugs that worked a bit, then a bit more, then some more, and so on. It is science's duty to expose this and make the virus electrocuting device available to the world as soon as possible.

(BB) But a very funny thing happened. Two years later a patent popped up. The U.S. Government patent office described the entire process. You can obtain patent #5188738 in which the same Dr. Kaali describes a process which will attenuate any bacteria or virus including aids/HIV, parasites and all fungi contained in the blood rendering them ineffective from infecting a normally healthy human cell.

This is in a government document. This was in 1990. Why haven't they told the public about it? I decided if there was a sure-fire cure for aids I had to find out about it.

I urge people to look up this patent for themselves. I also urge scientists to look it up and act upon it.

(BB) When I looked into Dr. Kaali's work I decided to go ahead and fund it. We found that it worked all of the time. For two and a half years we gave full credit for this invention to Dr. Kaali, whose name is on the patent.

Then I discovered that there was a long history of this technology. We followed a trail of these patents back 107 years. We also found a patent, #4665898, that cured all cancer dated May 19, 1987. Why has this been suppressed? Why hasn't your doctor told you about an absolutely proven, established cure for cancer?

This is a shocking indictment of science and one which you might like to raise with your MP (if in the UK) or your Congressman (if in the USA) or with your appropriate representative if in other parts of the world.

(BB) The answer is that doctors get $375,000 per patient for surgery, chemotherapy, x-ray, hospital stays, doctors and anaesthesiologists. This is the official statistic from the U.S. Department of Commerce. The medical patient cured is a customer lost.

And, inevitably, here is the answer. Doctors should be held accountable for this. In fact, you might like to seek out your local cancer specialist and present him with this information and gauge his reaction. If he looks shifty or attempts to change the subject then you can judge for yourself.

(LE) Can we talk about the technology?

(BB) Yes. The most important step in taking back your health and your power is through blood electrification. Research from Harvard, MIT and Albert Einstein College of Medicine has shown that microcurrents are known to eliminate all viruses, parasites, fungi, bacteria and pathogens in blood. I can prove that this research was lost or suppressed. There is no known cure for herpes, Epstein-Barre or Ebola. If it gets loose here, it will kill thousands. But not if you have colloids and a blood cleaner.

It is quite appalling that such research has been 'lost' or supressed. That Bob Beck can prove it is shocking and he must be encouraged to place his evidence into the public domain. Then, for the first time, the public will be aware of the scale of the possible cover-up which is preventing the cure of herpes, Epstein-Barre and Ebola.

(RT) Okay, and I will tell you how I became involved with this technology. I am from Vancouver, Canada, and my company, Sota instruments, is located there. My wife had chronic fatigue syndrome and we tried everything without success. She had been a comptroller for a multi-million dollar clothing company and she finally had to quit her job and stay at home just trying to survive.

I came across Bob at a lecture. I was absolutely intrigued by how much he had to give, not only through his knowledge, but through his kindness. Then I attended one of his lectures in Seattle. I have a nuclear and electronics background so many of the things Bob mentioned rang a bell. I recognized his information as basic and sound physical data that could be measured. My green lights were going off and I said this man's right on the money here. So I built one of the units.

It may come as a surprise that scientists weren't present at this lecture but it could also be the case that they were part of the suppression. That said, it is fortunate that someone with a nuclear and electronics background was present and that his green lights were going off.

(RT) We have one we call the silver pulser which is a two-in-one unit. It does the blood cleaning and also makes the colloidal silver. It is a unique device which runs off a single nine-volt battery but has a special constant voltage output of 27 volts. It is very lightweight, portable and easy to replace.

Medical science should move as quickly as possible to integrate the silver pulser into modern medical practice. As a matter of urgency, the silver pulser should be introduced into those areas of the world most affected by herpes, Epstein-Barre and above all Ebola. The World Health Organisation and Medicine Sans Frontières should equip their field operatives in Africa with the pulser so that the Ebola pandemic can be brought under control in the shortest possible time.

A company specialising in battery manufacture (such as Duracell) might like to make a humanitarian gesture at this point and supply the required nine volt batteries free of charge.

(RT) We also have a unit called the magnetic pulse generator which has a tremendous magnetic field. It has one huge output, like a gun, causing the micro-currents to occur deep within body tissue.

You need tremendous power to penetrate deep within the body. Our units have measured over eight inches of penetration so front and back you are covering 16 inches, which will cover most people.

This may be optimistic given the rising obesity rates in the West, but is still highly impressive. Science has never been able to replicate the tremendous power required to penetrate deep within the body despite millions of dollars of funding and years of research. Scientists I have contacted are also intrigued by the tri-field meter used in these experiments. Most research centres are equipped with the standard issue bi-field meter so again, there may be an advance on current techniques here.

(RT) The ozone unit is a brand new instrument we designed based on Bob's specifications. It is important that it is battery operated and portable and that the ozone gasses not escape from the unit. Ours has a unique ozone restrict mechanism that is a charcoal filter. As the ozone unit is actually giving up the gas it goes back through the charcoal filter and is destroyed.

At the same time, as you pour the water through the charcoal, it is purified, removing the organics and chlorine. So now the ozone

that is in the water doesn't have to work as hard to get rid of all the foreign stuff in the water. It is very important that if you do have a portable unit that you purchased at the local supply store, only use it outside in the fresh air.

Science has persisted in making ozone work far too hard to get rid of all the foreign stuff in the water. Experiments with charcoal filters have been sporadic but here, for the first time, we have evidence that they are effective in removing organics and chlorine as well as ozone.

(BB) Drinking two or three glasses a day is essential. This provides universal detoxification by oxidation of wastes and dead and neutralized pathogens. They are all anaerobic. In other words they can't live in oxygen. Cancer cells can't live in oxygen. Neither can most of the other disease cells known to humankind.

It is shocking that science has not yet discovered that cancer cells can't live in oxygen. One wonders where all the money allocated to cancer research has ended up. Even a layman would suggest that a primary area of research should involve the conditions in which cancer cells can and cannot live.

Perhaps it is true that scientists do indeed live in ivory towers and ignore the suggestions of ordinary people in favour of their own pet theories. Surely at least one scientist has suspected that pathogens are all anaerobic. If no-one has then we have a shocking situation on our hands.

(BB) The ozonated water puts oxygen into every cell in your body. You can prove that you are oxygenated by using a little spectrometer that attaches to your finger made by Nelcor company. It tells you on a meter the exact percentage of oxygen in your blood. The haemoglobin of the red blood cells carries oxygen to all of the tissues of the body. You may start at 93% to 95% oxygenation, which is low, and after two and a half to three minutes of drinking this water you will see 100%.

Science has been particularly lax in confirming that after two and a half to three minutes of drinking ozonated water every cell in the body will have had oxygen put into it. The fact that 100% can be achieved in so short a time is a very exciting result.

(BB) We have stressed over and over not to take any medicines. For example, if you have garlic in your system a little bit of it is tolerable. But garlic has sulfonhydroxyl in it which is a deadly poison. Gangsters and soldiers used to rub their bullets with it before they fired so if you were nicked it would kill you. Organic gardeners use garlic now that they can't get DDT. It kills everything.

Garlic also kills lots of brain cells which causes desynchronisation of the left and right brain hemispheres. Pilots are not permitted to ingest garlic before their flights because it slows down their reaction times. Do not touch garlic. It is not a health food. We tell people to eliminate everything that is known to be toxic in large quantities, including certain vitamins such vitamin A and niacin, and also garlic.

Brain surgeons attempting the difficult procedure of desynchronisation reversal offer anecdotal evidence that on opening the skull the smell of garlic was noted by theatre personnel.

(RT) So far, we have treated eight PCR-tested, HIV positive customers. After treatments ranging in length between two weeks and two and a half months they each went to zero, below detectable levels. We were absolutely astounded to get that type of documentation. These are separate from the Aids patients Bob has mentioned. We did our own testing. Many doctors, particularly naturopathic doctors, are trying out our technology and their patients are getting well.

Naturopathic doctors are key here. It is not often reported in the literature how effective these doctors are yet here we have solid, verified and incontrovertible proof that their patients are getting well. It is important they did their own testing. Independent testing cannot be trusted where this technology is concerned for obvious reasons.

(RT) The blood of an Aids patient is extraordinary to look at under a microscope. What you see will just shock you. There are life forms in their blood that look like octopuses with a hundred arms and there are things creeping around. Then we look at their blood after a few weeks and all of these things are disappearing. Their blood returns to the natural healthy state it was when they were born.

HIV specialists will confirm that the blood of an Aids patient is extraordinary to look at. They have been unable, as yet, to remove the life forms that look like octopuses with a hundred arms and that is what has been hindering a cure for all these years. Now the cure seems to be within our grasp.

Science should make pictures of the octopus forms and the creeping things available then provide later pictures which show them disappearing. In that way the general public could see for themselves the effectiveness of the cure and demand that money be diverted immediately from dead-ends such as retroviral research into octopus form removal research.

(BB) Let me summarize by reading a paragraph from my book. Why haven't doctors revealed this before now? Because it has been in the Patent Office. It has been in the journals. It has popped up from year to year. When actualized this data could interrupt HMO profits, disrupt medical pharmaceutical cartels, abort all biological schemes, eliminate most drugs, medicines, debilities and early deaths, wipe out hospitals and health care capital investments, minimize insurance mechanizations, dramatically abate sickness and suffering plus absolutely imperil social security futures with bankruptcies.

My book is soon to come out. I have personally taken time to look into this before I put my reputation on it. And now that I believe in it, I am evangelical about it.

And here is the real reason why this new technology has been kept from the public. The world's big pharmaceutical companies don't want you to know about it. In the final analysis they have little interest in making you well, they are more concerned with profit.

Fortunately, Bob's book will come out soon and the technology is already available for purchase.

I can attest to the effectiveness of Becks and Russ's devices. I ran across Beck's research by accident in 2007. It sent me on a wonderful 6 month discovery/fact-finding mission. After much research I decided the concept was sound and bought the silver pulser and later Beck's Braintuner the BT-7.

Science spends far too much time listening to other scientists in a closed circle of so-called specialists. Yet here we have someone who can attest to the effectiveness of Becks and Russ's devices, can back things up at the core level and after much research has decided that the concept was sound. This is the scientific method in a nutshell and this researcher's findings therefore have just as much validity as any other scientific investigator.

I must say in my and my girlfriend's case the bacteria die-off from the electrification manifested major flu symptoms for about 2.5 days. Extreme fatigue and muscle soreness being the primary effects and this from two healthy adults in their 40's. We did this for a week coupled with 8oz of colloidal silver per day. We began to feel in our

20's again. Completely energized, no more fatigue from doing relatively nothing and my g/f lost about 10 pounds in the process.

Science should also investigate the weight loss reported here as it could be the first step to a cure for the current obesity epidemic. It would not be the first time a spin-off had occurred from a cure initially directed at quite different symptoms, nor will it be the last.

After much hesitation I turned my parents on to the colloidal silver and they swear by the stuff now. Both experience revitalization when they keep up with the silver.

And here is further proof. Revitalization and weight loss observed in different subjects both using the same technology. Science should investigate as a priority if the same curative mechanism is involved or whether another, as yet undiscovered, principle is at work.

Quick notes: We have only been sick twice in about 2yr. First time was a cold. We drank some CS and were well within 24hrs. The second time was this past January '09, we both got something that completely floored us. Our voices were 2 octaves lower from soreness and we experienced a fatigue and whole body soreness like never in our lives.

Getting out of bed was practically impossible. Additionally, my girlfriend developed major congestion in the lungs. No fever for either of us though this went on for a week and a half before we kicked ourselves and remembered the gizmos. That's what we call them.

We first made some silver and drank 8oz 3x a day as well as putting some in a nasal sprayer and using it nasally as well as inhaling 5 to 6 shots from the inhaler through the mouth every 2 hours to get it into the lungs. In about 5 hours my g/f's lungs cleared up and in 12 hours the fatigue was half what it was. About 18 hours later when we woke up we both felt 100% better and our throats & voices were almost back to normal.

This evidence contains both the amount of silver used as well as the stages during which it was applied and the method of application. There is also clear evidence recorded concerning the time for the application to have an effect and quantifiable data concerning the halving time of fatigue as well as the completion period for 100% recovery.

It is exactly the type of data which a well-equipped medical research lab would collect while in the process of proving a new treatment using the standard double-blind protocol. This evidence can therefore be used as primary data.

About day and a half later the symptoms of whatever we had were about 90% gone. We then did the blood electrification to knock whatever was left out. We did not do it initially because we felt the die-off would have compromised us too much given our first experience with it.

This seems wise but science should investigate in order to show whether or not blood electrification, whilst knocking whatever was left out, would indeed produce sufficient die-off to compromise a person too much. It should be possible to determine whether genetic factors are at work here or whether the pathology is endemic and therefore insurmountable.

The BE brought us back to full health by 2.5 days with none of the symptoms from the first time we tried it manifesting. Several months prior to this event my g/f tested positive twice on a pap smear for cervical cancer. She has a history of cervical cancer and has had tumours frozen off in the past. It had been almost 15 years since she last had one frozen off and she did not want to go through it again. I suggested trying the electrification and silver before making an appointment for standard medicine.

Here we have the full medical history of the g/f and with this it will be possible to set a baseline against which either improvement or reversion can be measured. This is normal practice.

To cut a long story short she did the blood electrification about 1.5hrs twice a day, drank about 12oz silver roughly twice a day and douched twice a day with colloidal silver. She did this for a about 4 weeks. She missed a test at 4 weeks but had one the following week.

The pap test came back negative and much to her doctors surprise the scarring from prior freezing was gone. She said her cervix looked pristine. Needless to say, my experiences with this technology to date have been far better than I ever expected. I highly recommend it. This stuff works.

This is scientific proof. The baseline had been set and the medical history was known. The effects were better than expected in that the pap test came back negative (which was expected given the parameters of the treatment) but the scarring from prior freezing was also gone. This shows that the treatment, fully documented, obtained results beyond those expected. In scientific language the protocol exceeded regressive normalization. In layman's terms: 'This stuff works'.

This sounds promising and I want to try it out. But what's the deal with drinking colloidal silver? People seem to be turning blue

after taking it so can that be avoided? Is it inevitable that you will turn blue eventually?

Of course scientists, especially those employed by Big Pharma companies, are loath to lose their pre-eminence so they attempt to spread scare stories. The turning blue canard is an example of such disinformation and those scientists involved in continuing to disseminate the rumour should be ashamed of themselves.

Trust me, you've been reading too much silly big pharma-generated propaganda. True silver colloids, .0008 microns or less in a concentration of 7-10 ppm is an extremely potent immune system enhancer and has absolutely zero negative side effects. My wife and I have been making our own for over four years now and haven't had so much as a sniffle the whole time.

Which comes from a source which has no interest in placing false information into the public domain.

If the world's populace suddenly became aware of the incredible healing properties of CS, Big Pharma would literally collapse overnight just from that alone. . . And they bloody well know it.

It is appropriate to end with this cry from the heart. Ignoring the invective, born of frustration, it is easy to imagine the exasperation experienced by those who have direct knowledge of the benefits of blood electrification and colloidal silver because they have seen first-hand the effects they have on a wide variety of illnesses.

It may be true as they suspect that science is fighting shy of endorsing the treatment because it is under the influence of the world's large pharmaceutical companies and this seems the most likely explanation. On the other hand it could be that yet again science is sulking because it did not get there first. Either way, the time has now come to end such differences.

The facts are available for science to confirm and it should move to do so as soon as possible. Where the health of millions of people are at stake it should be a priority for one of sciences next steps to be the validation without delay of so-called 'alternative cures'. As can be seen from multiple testimony above there is nothing alternative about them.

They work, and it is only hidebound and reactionary science which is preventing them largely taking over from standard medicine.

Chapter 16: MindScience

What is MindScience? MindScience is a new Science developed by Dr. Baskaran Pillai to study the whole range of the human mind complex that include sensory mind, intellect, sleep consciousness, dream consciousness, the subconscious, personal unconscious, collective unconscious and the mystical states of the superconscious mind.

This new science appears to be a wide-ranging synthesis.

MindScience is not simply Western psychology or parapsychology. MindScience attempts to study the link between mind and matter at the fundamental level and draws upon principles from quantum mechanics and the wisdom traditions of ancient civilizations. In philosophical terms it is the merger of ontology and epistemology. MindSound technology is a branch of MindScience that explores the relationship between sound and mind.

And it already has sub-fields, also developed by Dr. Pillai. This is indicative of a mature science.

Science of thinking and probability living: Thinking is everything. Dr. Pillai teaches that we become what we think. When we think a thought over and over again and bring emotion to the thought we manifest that thought into reality.

Given the above, MindScience is obviously a full Quantum Theory based firmly on the Copenhagen Interpretation.

Other than typical human thoughts there are subtle sound frequencies and energies that we can use as a tool to re-program our thinking process to create expanded realities. At the highest level, there is no thought at all.

The analogy of a business may be useful here.

Infinite freedom, flexibility, opportunities and all probabilities exist in the domain of no thought. We need to learn the Science of thinking so that we can harness the tremendous power of our mind for living a life of probabilities and utter positivity.

The corollary being that it is only in the domain of no thought that we can escape a life of certainty and utter negativity. This seems intuitively correct.

Quantum physics has revealed an understanding of the subtle laws of nature. Power increases as we delve deep into smaller units of matter and energy. **Atoms are powerful, but more powerful than atoms are particles. Finally we have vacuum state and waves as ultimate levels of power and energy.**

This has never been disputed.

Our mind represented by thought forms is not a powerful mind. In other words, thinking based on logic and reason can be explained as molecular levels of intelligence. Atomic and particle levels of intelligence are superior to thoughts of logic and rationality. A quiet mind without thoughts is the quantum state of mind.

Scientists concerned with molecules are widely regarded as having inferior levels of intelligence to those working with atoms and particles. Dr. Pillai's theory clearly reflects reality here.

The human conceptual mind is based on thoughts that are linked to words and sentences. Dr. Pillai maintains that what weakens the minds is the sentence mind. Dr. Pillai humorously says that everyone is sentenced by the sentence mind.

How unusual to meet a Scientist with a sense of humour and how refreshing. Dr. Pillai has hit the nail on the head; modern linguistics should soon be in a position to confirm that we don't know any other way of communicating other than through sentences. Psychologists have measured the debilitating effects on the mind due to sentence over-use but so far their readings have lacked a theoretical underpinning. Dr. Pillai's excellent joke could well be used as an aide memoire to those working in the field.

To help us explore the inner journey to the quantum state of mind Dr. Pillai has done extensive research to develop simple and practical applications to replicate the significant results students have experienced during his 30 years of teaching. Just as the study of chemistry has theory which must be validated by individual lab work with specialized equipment, MindScience has a range of transformational tools for everyday personal practice.

These tools draw from several disciplines and are engineered for the modern world. The tools of MindScience include the following: Particle mind process: Identifying with the finest and tiniest unit of matter, such as electrons, neutrons, photons and bosons. Creates tremendous energy and power for heightened awareness and action.

Identification is an important first step. Only when people are familiar with electrons, neutrons, photons and bosons can progress be made. Particle physicists are already familiar but lack heightened awareness. This could be explained by their failure to *identify* with the units of matter.

Body dynamics mind process: Visualizations on inner body dynamics such as the circulatory system, nervous system and respiratory system. Creates rejuvenation and enlivening of the cells.

Enlivening of the cells has long been a dream of biology. Visualisation of inner body dynamics such as the circulatory system, nervous system and respiratory system could be just a starting point. Modern science should be able to extend this list to cover the excretory system. Visualisation of this can only help with appreciating MindScience.

Gamma light mind process: Revolutionary infusion of gamma light rays to deliver specific attributes of intelligence and compassion. Though gamma rays are incompatible with the physical body these rays are the primary vehicle for higher intelligence and interacting with archetypal energies. Gamma rays give the best results to help the mind and body evolve to advanced life forms.

As Dr. Pillai correctly notes gamma rays are incompatible with the physical body but science should make a greater effort to determine their relationship and interaction with archetypal energies. This should not be difficult provided suitable archetypal energy measuring instruments can be developed. It is a matter of priorities.

Health and beauty mind process: Communicating with our active DNA and accessing our inactive DNA with language and quantum sound frequencies to reprogram the body for vigorous health and ageless beauty.

DNA profiling and gene therapy are already mainstays of science but here we see frontiers being pushed back still further. Communicating with our active DNA is not so world-shaking but the idea that we could access our inactive DNA is a real breakthrough.

Quantum sound frequencies: Many ancient cultures discovered that the mind is made up of subtle energies in the form of sound frequencies. These sound vibrations make language possible. If there is no language we cannot think. Thinking is dependent on the mind, the mind is dependent on language, and language is dependent on sound vibration. The fundamental equation is mind equals sound.

This is the fundamental equation of MindScience and is as important as $E=mc^2$ is to Physics. As has been seen again and again it is the case that ancient cultures discovered many important things; the fact that they were aware the mind is made up of sound frequencies should humble modern science.

Dr. Pillai's concept of the mind as sound has led him to develop a science called mind-sound technology. Mind-sound technology employs the use of powerful phonemes, sound frequencies to enhance intelligence. Different sound frequencies create different kinds of intelligence.

This is confirmed by mainstream science. For example it is well known that bats use very different sound frequencies (ultrasound) compared to humans and there can be little doubt that human intelligence is a different kind of intelligence to that of a bat.

Fire labs: Technology from Vedic sciences based on invoking gamma light using the medium of fire. Fire is a powerful means of communication with gamma light and archetypal energies of higher intelligence. In the fire labs we raise fire with specific herbs that produce electromagnetic fields that support wealth, health or relationship.

Vedic Sciences are derived directly from ancient cultures. Science has so far failed in its attempts to invoke gamma light via the medium of herbs but trials continue.

Rituals: Accessing higher intelligence requires the activation of powerful sound vibrations and archetypes in our mind. Rituals are a powerful tool of activation. Vortexes built by ancient civilizations were laboratories designed to interact with the archetypal forces and the rituals performed there were designed to access their intelligence and energy.

This passage is so rich in insight that it threatens to keep a well-resourced research laboratory in work for years to come. To start with, the exact correlation between powerful sound vibrations and archetypes in our mind needs to be studied and, if suitably reinforced areas are available, vortexes could be analysed and their interaction with the archetypal forces determined. With help from Anthropologists, the rituals could be performed again in order to make tests on how they accessed their intelligence and energy.

Archetypal symbols: Scholars Carl Jung and Mircea Eliade both acknowledged that archetypes are endowed with special forms of intelligence. Jung described the gods, goddesses and angels as archetypes in the collective unconscious.

Eliade claimed these archetypes were actual energies or entities that reside in invisible, subtle realms. Higher intelligence resides

with the archetypes. **These archetypes can be accessed through mind processes, quantum sound frequencies and also symbolic devices**

These scholars are both well respected. Accessing of the archetypes which reside in invisible, subtle realms is not straightforward but is obviously an important next step for science. Fortunately help is at hand:

Particles Of Power is a monthly download subscription program. Price: $13.00

This seems a small price to pay.

9 planet gemstone bracelet: Planets are agents of our karma. This 9-planet bracelet holds stones compatible with each. Price: $18.00

Again, a very modest fee to have access to stones which are compatible with planets.

Siva Shakti mala with Rudraksha beads and crystal beads: Mala beads are a special rosary used for counting recitations of 108 sounds. Price: $35.00.

Science should really get hold of some Rudraksha beads and use them to investigate the collection of 108 sounds. It works out at less than 33 cents per sound.

9 planet incense pack of 12 boxes: This 9 planet incense 12 boxes is also known as Navagraha incense. Price: $40.00

Again, very reasonable. But not content with his achievements so far Dr. Pillai is now looking to move forward in his journey.

Message from Dr. Pillai: I'm at a crossroads in my life and in my understanding of human nature. I have considered my teaching, my interaction with people and its impact on the general public. I want to make some important changes based on my current understanding of life. I'm expanding my identity beyond the limited identity of an Indian Guru.

This identity has been too confining and has raised the eyebrows of many people who would have otherwise join my movement to transform the world. Many 99.9999% are standing outside the gate. They are afraid of the idea of Guru, it could be a dangerous cult. I am now joining the respectable cult of science, technology and academia. This is not new to me. I was invited to the United States for my PhD in 1983 based on my publications and research at the University of Madurai.

Dr. Pillai echoes postmodernist theory here in his identification of science as a cult. There are a number of French postmodernists who would wholly agree with his analysis. Science is merely one way of seeing the world and one which is culturally determined. The former way of Guru is not invalidated, the shift in cultural viewpoint is merely a way of accessing those individuals who have not shared his previous, equally rich yet culturally different, view of the world.

The trip to Egypt in February reinforced this decision. 5,000 years ago the Egyptians had a better knowledge, an understanding that was not simply linear scientific knowledge, but a holistic understanding in which they had access to the knowledge from all the galaxies. I want to incorporate this holistic knowledge into my teaching. To do this I need to renounce my limited identity. It's not that what I've given you so far is unsound but I'm going to expand the horizon.

Linear understanding is becoming increasingly seen as impoverished. The holistic world-view is more suited to the multicultural aspect of today's modern synthesis and science can only be the beneficiary of a more enlightened viewpoint. It is something of a shame that science, at least in the West, should be so, but that is the world in which Dr. Pillai finds himself.

As an aside, it is to be hoped that NASA will allow time on the new James Webb Space Telescope (the successor to the successful Hubble Space Telescope) in order to access knowledge from all the galaxies. The first aim is to regain the better knowledge the Egyptians had 5,000 years ago.

First, we need to end our current way of thinking thoughts. When we think thoughts we are in the molecular mind. Molecules are gross forms of matter with limited power.

Nuclear Physicists will confirm that molecules have limited power. This is why we have atom bombs, not molecule bombs.

We give up limited thinking when we are in the particle mind. In this realm, thoughts disappear and we are in the domain of infinite knowledge. In that domain of infinite knowledge, we access intuition, the power of creative action, and feel unbounded by time or space.

I can't actually recall when intuition last played a part in a scientific discovery. Today's science seems to be all about committees and how many millions of dollars will be spent splicing yet another artificial gene into an unsuspecting plant.

As you go deeper into the particle level, and that's why I use the expression the particle mind, the mind becomes as powerful as the subatomic particle. You begin to function from that level.

All areas of your life begin to change. You are able to do well at school because you are using the particle mind as opposed to the molecular logical mind. At work, you are able to perform work better than other people because you are using the particle mind.

Doing well at school is increasingly being linked to not thinking logically. Indeed, not thinking logically has long been established practice in University Education Departments. It is therefore unsurprising that the technique is finally taking root in schools.

So the particle mind has application for everyone because it is primarily dealing with unlimited intelligence. A benefit of particle mind is not just intelligence only, but also unlimited energy.

Dr. Pillai is perhaps exaggerating when he speaks of unlimited intelligence and unlimited energy but he is to be forgiven his natural exuberance as he has just invented a whole new science.

This is one technique that is going to improve all areas of your life including your health. Diseases are due to ignorant behaviour of the cells.

Biologists will confirm that well-informed cells are much less disease-prone.

Ignorant behaviour of the cells is due to the ignorant mind.

It has long been suspected that people with few GCSE's are most likely to have poorly behaved cells.

Once the mind consciousness becomes more intelligent there is no room for diseases in the body or diseases in the mind. In general, unhappiness will be wiped out.

Clever people have less diseases. It is that simple.

Oct 8 - 11: Goddess Retreat: Embody the power of the divine feminine: Join Dr. Pillai for a 4 day retreat in the Poconos, Pa, during Navaratri, the most powerful time of the year to commune with the goddess. Experience secret sounds, sacred geometry, lecture and fire labs to invoke the divine feminine archetype. She controls power, joy, beauty, expansion, abundance, prosperity.

It is good to see Dr. Pillai promoting the advancement of women in science.

Nov 14 – 21: Commune with the Virgin Goddess at Kanyakumari, India. Receive profound initiations into the mysteries of goddess worship from Dr. Pillai at this great power spot in Kanyakumari, land of ancient Lemuria.

Archaeologists will be pleased that Dr. Pillai has located the ancient land of Lemuria which they have been seeking for some time now.

Spring, 2011: Journey to Greece with Dr. Pillai during Spring Navaratri, the 9 nights of the goddess, to participate in a history-making revival of the great Greek archetypal beings at their special vortexes of power.

It shows a modern and enlightened approach for Dr. Pillai to open his expedition to members of the public who will, as a result, experience scientific field-work first-hand. Scientists of all disciplines should follow suit.

What is an energy vortex? An energy vortex is formed in a place where focused thoughts and feelings have gathered in unison of common vibration. Such energy vortexes can often be found at the sites of schools and universities, large companies and organizations, temples, churches and mosques. Wherever people have gathered for an extended period of time in common purpose with powerfully focused intention you will find an energy vortex.

For example, the queue for the toilets at the Monsters of Rock Festival, Castle Donington.

Dr. Pillai often organizes special trips in the fall of each year to such energy vortexes. This year's trip will be in the south of India to energy vortexes that include water, forests, caves and temples.

Dr. Pillai would hopefully be open to scientists joining him. Research laboratories have an incomplete map of energy vortexes and although progress has been made with whirlpools, forest, cave and temple vortexes remain elusive.

Particle mind vs. Molecular mind: I am excited by the responses I have been receiving to the 90 day quantum sound frequency (QSF) challenge. I have always pondered the origin of thought, sound and vibration – the bridge between knowledge and experience.

After years of experimenting with the principles shared in such works as Shakti Gawain's 'Creative Visualization', Rhonda Burn's 'The Secret' and Wayne Dyer's 'Change Your Thoughts – Change Your Life' I find that the information shared by Dr. Pillai has really helped me to connect the dots on a more concrete level.

Quantum scientists as well as sound frequency scientists are equally excited by the QSF challenge. Moves are afoot to ask Shakti, Rhonda and Wayne to head up a research group giving them access to and control of a modern facility and a team of postdocs. The whisper is that Fermilab could be involved.

Regardless of whatever your situation in life is, whether you have money or don't have money, whether you have health or you don't have health, whether you have an education or you don't have one, all of these are relative. The probabilities are infinite. It is all about where you are going to put your mind to, or look at. So when you look at it, it is there, when you don't look at it, it is not there.

Dr. Pillai has hit the nail on the head with this one statement.

This is such an exciting time that we are living in, where we are actually realizing how to consciously co-create our life experiences. You literally become what you think. This is science.

Anyone who doubts this is science should first take the time to try it. The results will be highly instructive.

I am Higgs Boson, by Bilva, Ann Arbor, Michigan.
Dr. Pillai has been my Sadguru since June, 2007 and every day is a miracle. Lately with Om Higgs Boson Aadhi Sivaya Namaha, I have found an even deeper connection with Dr. Pillai in an electric energy field that I receive that is his energy field. I feel this energy field within my body.

The formerly theoretical particle has, of course, now been discovered in a series of experiments at the Large Hadron Collider (LHC) thus giving credence to Dr. Pillai in his chanting of 'Higgs Boson'.

Peter Higgs originally proposed his particle as a theoretical explanation of mass; to have Bilva potentially extend his theory to cover the electric energy field is an exciting new development.

Ancient wisdom Science, by Singh, Massachusetts, USA.
I am an intransigent left brain personality believing in logic and factual reasoning. With sheer disgust and utter defiance to so called cult personalities I unknowingly came across a Dattatreya video recording. To my utter disbelief and astonishment I received favourable reply from my associates after almost 60 years, 5 solar cycles of back breaking drudgery and relentless struggle for a favourable business deal.
My colleagues at Harvard Medical School have calculated the positive power of meditation to be significant to the order of praised to the power -30. This study is done by brain scanning techniques. After experiencing Dattatreya grace I personally came to a simple

conclusion. The power of meditation and mantras is beyond the grasp of mathematics.

There can be little argument. Mathematicians, however, are disputing that the power of meditation is beyond their grasp, though they have yet to offer convincing proofs.

Below is Heather's miracle story in its entirety. Heather just only started practicing baba's mantras Shreem Brzee for wealth and abundance and Om Higgs Boson Aadhi Shivaya Namaha to get into the particle mind.

Heather's experience is particularly impressive. Science often loses the individual experience with its habit of averaging out results. This may well come to be seen as a mistake. Averaging out results is all very well when one is considering the effects of a drug on thousands of patients for example but of what use could averaging have been in the discovery of television? There were no other candidates, there was either a television or no television.

So Heather's story should be taken by science not as a miracle but as something which happened to an individual which cannot be explained within the presently understood scientific paradigm. This makes it a natural subject for scientific study as science progresses by looking carefully at the anomaly, not the routine. And, yes, science should admit that anomalies happen to individuals, rarely groups.

Since last week I have been chanting Shreem Brzee as often as I can, either out loud or in my head and things have started to really happen for me. I was inspired by the gentleman who shared his story last week about Shreem. He had gone from losing his job in the financial sector to creating huge personal wealth by immersing himself in Shreem.

Perhaps those scientists who complain most loudly about cuts in research grants might like to try it! (joke).

So the day after the last event, I chanted Shreem Brzee a lot. I also have the words written on my water bottle to embed the concept into the water particles and make sure I'm receiving the idea of Shreem in every cell of my body. And quite a few positive financial things have happened since then:

Writing words on a water bottle to embed the concept into the water particles thereby ensuring that the idea is received into every cell of the body may sound unusual but stranger things have happened.

Radio was deemed impossible by science, after all how could information possibly be transmitted by invisible waves?

Train travel at above 10 miles per hour was deemed impossible by science, after all how could people breathe if the atmosphere was moving past them so fast?

The world market for computers was estimated by science to be less than five for the whole planet. After all, what would people use them for?

The history of science is littered with scientists claiming that one idea or another is either impossible or useless. Water imprintation looks likely to be one of these 'impossible' ideas – right up until it is shown to be true.

Within one day I received a cheque which I was certain would not be coming to me for at least another month. I freelance, so payment can be erratic. I really needed it and had focused on it the night before. On the second day I was given an offer of a small freelance job with someone I haven't heard from in years. An old boss in fact. It turned out to be a great fun experience, 3 days with a lovely team and happy client and I get more unexpected money. And he's offered me more long term freelance projects now.

And if scientists are looking for verification then here is powerful evidence indeed:

After the first event the week before I had focused on Higgs Boson intently, and as I shared with Vijay last week, so many things began to spring into action right away. My jewellery design took a huge leap with the generous offer from a producer friend to create a free viral video for us featuring an interview with me and footage of the designs.

This is now complete and we'll be putting it out on YouTube soon. I recognise this as a wonderful gift because we will now be able to reach endless numbers of people with our concept... something that would not have been possible on our limited marketing budget. I believe this is going to open the doors to amazing opportunities.

Solid evidence. The kind that science demands.

In addition a few months back I had come up with a concept for a TV show featuring a friend of mine who is undertaking a tattoo apprenticeship, an unusual career for a woman. I really wanted to help her get her new career off to a great start and so I came up with an idea for a reality TV show. I somehow managed to get a production company interested but there had been no further

response in over a month and I had quite given up on the idea of it ever happening.

But... Since chanting Higgs Boson, their producer, not the same one as above, came back out of the blue and everything has moved very quickly. He met the team last weekend, we spent 4 hours together and by the end of it he was convinced it was worth pursuing. This Sunday, tomorrow, he is bringing a film crew to shoot footage so he can approach TV channels. I believe this last week's development has been a combination of Higgs Boson and Shreem Brzee.

There can be few doubts that this is the case as the 'cause and effect' paradigm demanded by science has been observed.

And there is a larger lesson to be learned. Science may not be in the business of coming up with an original format for a reality TV show but the methodology behind the idea deserves further investigation.

Scientists are often stumped when it comes to thinking up new ideas and they fall back on tried and trusted methods. These include reviewing the literature, analysing data, changing variables or, if desperate, inviting new members onto their research team in the hope that they will bring something new to the endeavour. But there is another possibility, as outlined above. Perhaps they should put into place a protocol where the chanting of Higgs Boson is prominent if not primary.

Looking at the evidence some points are immediately clear. First, the Higgs Boson is now a scientific fact so they are hardly leaving themselves open to ridicule. Second, when Higgs Boson is chanted things can happen out of the blue and everything can move very quickly.

There are other subatomic particles whose properties have not yet been investigated in this manner. I don't think it is giving too much away to say that in the near future we may be hearing of a quark in this context and I am reliably informed that at least three groups worldwide are in final pre-preparation before publication in referred journals concerning the chanting of 'Positron'.

A splinter group promoting the chanting of 'Anti-Electron Neutrino' has apparently met with less success.

This is one of those incidences where a man of science from outside of the mainstream has opened the way to a new discipline. Initial resistance has occurred but this is only to be expected; science has its entrenched views similar to any profession.

However, science now has the opportunity to work with Dr. Pillai and fully develop his theory into something both practical and effective.

Investigation and confirmation of MindScience should certainly be one of sciences next steps.

Chapter 17: Numerology

Numbers are vibration energy.

No serious scientist disputes this.

We can use numerology to determine our 'number' or our vibration by appointing numbers to the letters of our name, totalling them up until they become a single digit. We can do the same with our birth dates, such as 01/01/59, which could be totalled as $1 + 1 + 5 + 9 = 16$ and $1 + 6 = 7$. Here, we would be a seven vibration.

We could then find out how this might be significant in relation to our personality, character and so forth. A very interesting point about the number nine is it always becomes itself. It cannot be anything else.

Mathematicians have noticed this about the number nine. It has now been proved that it cannot be anything else.

Multiply nine by anything and add up the digits and they will always come back to nine.

$9 \times 9 = 81$.

$8 + 1 =$ nine.

Or take $9 \times 219 = 1971$.

$1 + 9 + 7 + 1 = 18$

$1 + 8 =$ nine and so on.

The number nine will always come back to itself. Just like a circle. What is even more interesting about numbers is when we add together one through eight, we get 36.

$3 + 6 = 9$, nine again.

We then add one through nine $1 + 2 + 3 + 4 + 5 + 6 + 7 + 8 + 9$ together and we get 45.

$4 + 5 = 9$, yet again.

So it would appear that nine is the culmination of all the other numbers one through eight.

In mathematical circles this is known as the Culmination Proof of Nine.

In ancient times the cycle of the Earth around the Sun was a perfect 360 days. They therefore calculated the year as 360 days - 12 months of 30 days [and/or 36 weeks of 10 days].

Until the seventh and eighth centuries BCE calendars from around the globe recorded the 360 day cycle but suddenly calendars

began adding days to both the length of the year and to the months to try to re-synchronise the calendar with the solar and lunar cycles detailed in the fascinating book Worlds in Collision by I. Velikovsky.

Worlds in Collision is a source book for Geologists, Astrophysicists and others. It is rare that one book has such an impact across so many disciplines.

It has been postulated that the misalignment may have been due to one or more cataclysmic events such as the Earth passing through the tail of a comet which caused the path curve of the Earth and/or its velocity to alter. This could be the reason why the number nine in numerology has also been associated with natural disasters such as floods.

Modern-day Cosmologists are only now coming to terms with the idea that the orbit of the Earth could be altered by its passage through the tail of a comet.

However, we do retain this ancient order in geometry because we still divide a circle, the ancient symbol for the Sun, into 360 degrees.

360° numerologically is 3 + 6 + 0 = 9. Therefore, this is a nine number.

This is good evidence for Velikovsky's theory.

If we apply this method to other geometric shapes we get interesting results.

The diamond is four lots of 90°, which equals 360°; nine again.

The square is the same, nine.

The equilateral triangle is 60° + 60° + 60°, = 180°: 1 + 8 + 0 = 9, another nine.

If we look at the pentagram, five lots of 36°: = 180°; another nine.

The hexagram star of David = 360°: the same story, a nine. We could go on and on.

This cannot be coincidence.

The hexagram is particularly interesting in relation to the circle, the Sun and nine. In ancient times the hexagram within the circle was used as a symbol of the zodiac cycle. The Sun was seen as the centre of the circle which, due to the six-pointed star pattern, is surrounded by 12 divisions - the 12 houses of the zodiac. In Sumerian times the zodiac was based on the Earth's precessional cycle of 25,920 years - again a nine.

The precessional cycle is the length of time the Earth takes to 'wobble' around its axis. Just like a spinning top which is tilted to one side, the Earth's central axis is tilted towards or away from the Sun and it takes that length of time for one revolution of this tilt to return to its former position. During which, the constellations of the heavens will slowly move backward across the sky to new positions according to the view from the Earth.

Modern day Astronomy is largely based upon knowledge from ancient times. The Sumerian people are of primary importance here.

1080 is the average number of human breaths drawn per hour, another nine. 1080 is also linked with the energy of the mother in her various forms as Diana etc. and the lunar energy. Interestingly the radius of the Moon is 1080 miles.

Science has yet to confirm that the 1080 is linked with the energy of the mother though research continues.

Amazingly, the Moon has always been associated with the colour of silver, whose atomic weight is 108 - nine yet again.

Incredibly, this has proved to be the case.

If we look at the number 666 - the sum of the numbers 1-36 (nine) - to some people it represents the devil, the mark of the beast. The book of revelation relates this number to the apocalypse, which is again linked to natural disasters.

Could this be what is encoded in the book of revelation? A numerological memory of the instability of the Earth and its vulnerability to outside forces such as may have altered the ancient 360 day cycle in the pre-Judaic times?

Anthropologists are aware of this and have made it a target of research.

To other people, however, this number represents the energy of the father, the solar energy. If we add 1080 and 666 together they equal 1746, which, to some represents the grail, the seed and the vessel, the totality of creation. A nine number. 1746 is also deemed by some as being related to the earth's electromagnetic grid.

This passage is highly technical and science is yet to come fully to grips with it. However, preliminary results do tend to confirm that 1746 could well be construed as representing the grail, the seed and the vessel. On the other hand there are many experiments which are measuring the Earth's electromagnetic grid and, fascinatingly, 1746 may have a part to play.

186,282 is the maximum speed of light in miles per second which represents the fastest speed achievable in the physical Universe. There are 4,320,000 years in the Mahayuga, the Hindu cosmic time cycle. Both nine numbers. We have a gestation period of nine months for a human birth, surely a clear indication of nine being symbolically representative of the physical aspect of creation. Moreover, within our solar system there are nine physical planets in orbit around their spiritual centre - the Sun.

None of this is disputed although we may note here that science has recently demoted Pluto from being a planet to not being a planet. If one were being uncharitable it might be thought that scientists were seeking to deny ancient knowledge only to suddenly 'discover' it for themselves in the future and reinstate Pluto. It wouldn't be the first time such a thing has happened.

Nine is made up of and contains all the other numbers or all the other vibrations and would appear to be represented in many geometrical shapes. As mentioned, on an energy level of creation the level of the Ahnu ether/spirit, nature does not work in angles. Only in the physical world where the ether/energy/spirit animates matter do we see these geometric shapes form.

Geometric shapes are what man sees and perceives with his intellect as being the source of creation when they merely represent the end result. Can we assume for simplicity that all the numbers one to nine are different vibrations as numerology would have us believe?

Can we further assume that all the numbers, or vibrations, which make up the nine are therefore all the various vibrations of the Universe in a physical sense? That is the inert part – matter - the 0.0001% of who and what everything is, as Bruce Cathie speculated?

With further investigation I think it is reasonable to say that science will recognize Bruce's theories.

Accepting this, we can also accept that the physical and the intellect can go no further because once it becomes nine we know it will only keep on repeating itself. The vibration of nine is the limit of physical creation just as the speed of light - a nine - is the limit of speed in the physical Universe. Could this be therefore the number of man in a physical and intellectual sense?

This is what the book of revelation tells us - that 666 (nine) is the number of a man. Therefore the number of the beast is the number of man, that is, man is confined by the number to being a beast - i.e. to expressing the baser aspects of the self. Nine is the limit of the ego, the limit of the intellect.

Nine is a vibrational prison, as it can be no other than what it is, a vibration which keeps repeating itself and like the speed of light cannot be physically or intellectually transcended.

Science is coming, slowly as always, around to thinking that nine is indeed the limit of the intellect.

So what feeds the nine matter? We know from open-minded physicists that ether spirit animates matter.

Closed-minded physicists dispute this but they are best ignored.

If nine including the numbers 9 down to 1 is representative of the whole of the physical aspect of creation something has to be responsible for creating and animating it. For the answer to this we need to go back to the circle and it is here we can discover man's 'missing link'. We know the circle is a nine number and that nine can only ever be itself so nine must be represented somewhere within the circle. This leaves us with two possibilities for looking at how the circle is made up.

We can divide the circle by eight, which would give us eight segments of 45° degrees, preserving the nine. However, eight is a lower vibration than nine so I believe we need to divide the circle by the number greater than nine - ten - to give ten segments of 36°, 3+6=9. We now have ten aspects of nine making up the circle which is also nine. So the aspect of nine repeating itself is preserved.

10 x 9 = 90

9 + 0 = 9.

This introduces us to another number, the number ten. If we now add one through nine and finally add ten we have

45 + 10 = 55

5 + 5 = 10.

It appears that nine cannot exist without ten.

Every mathematician I have spoken to has never offered a convincing proof that this is untrue.

Because we know that ether feeds matter then ether feeds nine from within, from beyond nine. Numerically speaking, here we see that 10 comes from 'above' nine and would seem to be the transcendent number of 9. Could this be where the axiom 'as above so below' originates from? Is this why we have had this simple inversion of truth expressed upon us because once again we have taken this symbolic representation too literally?

From this example we need ten for the nine and all it is composed of to exist in the first place. It would appear ten creates and animates all of the other vibrations, one to nine, the physical aspect of creation. Without ten we cannot have nine or anything below it.

Advanced mathematics deals with transcendental numbers and the reader is invited to explore this concept for himself before coming to a definitive conclusion. For physicists, the idea that ten creates and animates all other vibrations could be explored in the Large Hadron Collider. Here, ten could be tested in billions of events to see if nine or anything below it can be created.

If we ask the question 'what feeds the nine?' in relation to our solar system we see that the Sun feeds the nine planets with energy. The Sun is the 10th body at the centre of our solar system. The 10 feeds the 9. In some ancient cultures 10 was held as the sacred number which should not be written down.

It was therefore often written as two lots of 5. This indicates that far back in antiquity our ancestors had a profound understanding of the nature of the interconnected order of the Universe and the role numbers play in that cosmic order.

There can be no other explanation.

Due to our tendency towards the externalisation of perception we have intellectualised over the years that the symbol for the Sun must therefore refer to the physical Sun, which we see with our physical eyes, and so we have continued to look outside into this physical reality for the answers to our spirituality. We know this is an error because we know that spirit resides within. Also, because as physicists will tell us, everything comes from within not from without.

This is still controversial but it is true to say that a consensus is developing amongst physicists that not much at all comes from without. If present trends continue physicists will reduce that figure to exactly zero by the beginning of the next decade.

The natural flow of creation works from in to out and it is the ether/spirit/fire which animates matter - from within. The ten, the Sun/fire/spirit is the 99.999% of who and what we are. This symbol has always been a symbolic representation of the inner Sun, our inner Sun, our fire within, the inner flame, the spirit, the energy source which supports all life.

Biology is based on the idea that the natural flow of creation works from in to out - Babies start in and come out for example.

It is no coincidence that both the ten and the Sun/fire/spirit are symbolically represented as being one and the same. Numerologically speaking the vibration of the inner ten which creates everything is the spirit, ether, the inner Sun and the inner fire which creates life and animates matter and animates all the other number vibrations of matter so we can experience the physical reality we have created.

The source of the animation of matter has been a long term goal of scientific research right up to the present day. What is it that animates matter? Matter is usually inanimate except where life is involved. Here, for the first time, we have a comprehensive and straightforward account of life which is as persuasive as it is compelling.

Ten, being the source of all there is, the oneness of everything, must therefore be represented in the physical aspect of creation in order for it to experience itself. After all, what exists outside of us is merely a reflection of what lies within. Ten/spirit must be represented somewhere within physical reality otherwise there would be no physical reality at all.

And it is. Symbolically the male energy is a linear energy and it works in straight lines like this: — or like this: l, and when viewed from its end this energy becomes a dot ·, whereas the female energy is circular, 0, or three dimensionally a globe or perfectly round ball.

It makes sense that the male energy is linear for reasons which need not be elucidated here. Likewise, the female energy. In order to see this it is only necessary to imagine it the other way around. It just doesn't make sense that male energy should be circular. And no-one would intuit that female energy could possibly be linear.

On an energy level, balance between the male and female aspect results in the perfect vortex of energy, energy spinning both ways and exploding and imploding all at the same time. The female energy – the 'globe or ball' – is 'pulled' in towards its horizontal centre by the male linear energy forming what we would recognise today as the 'infinity symbol' ∞.

Hence, as we have discovered, we have the ancient symbol for spirit, the light we are and also the symbol for the vibration of ten, which can also be shown as 10.

Many people are familiar with the idea of female energy being pulled towards its horizontal centre by the male linear energy so this seems a sensible account of the origin of the infinity symbol. The origin had been previously shrouded in mystery.

All we need to do is to stop looking up and out into the stars and the heavens for a moment and look down. Look down at our hands and what do we see? Ten fingers. If we look further down we see ten toes. Realise we are the ten. We have always been the spirit/ten. We have just forgotten that's all.

It is the task of modern science to enable us to remember. Astronomers are particularly at fault for insisting on looking upwards and outwards to the stars. Surely some of their research time could be spent on looking down. Telescopes have a high magnification factor and while seeing ten fingers would be simplicity itself who knows what else could be discovered?

And so we focus 99.999% of our attention on the 0.0001% of who and what we are, the physical part, the nine of man, the geometrical and mathematical part.

Interestingly, the sum of these two figures yields 99.9991% meaning there is 0.0009% (another nine!) missing

.

I quote Deepak Chopra M.D. from his book Quantum Healing: 'It is from this energy structure that all geometric shapes evolve. For example: if we were to place a dot in the centre of all the outer circles, or where the outer circles join the main circle and join selective dots together, it is possible to create almost any geometric shape we desire.'

The importance of this is impossible to overestimate.

If we look at a rainbow we see light refracted into the seven colours. The top colour is red, then orange and so forth down to the bottom of the rainbow where we find the spiritual colours or vibrations. If we have been fortunate to observe a rainbow from an aeroplane we would see the rainbow forms a circle and the spiritual colours are in the centre.

This complements how nature works, from within to without, with the higher vibrations closer to source and the lower vibrations forming the further from source they are. The lower physical vibrations red, orange etc. are on the outside, not on the bottom or the inside. The vibrational source of spirituality and therefore truth always lies in the centre.

This is a rather technical passage and as such may be fully accessible to experts in the field only. However, the layman can get a flavour of the argument by considering any rainbow he or she has observed directly themselves. If one has a vibration meter to hand one could easily confirm the higher vibrations closer to source and the reverse also being true. The vibrational source of truth has been a little more challenging for science to measure but initial results are tending to confirm that it always lies at the centre and if not, then it is very, very close.

When we look at the recent 'prophecies' regarding the May 2000 planetary alignment, we have been led to believe by many prophets, seers, guru's and the like that during the three-day period 3rd to 5th May inclusive, the height of the planetary alignment, that the Earth's magnetic and biophysical fields would disappear.

Therefore, in order that we avoid the consequence of 'losing' our biophysical and magnetic fields, along with the memory of who and what we are - and the possibly of being sucked out into space somewhere - we have been encouraged by these 'prophets of doom' to 'build' the merkaba - an energy body based upon the star tetrahedron - around our physical body to protect us.

Prophets, seers and gurus are not often wrong when it comes to predicting the future of the Earth's magnetic field so there must have been another influence causing their predictions to be a little off this time. It would be interesting to know what caused this and science should be at the forefront of the research effort.

The planets will realign again at some point. When that happens the prospect of losing the memory of who and what we are as well as the possibility of being sucked out into space somewhere, will disappoint many. It is sciences responsibility to avoid this happening, or if not to prepare the world for bad news. Perhaps building the merkaba *is* the answer.

Science has a duty to investigate and report back on alternative responses to the imminent danger of planetary alignment and to do so with all possible urgency.

Chapter 18: Orgone Energy and Tesla

What is Orgone energy? What is a human being? What is life? Can science give us reliable answers to such questions? The electricity of life. The meaning of human consciousness. Are we alone? Are the traditional contests between science and religion still relevant? Does the word spirit still hold meaning today?

These are important questions.

I know it has been discussed before but no one really talked about it scientifically. What is it? An electrically charged particle? I have no idea why can't modern technology see this orgone, this is what I do not understand? Is it real? Made up? Can someone help me explain?

A cry from the heart such as this should be the signal for science to supply answers.

Orgone is a subtle energy. I'd call it an emotional energy. It is not detectable nor measurable by current science and therefore 'does not exist'. It requires a holistic approach to recognize orgone.

Unfortunately science is not noted for an holistic approach. It is one of the limitations of current scientific methodology. But, as orgone energy demonstrates, this methodology may have to undergo a fundamental transformation if phenomena such as subtle energy is to be fully understood.

From what I heard orgone might be the bluish tone over the atmosphere but supposedly that is due to Rayleigh scattering.

Rayleigh scattering is the current paradigm to explain the bluish tone over the atmosphere but just because a paradigm is widely accepted does not mean that it is immutable. The Earth was widely considered to be flat and that was the accepted paradigm for thousands of years. A revolution occurred when the Earth was found not to be flat and the new paradigm became accepted after a remarkably short time.

The same thing can happen to any currently accepted scientific 'truth'. That is the whole point of science. Science is our latest best guess. There is every reason to suspect that Rayleigh scattering is merely the latest best guess as to why the sky is blue.

Wilhelm Reich's theory of Orgone energy: Reich's approach laid the foundations for a microfunctionalist treatment of physico-mathematical quantities and processes but failed to generate a

consistent method capable of successfully distinguishing gravitational and electromagnetic interactions and properties from orgonotic and mass free interactions.

This paragraph is for the specialist but the layman can gain some understanding of the advanced scientific nature of Wilhelm Reich's theory purely from the sophisticated scientific terms employed.

Reich's exclusive assimilation of mass free properties to orgone energy prevented him from realizing the difference between electric and nonelectric manifestations of the aether as a primary form of mass free energy. This left his followers mired in the premature identification of aether with orgone.

Moreover, only late in his investigation did Reich begin to realize that what he called orgone energy was no different from what Tesla thought was the aether electric radiation. The theory of the orgone remained prisoner of these shortcomings, and the premature death or murder of W. Reich damned its continuation and consistent development.

Again, this passage is comprehensible only to those with an advanced degree in Physics. However, it should be investigated as a matter of priority, as should the possible foul play indicated.

This second volume of experimental Aetherometry was a very special experiment. It attempted to bridge on solidly empirical grounds the incompleteness of Nikola Tesla's discoveries and his errors in understanding with the incompleteness of Reich's discoveries along with their errors and limitations. In the process we came to discover that Tesla radiation consisted not in the production of 'electromagnetic radiation' or photons but in the generation of mass free ambipolar electric radiation and that this radiation - or its spectrum - encompassed precisely the true physical sense of Reich's discovery of what he termed orgone energy (OR) and its functional opposite, deadly orgone (DOR).

These passages are daunting even to a professional. There is clearly something important under discussion here.

Concerning the azure blue sky, OR and DOR radiation & Cherenkov radiation. Other authors have spoken of a reality akin in some respects to the aetherometric claim of the existence of mass free ambipolar charges.

Tesla spoke of non-ordinary electricity, primary electricity, ether electricity, longitudinal electric waves distinct from electromagnetic radiation, manifestations that have fallen under the rubric of Tesla

waves or Tesla radiation. Reich spoke of his mass free orgone energy and orgone charges.

Cerenkov spoke of a pilot or phase wave that transmitted potential, or its envelope, at speeds greater than c but did not transport electromagnetic energy. Maximo Aucci and Thomas Bearden have described massless electrons associated with longitudinal electric field propagation. Harold Aspden has described cosmological charges that escape the constraints of mass-based relativity as elements of a dynamic aether of space.

Professors I have spoken to have difficulty with this passage, such is the richness of new insight.

This leads us directly to propose a basic ambipolar radiation structure to the photon signatures characteristic of the basic layers of the atmosphere and correlate these layers to critical transition processes in the allotropic cycle. Our findings indicate that the aetherochemical and physical processes underlying this cycle play the most fundamental role in structuring the layers of terrestrial atmosphere.

This paragraph is also for experts in the field but the layman will note the impressive use of scientific terminology. That alone makes it impossible for mainstream scientists to simply dismiss. As is the case in all scientific discussion, equally important scientific terms need to be used as a matter of course in any counterargument. Since opposing views have not, up to this point, deployed equally imposing scientific terms then the hypothesis proposed cannot be considered to have been countered in any objective scientific sense. It is simply not good enough to say 'that's rubbish'.

Finally, as the most poignant example of the previous statement, we note that the blue light emanating from the last chemical step in the specific formation of water occurs very near the ambipolar solar radiation mode we have discovered and indicates how its dominant atmospheric role in producing a blue sky is the result of an aether electric resonance in the process of water formation – determination of the OR and DOR energies, frequencies and wavelengths driving the atmospheric allotropic cycle of oxygen, ozone and water.

Put this way, it is easy to see how the blue sky could be the result of a chemical reaction. We are of course left with the problem of a red sunset. Perhaps the reaction is reversible.

Indeed, the interaction of orgone or ambipolar electric radiation with the atmosphere is not limited to the release of 'latent heat'. It is also the same interaction that generates blackbody photon spectra - one for electrons and one for baryons, beginning with protons. These spectra merge, one with the other, which is why physicists have still not managed to distinguish them as distinct blackbody spectra, but the photon distributions produced are as distinct as a baryon is distinct from a lepton.

As things stand Physics cannot deny that a baryon is as about as distinct as it is possible to be from a lepton, so it follows that the photon distributions as described above will indeed be totally distinct. This should give Physicists considerable pause for thought.

Once physics grabbed hold of electricity all knowledge of it ceased. Electrons have nothing to do with the flow of electricity. Electrons are the rate at which electricity is destroyed. Electrons are the resistance. Eric Dollard.

It takes a visionary such as Eric Dollard to introduce into science that which Philosopher of Science Thomas Kuhn refers to as a paradigm shift. In Kuhn's view science normally proceeds within an accepted world-view and advances incrementally. However, eventually a crisis is reached when experimental results depart so far from the accepted world-view that the established paradigm is no longer sufficient to account for new results no matter how much tinkering is done. The stage is then set for a paradigm shift where new theories are advanced and eventually oust the previous paradigm.

It usually requires a genius to instigate such a paradigm shift. Einstein was widely recognized as a genius when he advanced his theory of relativity. It would not be too much of an exaggeration to claim Eric Dollard as another genius in respect of his theory of electricity. Just as Einstein completely overturned accepted theories of space and time so Dollard overturns accepted theories of electricity.

Einstein was tempted to say that once physics grabbed hold of space and time all knowledge of it ceased but he lived in an era where he couldn't phrase it like for fear of upsetting sensibilities. Dollard lives in a more forthright age where candour is commonplace. But that should not detract from his central argument: Electrons have nothing to do with the flow of electricity.

Eric Dollard is the only man known to be able to accurately reproduce many of Tesla's experiments with radiant energy and wireless transmission of power. This is because he understands that conventional electrical theory only includes half of the story.

It is the mark of the true genius that he (or she) recognizes when a conventional theory is incomplete. Most scientists simply accept that the theories they work with are complete and therefore do nothing to attempt to extend them.

The typical Hertzian electromagnetic field of transverse waves is the gross by-product of a much more powerful but hidden energy envelope which is manifested as longitudinal standing waves in a scalar nodal matrix, not propagated in the up and down ocean wave fashion of transverse waves.

Many Physicists are aware of the nascent paradigm shift of the hidden energy envelope and talk about it amongst themselves but as yet no-one has been brave enough to put their head above the parapet and announce it publicly. However, the truth will out and no doubt within the next few years we will have a fully comprehensive theory of the electromagnetic field which will relegate the transverse wave paradigm to a by-product of longitudinal standing waves in a scalar nodal matrix.

This is yet another example of science holding the advancement of knowledge back for decades while visionaries such as Tesla and Dollard find themselves languishing. Science as a whole and individual scientists in particular should be ashamed of themselves and when the paradigm shift arrives should feel it incumbent on themselves to offer a full apology.

Tesla magnifying transmitter: The Tesla magnifying transmitter is a converter which converts electromagnetic energy into what is called magneto-dielectric energy.

Magneto-dielectric energy link to orgone energy: If you take a low pressure gas in a bulb and place it in two superimposed dielectric fields then you get spiral formations such as Reich wrote about in his book Cosmic Superimposition. These formations appear as spheres, galaxies and other cosmic forms.

This is good evidence that Tesla and Reich were correct.

The only successful demonstration of Tesla's longitudinal dielectricity ever made available to the public: Tesla's longitudinal electricity: 60 minutes: A laboratory demonstration video with Eric P. Dollard & Peter Lindemann.

If you've ever wondered if there is more to a Tesla coil than just making big sparks then watch this video. A series of experiments providing you with factual data on the reality of Tesla's theories. You will see experiments on: The one-wire electrical transmission

system: The wireless power transmission system: Transmission of direct current through space and a novel form of electric light which attracts material objects but repels a human hand.

The latter is particularly impressive and threatens to overturn accepted theory by itself. Who would have thought a human hand is not a material object?

Transverse & longitudinal electric waves: 50 minutes: Eric demonstrates the reality of longitudinal waves and their application to the natural transmission of electricity. Analogue computers, networks of coils and capacitors, are presented opening up an entire new field of electrical researches into the wave forms discovered and used by Tesla in his wireless power transmissions. The longitudinal magneto-dielectric wave is shown to have a propagating velocity greater than the speed of light.

One would expect Physicists to show a little more interest in a wave which has a propagating velocity greater than the speed of light. Physicists seem unwilling to investigate anything which threatens to overturn Einstein's paradigm that *nothing* can travel faster than the speed of light. Perhaps they should be more forward thinking. Or perhaps they are worried they will have to rewrite their lecture notes.

The transverse electromagnetic wave is shown to be a retarded, unnatural form of energy transmission. A multi-pactor orgone detector is demonstrated indicating a relationship between Dielectricity and orgone. A competent researcher can duplicate the experiments on this video from the information given.

So here is the ideal opportunity. For a small outlay physicists can purchase Eric's video and duplicate the experiments. However, Nobel Prizes may be at stake so one hopes they will acknowledge Eric in their papers and make sure he as well as Tesla are given recognition for their contribution.

Four quadrant theory of electricity by Eric Dollard. If we take Tesla's three phase electricity or rotating magnetic field we find it is based on the archetypal form known as the solar cross or by various other names, Mandalas, medicine wheels. These are four quadrant types of forms, a balanced cross as opposed to an unbalanced cross.

This is where you get the four quadrant theory of electricity. electricity has to be viewed from a four quadrant type of situation. The right angle plays an extremely fundamental role in electricity. It is generally a right angle phenomenon.

Many Physicists will argue that electricity is, in general, a right angle phenomenon though a few heretics hold out for 45 degrees. There is widespread agreement that electricity has to be viewed from a four quadrant type of situation. This is key.

Tesla experimented with impulse current and oscillating current. Our electricity is direct current and alternating current. The four quadrant theory of electricity is impulse current, oscillating current, direct current, alternating current. Alternating current plus direct current are transverse electromagnetic. Impulse and oscillating current are longitudinal di-electric.

It is surprising that science has not promulgated this general theory of electricity more widely. Everyone is familiar with direct and alternating current but everyone automatically assumes that there is no more to electricity than that. As Tesla showed there is also impulse current and oscillating current.

Some Physicists contend that electricity theory already takes account of these because impulse current is just a relatively brief spike of direct current and oscillating current is just another name for alternating current. But if that were true how can they account for the four quadrant theory of electricity?

Eric P. Dollard - Symbolic Representation of the Generalized Electric Wave - 86 pages: Extension of the theory of versor operators and imaginary numbers to represent complex oscillating waves such as those encountered in the researches of Nikola Tesla and everywhere in nature. Theory of free electricity produced by rotating apparatus such as variable reluctance devices. Waves flowing backwards in time are explored.

There is plenty to keep not only mathematicians involved in complex analysis occupied but also those physicists involved in investigating second law of thermodynamics anomalies (free electricity) and temporal irregularities (waves flowing backwards in time). Research has stalled in both areas but for the small outlay of purchasing Eric's book both fields could receive the kick start they so urgently require.

Longitudinal energy in other videos: Dr. Konstantin Meyl: Power Engineering Scalar Field Theory: This short but highly informative presentation is probably the only one in existence for English-speaking audiences to correctly expand the classical electromagnetic field theory to include longitudinal/Tesla waves.

The impact of such an expansion is immense and requires a complete change in thinking and a revision of the very foundations of physics in general. Do scalar waves exist or not?

As usual, Physicists are content to ignore Dr. Meyl just as they are content to ignore Eric. It is difficult not to imagine physicists sitting comfortably in armchairs in their favourite club sipping sherry and congratulating themselves on their tenure while the real work is being done by others. It seems obvious from the outside that modern physicists are missing so much which is fundamental to their discipline. In any other walk of life this would be cause for alarm. In science it seems to be a commonplace.

In Meyl's extended field theory they come about naturally eliminating the need for any exotic theories such as superstrings, dark matter, revising even the theory of relativity. Meyl explains how Faraday's experiments were actually describing a much broader electromagnetics than was later derived by Maxwell and curtailed to an even more primitive state by others so that today's engineers are literally handicapped when using it.

One would think even a Physicist would take an interest when the theory of relativity was up for revision.

Meyl's extended field theory does have an important scientific principle in its favour, that of Occam's Razor, which says that in any explanation it is wise to refrain from referencing ideas and concepts which may turn out to have no bearing on the question and may not even exist. Meyl's extended field theory eliminates the need for superstrings and dark matter and therefore passes the Occam's Razor test. Scientists have no idea what a superstring is or whether dark matter exists and so they are guilty of multiplying entities unnecessarily in contradiction to the principle. On that basis alone, Meyl's theory appears superior.

The missing experimental pieces eventually came from the work of Tesla and a small version of his wireless transmission of energy is demonstrated here, carrying electrical energy without losses to the other side of the table with an impossible 1.5-times the speed of light.

One might think that governments throughout the world might be interested in the transmission of electrical energy without losses. Yes, it has so far been achieved from one side of a table to the other but this is proof of concept. Scaling it up is a matter of engineering.

I can only assume Governments are listening with deference to the very Physicists who are too indolent to pursue anything which threatens to upset their scientific applecart. Either that or they are influenced by global energy companies whose profits stand to be diminished. But even that pales into insignificance with the lack of interest in investigating energy transfer faster than the speed of light. It seems that Physicists will not go against the teachings of Albert Einstein even when the proof is right there in front of them. This makes science like a religion with Einstein as its pope whose infallibility is not to be questioned.

Today's Science has moved a long way from its principles rooted in the Enlightenment and is all the poorer for it.

So are we, paying so much for our wasteful forms of energy, heating up the planet as we do, and finding ourselves saddled with an authority figure whose pronouncements dare not be questioned. We live in a diminished age.

During this amazing lecture Meyl takes you through the forgotten ideas of Nikola Tesla and explains how they are the missing pieces which prevented scientists from reaching the goal of a comprehensible Unified Field theory. He also discusses many new applications which can now be not only better understood but also built and used in our everyday life.

Those Physicists who have been seeking the holy grail of physics, a unified field theory, the search for which fruitlessly occupied Einstein's later years, will find much food for thought here. Physics took a wrong turn when it ignored the ideas of Tesla but Professor Meyl appears to be back on the correct track and his elucidation of Tesla's theories plus his own thoughts on new applications should form one of science's urgent next steps. A 100 year detour in the wrong direction, mistakenly signposted by Albert Einstein, now stands ready to be corrected by Professor Meyl and Eric Dollard. All it will require is Physicists, willing to take the first step and challenge prevailing orthodoxy. I wonder who will be brave enough to be the first?

Chapter 19: ORMES

In the late 1970s an Arizona farmer named David Hudson noticed some very strange materials as he was doing some soil restoration on his land. Hudson spent several million dollars over the following decade figuring out how to obtain and work with these strange materials.

It is lucky for us that David decided to go through with this. As is usual in the history of science the main advances have been driven by curious individuals motivated by the their thirst for knowledge. This contrasts sharply with the modern figure of the scientist, cosseted by his research grant and increasingly anonymous; blending into and becoming indistinguishable from his research group.

Hudson's curiosity led him to work with spectroscopists at Cornell University and other labs to discover the elemental ingredients of this powder. Initial findings of the sample yielded iron, silica, and aluminum. Extraction of these elements left 98% of the powder intact. The surprise was that this 98% consisted of nothing which could be identified through normal spectroscopic analysis. Hudson knew that this 'nothing' could be seen, felt, tasted and weighed. It had to be SOMETHING and he was determined to find out what IT was.

It is worth noting the spectroscopists at Cornell University and other labs seemed relaxed by finding 98% of the powder they were analysing consisted of nothing which they could identify. One would expect a little more adventure on their behalf.

In the course of his research he found a paper from the Soviet Academy of Sciences stating that proper spectroscopic analysis requires a 300 second burn instead of the 15 second burn used in the United States.

If David Hudson could unearth this paper then the question must be asked why couldn't, indeed why hadn't it been read by those in Cornell and the other labs? The Soviet Academy of Sciences is hardly the world's most obscure organization. One can only refer yet again to the inertia demonstrated by self-satisfied interest groups such as the spectroscopists at Cornell University who know they will be funded year in year out no matter what.

Utilizing the Soviet technique of fractional vaporization Hudson discovered his sample contained the elements palladium, platinum, ruthenium, rhodium, iridium and osmium. Even more astounding was his discovery that each of the non-metallic elemental forms was a superconductor, a substance that allows an electric current to flow without resistance in the absence of a continually applied potential.

Look what they missed!

Hudson continued his research and found four papers by the U.S. Naval Research Facility showing that cells of living tissues communicate with each other by a process identical to superconductivity but the nature of the superconducting substance was unknown.

The U.S. Naval Research Facility may seem an odd place for that discovery but there it is. A pattern is beginning to develop here and it is not a pretty one. What were Biological research labs doing in the meantime?

On a hunch Hudson analyzed the brain tissue of pigs and cattle and found the brain dry matter weight was 5% rhodium and iridium! As a result of Hudson's research he knew that the electrons flowing through a superconductor pair off and are converted to a light frequency in the process. He theorized that this might be the same process occurring in human cells.

One throws up ones hands in despair at this point. Why should it take an Arizona farmer to show that the brain tissue of pigs and cattle were 5% rhodium and iridium?

Hudson discovered papers which indicated intense experimentation was being done utilizing precious elements in the treatment of cancer. These elements were shown to interact with the cell by a vibration frequency or by light transfer to correct the mutant DNA.

It is futile to complain once more concerning science missing these papers. The point has been made. One must move on and now question why science has not attempted to utilize precious elements to treat cancer. Could it be that today's pharmaceutical companies simply do not want the truth regarding vibration frequencies interacting with cells to be made public because once the treatment became known it would affect sales of their own, rather dubious, chemical attempts at a cure?

This appears likely. Light transfer correcting mutant DNA has only been mooted in the scientific press so far yet here it is, a big step forward, and one which science should have been in the forefront of investigating.

When Hudson applied his proprietary method of analysis to Essiac Tea (an alternative treatment for cancer) he discovered high levels of rhodium and iridium. He also found that Acemannan, a derivative of the Aloe Vera plant currently being tested on AIDS patients, is 90% rhodium.

As a result it was suggested the botanical name of Aloe Vera should be changed to *herba metallica* (metal plant) .

David Hudson re-discovered the existence of ORMEs (Orbitally Rearranged Monoatomic Elements) which are virtually undetectable by conventional means (except for a distinguishing IR doublet located between about 1400 and 1600 cm^{-1}) because they lack a d-orbital electron.

Hudson and associates developed a method to recover ORMEs and convert them into their metallic forms. While it is not a transmutation of one element into another (but rather, the conversion of an allotrope into the common visible form of the element) the extraction and conversion of Ormes to metal may explain the claims of some other experimenters.

The clues were there concerning ORME's, allotropes, IR doublets and so on but science seems stuck in an inertia which is largely of their own making and from which individual scientists and research groups seem incapable of breaking. Scientists have lost their 'mojo' is one complaint which I have heard several times.

Certain ores, particularly sodic and calcidic plagioclase, contain large amounts of Ormes which can be extracted.

No geologist can be found on record who disputes this.

The precious metals known to exist in this 'Ormes' state (known as the 'light platinum group') are cobalt, nickel, copper, ruthenium, rhodium, palladium and silver together with osmium, iridium, platinum and gold (known as the 'heavy platinum group'). These minerals are known as superconductors and can in a monatomic, super deformed, high spin and low energy state lose their chemical reactivity and metallic nature thereby resulting in a state of superconductivity.

Meta-analysis of the data suggests that scientists have failed to reproduce these results because they only experiment with low spin states.

These precious metals have the unique ability to remain stable in the monatomic form which can then lead to effects ranging from Levitation (weight losses) to Zero-Point Energy applications and to fundamental biological and/or human physiological effects.

One wishes that modern-day scientists were as curious about the world as their predecessors. It is difficult to imagine Thomas Edison, for example, remaining blasé if he caught wind of such discoveries.

The Future of Medicine: The discovery of ORMES elements is today's equivalent of the discovery of $E=mc^2$, it is that significant and will be recognized so as more applications evolve. We predict it will be a major part of our future medicine.

When Einstein discovered his famous equation it would not be too much of an exaggeration to say that it led to the modern world as we find it today. Zero Point Energy applications in turn have the prospect of renewing the world yet again. When one adds in the benefits of levitation there could be a real game changing opportunity just around the corner. To take just one obvious example, NASA would be able to do away with expensive and unreliable rockets and instead design spaceships to levitate astronauts to the Moon.

ORMES or m-state materials are thought to be the precious metal elements in a different atomic state. Monoatomics seem to have a quirky quantum mechanical state where each individual atom 'blurs' its identity across an array, creating properties similar to Einstein-Bose condensates that exist fractions of a degree above absolute zero.

When David Hudson's colleagues experimented with the material it would disappear when exposed to sunlight, or any kind of radiation, changing state into something slightly out of phase with our quantum reality and reappearing after a time, apparently losing its energized state from wherever it went upon dematerializing.

Science has only recently appeared ready to fully engage with materials slightly out of phase with our quantum reality. For too long it has tested for materials fully out of phase but this has proved to be a dead end and the best minds are moving away from the field. They are now designing experiments to determine just how long it takes for materials to reappear having lost their energized state.

Ormes elements have the same number of protons and electrons as their mineral counterparts that are found on the Periodic Table of Elements. The difference is in how the elements spin. Some rapidly

spinning elements form what are known as 'Cooper pairs'-electron pairs that become bound together as they spin.

Ormes elements are highly stabilized and chemically inert-they are no longer capable of interacting with other elements. In this state the Ormes elements are extremely bioavailable. They can be absorbed easily and rapidly by plants and other organisms.

Biologists have recently floated the view that chemically inert elements are those best placed to take part in the complex chemical reactions necessary in living organisms.

ORMES-Anti-aging minerals, the cutting edge of superfood nutrition: As monatomics (single, isolated atoms) and proven to being a superconductor they exhibit many 'supernatural' properties such as existing simultaneously in two or more places at once, or under certain circumstances, disappearing from our 3D physical reality into other 'dimensions' of reality, then reappearing.

It is obviously a benefit that they can exist simultaneously in two or more places at once. One can imagine the usefulness of a mineral promoting or actually taking part in reactions in several cells at once – the dosage required is automatically reduced for a start.

Monatomic elements also have substantial healing properties. These elements exist in the body, particularly in the nervous system. Many plants contain monatomic elements as well, interwoven within the botanical molecules. ORMEs elements are in the human brain and comprise up to 5-6 percent of physical matter.

David Hudson, the modern day re-discoverer of Ormes minerals, estimated that as much as 5% of the dry matter weight of our nervous system could be Ormes minerals (formerly mistaken as carbon, iron, and other minerals).

Biologists, not being Chemists, can be forgiven for not noticing what they thought was carbon, iron and other minerals was actually something else.

A quote from David Hudson: 'I'm not a doctor, so I can't practice medicine. Anything that is administered to someone for the purpose of curing a disease is medicine ... My purpose in this was not to cure diseases and illness, but I did want to know: Does it work? ... I can tell you, it's been used on Lou Gehrig's disease; it's been used on MS; it's been used on MD; it's been used on arthritis ... I can tell you that at 2 mg. a day it totally has gotten rid of Kaposi Sarcomas (KS) on AIDS patients (there's 32,000mg. in an ounce, 2mg. is nothing).

And it gets rid of KS. I can tell you that for people who have taken it at 2mg. injections within 2 hours their white blood cell count goes from 2,500 to 6,500. I can tell you that Stage-4 cancer patients have taken it orally and after 45 days have no cancer any place in their body.'

Scientific evidence does not always have to come from medical trials which can take years to complete. David, fully admitting that he is not a doctor, provides persuasive evidence.

The Ormes materials can be extracted from the air, water, rock, soil, food and the refinement of certain metals. Ormes are found in every organism including the human body. Essiac tea, sheep sorrel, aloe vera and slippery elm bark are high in rhodium and iridium if grown in volcanic soil. Magnesium, calcium and other minerals are also found in trace form. Rhodium ormes is non-toxic even at high concentrations. Rhodium and ruthenium play an important role in our DNA.

Crick and Watson gained fame as the discoverers of the structure of DNA but analytical methods were not as advanced in the 1950's as they are today and while their structure was perfectly adequate for those times it has become increasingly apparent that something was missing. Now it seems likely that the missing piece of the DNA jigsaw is rhodium and ruthenium.

The ormes iridium and gold speed up the metabolism of the body possibly by up to 40% and they do not enter the bloodstream nor the kidneys but rather enter the acupuncture system of channels and meridians, spinal cord and thymus.

Spinal cord scientists have so far ignored the importance of iridium and gold. This is largely because there is so little collaboration with metabolism scientists.

Science should move towards a collegiate model where specialists in other fields are invited to conferences outside their speciality. Initially there will be resistance and scientists will claim they are too busy or that they are not interested but such objections should be disregarded as examples of self-interest. A measure of compulsion where scientists are forced to attend meetings outside their own area of study should solve the problem. The potential benefits of such a system far outweigh any objections and in the future will come to be regarded as the norm.

Ormes is concentrated in volcanic soils, seawater and many rivers. All of these m-state elements are quite abundant in seawater. They also seem to be present in most rock, fresh water and in the air.

According to David Hudson's research these elements in their m-state may be as much as 10,000 times more abundant than their metallic counterparts.

Even if David's research is out by as much as 10% that still means that the m-state elements are 1,000 times more abundant than their metallic counterparts. Science should take a really hard look at itself and ask the question: How is it that we have missed so much?

There also may be other elements that occur naturally in the m-state. Various researchers, working independently, have identified these materials in this different state of matter. They have arrived at many of the same observations. These m-state elements have been observed to exhibit the quantum physical behaviours of superconductivity, superfluidity, Josephson tunneling and magnetic levitation. It looks like these are an entirely new class of materials.

Material scientists appear to be culpable here. How, with the most modern research tools at their disposal, they can completely miss an entirely new class of materials is beyond belief. This is frankly shocking and the people involved should be severely censured by their grant awarding committees and sent on a refresher course.

Plants grown in these soils absorb it growing healthier with substantially more nutrients. Food with this energy supports proper nutrient absorption & toxin elimination, balances your body's pH and repairs & regenerates cells. Ormes energy may be how our cells communicate with our consciousness.

Consciousness exists and so, obviously, do cells yet the link remains unexplained. This is known as the 'hard problem' of consciousness. Since cells produce consciousness there must be a causal link between the two but science has repeatedly failed to identify it and in recent times has given up and left the field to philosophers.

Summary: Ormes discovery is now a cutting-edge subject at the forefront of pharmaceutical and industrial technologies based on high spin state mono-atomic or single atom elements of iridium, rhodium and gold. Even better, these formulas can be scaled-up from a laboratory to a chemical plant.

Ormes technologies have developed high spin mono-atomic states for many of the transition elements including gold. White gold is unique among the monatomic elements because it is able to correct DNA deficiencies by a denaturing process. We hypothesizes that consuming doses of monatomic elements may decelerate the aging process and perhaps even reverse it.

Modern day Chemistry is in danger of being left behind by these new and highly significant discoveries. Even the best equipped laboratories have been unable to replicate high spin mono-atomic states of transition elements and frankly, given the funding they have received, this is simply not good enough.

Chemists should take a hard look at the paradigm they are following. How the discovery that white gold is unique among the monatomic elements eluded them should be the subject of a major enquiry. Why it was not linked to the discovery that white gold is able to correct DNA deficiencies by a denaturing process is something that only they will be able to answer. As far as the rest of us are concerned this appears to be a fault in the very method and practice of science.

Science should now be researching the likely probability that consuming doses of monatomic elements may decelerate the aging process and perhaps even reverse it. From a standing start it has the opportunity to make a real difference in people's lives.

The validation of Ormes materials should therefore be one of sciences most urgent next steps.

Chapter 20: Physics

The following is from an Internet Discussion Group dedicated to Physics. There are several people in the discussion and the topics are wide-ranging and have implications not only for Physics but for us all.

The major role played by electricity in our everyday life is obvious. Putting aside our appliances and tools our bodies can only function through electric impulses, nerves being the conductors, the brain the origin as well as the producer, the muscles the receptors. The basics of our body composition, the atoms, are electrically energized particles. The molecular activity is therefore electrical in nature. This leads us to the cellular model where electricity has to play a role or our bodies would be useless masses of cells.

Modern day Biology is increasingly coming to the view that atoms are indeed electrically energized particles. Physicists are the first to agree.

Now allow me to make a very wide assumption. If we live in an Electric Universe, wherever we go in this Universe we will be in a plasma environment, our electrified bodies are or will become immersed in this environment. Our thoughts are electrical impulses generated and formatted by our brains, in my case right now transmitted to this computer. They are the products of learned, processed and formatted information acquired over time and synthesized to accommodate my present state of mind.

Psychologists would agree that our thoughts are electrical impulses generated and formatted by our brains. The assumption is not all that wide however since if the Universe is electric we would undoubtedly be in a plasma environment and our bodies could therefore fairly be described as electrified.

The military in the US and likely other countries have, according to some reports, developed some mind control system that would change the behaviour of a crowd, the entire population of a country or the armies of an enemy country. True or not the only way they could achieve this deed would be through electrical impulses. The HAARP experiment in Alaska was supposedly designed to achieve this among other things like climate and weather control.

HAARP is the High Frequency Active Auroral Research Program based in Alaska.

It's not only electricity. Anything existing has impact on our mind and consciousness starting with the water you drink, the food you eat and the light you expose your body to, the environment you are in, including any waves available there. From magnetism to the vibrations of other living organisms around you.

The weather, wind and air pressure and all visible and invisible heavenly bodies. Anything leaves an impression in your mind. While most of the average people are not aware of them the impressions are there. Some don't sense anything at all; others sense some sort of white noise in their mind and a few are able to tune into many, many of the single signals within the white mind-noise.

Although it is well known how all visible heavenly bodies leave an impression in the mind the effect of invisible heavenly bodies is less well known. Science should address this gap in our knowledge as soon as resources allow.

As for mind control, if such exists, it's always about a sender and a receiver. And a receiver is always about an antenna. The human antenna is from your feet to the top of your skull. Feet is the grounding, top of skull the part catching the waves. The throat is similar to a diode and the heart similar to a transformer coil. The techies here will recognize the detector radio.

There is no, I repeat, no technology to play with your mind as long as your feet are mentally on the ground and your coil... erm, heart is tuned to the higher ideals. Properly grounded you will always receive the true signals. Just when you turn into a high-flyer you become vulnerable to artificial signals.

Specialists in the new field of bio-electronics will be familiar with modelling the throat as a diode and the heart as a transformer coil. When looked at in the correct way the comparison is obvious.

Personally my preferred reception environment would be in water. It is clear, somewhat pure and always flowing and contrary to what most people believe it is not always flowing downhill. And it is a good conductor of energies.

It seems intuitive that one would be less susceptible to mind control if one was taking exercises in the local swimming baths, for instance. .

I have considered sewing hats and capes for family and friends out of electromagnetic-resistant fabric to off-set the ocean of

electromagnetic we live in today; one might even fashion something like a 4 poster bed cover for sleeping but as I spend most of my time at the computer a PVC pipe frame over the computer and chair seems more to the point for me.

Hats and capes are a start but a PVC pipe frame over the PC and chair offer clear advantages.

The fabric is light weight and actually attractive, a light golden brown colour with a nice texture. Very light-weight. The company also make electromagnetic proof tents and yurts.

Yurts have been mooted to be superior to tents in this respect.

Additionally, when I discovered the Electric Universe theory it blew me away. At my core, I knew it was right.

This is how most science actually proceeds though it may come as a surprise to the layman. Scientists often know at their core that their theory is right and it blows them away.

The core of everything is not only electrical in nature, i.e. protons and electrons but in keeping with the atom, everything also has resonant component as well.

At this point new Physics begins to emerge. No scientist will doubt that the core of everything is protons and electrons but few have considered that everything also has resonant component. The occasional research paper has hinted at this but no consensus has emerged.

Much to be learned from the Russians here. Their science has been largely untouched / uncorrupted by politics -for instance back in the 50's the Russians discovered that they could cause the brain to produce any neurotransmitter by stimulating it with low voltage micro currents modulated at different frequencies, much to the chagrin of big pharma these days, I'm sure. Dr. Beck's Braintuner Sota's BT-7 works on a similar principle and quite well.

That Russian science was largely uncorrupted by politics is a matter of record.

This almost directly relates to the operations of Electric Universe theory, only on a micro scale, currents, voltage, resonance etc.

This is no longer a problem as modern scientific equipment is easily capable of measuring on a micro scale.

I'm working on how everything is connected and recently I have realized how utterly futile and doomed to failure the current modern science search of a maths formula for the theory of everything is.

Most scientists working on the theory of everything would agree. Progress has been slow since string theory was announced as the next great step forward, but it has failed to live up to expectations.

For now let's just say all you need to know is how Plasma Physics, resonance, electricity, electromagnetics and perhaps the nature of crystalline structures applies to and interacts with everything and you could very well understand and manipulate anything in any way you wish.

Physicists are slowly coming round to this interpretation.

Structure and function cannot be separated. The physical form of charge is a sphere. The folding insideout of the distributed charge is indeed the source of all forms based on its relationship to the RMF quantum spin of the surrounding aether. The flower of life is overlapping spheres, so it is with the electron cloud.

This passage is highly technical and is for experts only.

I stumbled across something this weekend. Regarding electricity and body it might be a good idea to read about ozone.

But here science could have a striking new direction to explore .

Were the natives' right when they said the astronauts should not bring back anything from the moon?

I think we can all agree it is not the first time that the wisdom of natives could prove invaluable.

I guess I feel that the Electric Universe has unlimited potential for real world applications. Those applications are not kosher if they are Skunkworks projects. All the while Tesla is not to be found in college textbooks in electronics and his work and real world applications are certainly evident and quite possibly still being explored in Skunkworks projects.

To not to be able to speculate on what the military is up to is leaving us to dabble in only the public domain. That's not going anywhere fast. To divorce the reasons for the foot dragging at so many levels is turning a blind eye to the truth and the facts.

From Wikipedia: 'Skunkworks is an official alias for Lockheed Martin's Advanced Development Programs (ADP), formerly called Lockheed Advanced Development Projects. Skunkworks is responsible for a number of famous aircraft designs including the U-2, the SR-71 Blackbird, the F-117 Nighthawk, and the F-22 Raptor. Currently its main project is the F-35 Lightning II, which will be used in the air forces of several countries.

The designation 'skunkworks' is widely used in business, engineering and technical fields to describe a group within an organization given a high degree of autonomy and, unhampered by bureaucracy, tasked with working on advanced or secret projects.'

I heard a telling commercial two days ago from an oil company and I quote 'If one day scientists find a magic ball of electricity that never stops then that day we can give up our dependence on fossil fuels; until that day we search for new and cleaner ways to keep you going.' Guess what big oil, I found the magic ball of electricity...

Science should make it a priority to investigate the magic ball of electricity as soon as possible. The implications could be profound.

A look into the industrial military complex leaves no doubt as to the control of Media and Government. The application of Electric Universe technology is in reality Tesla technology. That is hard to separate from public domain and military domain especially if we are to get to the heart of the matter and the meat of the topic. Yet somehow we must walk that fine line of imaginary divisions. Where only public information/disinformation is valid for evaluation yet military technology is an Area 51 no-man's land.

Ben Rich of Skunkworks and Boyd Bushman of Skunkworks have made very telling public confessions to the extent they could. Unless it was disinformation on their part then the limits on secret technology is honourable for national security but will forever be the cat they don't want out of the bag. I fail to see how one can intelligently discuss the Electric Universe without at some point in time crossing that line.

It is unlikely that Ben Rich and Boyd Bushman are engaged in spreading disinformation.

The rotating disk dynamo has mystified every scientist since Faraday's 1831 discovery.

The mystification apparently continues to this day. It is high time that scientists came to understand the rotating disk dynamo and it is surprising that research efforts have not been made in this direction.

Also called a unipolar generator or n-machine by Bruce DePalma its efficiency is often known to be above 95% in commercial models. Nikola Tesla's notes on a unipolar dynamo, Einstein and Laub's article on a rotating magnetic dielectric, Inomata's new paradigm and n-machine, a list of homopolar patents

and more are included. Can the homopolar generator become a self-running free energy machine?

This is important to know as, if true, it would finally prove the Second Law of Thermodynamics to be wrong. The second law has been on shaky ground for some time. An efficiency above 95% is a strong indication that the 'law' may well be on its last legs. An increase of as little as 5.1% would be enough to put the final nail in the coffin.

Investigating the Paulsen UFO story and the DePalma claims of overunity the author began an earnest scientific endeavour in 1980 to build a homopolar generator and test for the elusive back torque which had never been measured before.

The project helped complete his Master's degree in physics at SUNY in Buffalo. Only afterwards did the connection to John R. Searl's energy and propulsion invention become apparent. Each roller magnet in the Searl device is a small homopolar generator and the entire set of rollers create a radial Lorentz force too.

If modern day Physics is to be believed the back torque remains elusive. However, the scientists involved may well have not tested enough and so should be encouraged to revisit their experiments. The fact that the entire set of rollers in the Searl device created a radial Lorentz force should add impetus to this research.

Does the Skunkworks team have electro-gravity? Boyd Bushman and Ben Rich say an emphatic yes. So I fail to see why it's taboo or not kosher or intelligent to review the most advanced technology on Earth, that would be the Skunkworks team of Lockheed. Especially if they cracked gravity. That would be a milestone and we all know it. Would they hide it? Yes. Can we discuss gravity and not get into antigravity? I do not see how.

This seems undeniable. Physicists through the ages have often wanted to discuss gravity and antigravity at the same time.

Can we discuss gravity and discuss electricity with full freedom? Can gravity be an inverse square law function and electricity be an inverse square function and not be related? I fail to see how. I am sure even Wal has stated that he believes gravity is an electro-gravitic dipole function longitudinal field effect.

If this is so Wal is pointing the way towards a radical extension of modern day understanding.

This simple electrical model of matter has the great virtue of reducing all known forces to a single one, the electric force.

This is good evidence in favour of the electrical model. The history of Physics has largely been a story of simplification and unification of forces. This could be the final step of unification, something which Einstein spent the last years of his life investigating and which eluded him.

However, it has a price. We must abandon our peculiar phobia against force acting at a distance. And we must give up the notion that the speed of light is a real speed barrier. It may seem fast to us but on a cosmic scale it is glacial.

This is undeniably true. Given the vast distances in the Universe the speed of light is slow. Light takes over four years to reach us from even the nearest star.

Imposing such a speed limit and requiring force to be transmitted by particles would render the Universe completely incoherent.

And this is the problem with today's Physics. It is a matter of observation that the Universe is *not* incoherent.

If an electron is composed of smaller subunits of charge orbiting within the classical radius of an electron then the electric force must operate at a speed far in excess of the speed of light for the electron to remain a coherent object. In fact, it has been calculated that if released, the subunits of charge in the electron could travel from here to the far side of the Andromeda galaxy in one second.

A speed which is hardly glacial. Physics should therefore mount a research effort to confirm that the subunits of charge could travel that far that quickly.

Obviously we cannot have measuring devices on the far side of the Andromeda galaxy at our present stage of technological development but we have atomic clocks which could measure the time for the subunits to travel from one research establishment (say CERN) to another (say Fermilab) to the required accuracy. It is unusual that such a straightforward experiment has not yet been performed. Perhaps Einstein still influences physicists to an unwarranted degree.

We have direct evidence of the superluminal action of the electric force given that gravity is a longitudinal electric force. Indeed, Newton's celebrated equation requires that gravity acts instantly on the scale of the solar system. It has been calculated that gravity must operate at a speed of at least $2x10^{10}$ times the speed of light otherwise

closely orbiting stars would experience a torque that would sling them apart in mere hundreds of years.

Similarly, the Earth responds to the gravitational pull of the Sun, where it is at the moment not where the Sun was 8 minutes ago. If this were not so the Earth and all other planets in the solar system would be slung into deep space within a few thousand years. Gravity is therefore an electrical property of matter not a geometrical property of space.

Had Einstein lived he would have amended his theory accordingly.

What is the nature of light? Einstein's special theory of relativity was disconfirmed right at the start by the Michelson-Morley experiment which showed a residual due to the aether. This was later confirmed by far more rigorous repeats of the experiment by Dayton Miller.

But by then popular delusion and the madness of crowds had taken hold and contrary evidence would not be tolerated. The Dayton Miller story makes interesting reading. If it weren't for the extraordinary power of self-delusion, commonsense would tell us that a wave cannot exist in nothing. So Maxwell was right, light is a transverse electromagnetic wave moving through a medium, the aether.

Anyone who is familiar with the special theory of relativity will realise how much it goes against commonsense. It should have been doubted on those grounds alone right from the start.

But what is the aether? In the vacuum of space each cubic centimetre is teeming with neutrinos. And since neutrinos are resonant orbiting systems of charge, like all matter, they will respond to the electric force by distorting to form a weak electric dipole aligned with the electric field. The speed of light in a vacuum is therefore a measure of the delay in response of the neutrino to the electric force.

This passage, though somewhat technical, does demonstrate common-sense.

What about the bending of starlight by the sun, which discovery raised Einstein to megastar status? The residual found in the Michelson-Morley experiments shows that the Earth and all ponderable bodies drag the aether along with them. The bending of starlight near the sun is simply the effect expected of an extensive neutrino atmosphere held to the sun by gravity. Light will be slowed in the denser medium, causing normal refraction or bending of light.

Ponderable bodies could easily do this. Astronomers should mount a research effort to test neutrino atmosphere theory as soon as possible. Surely the carrot of overturning Einstein must appeal to someone?

What about time? With all bodies in the Milky Way galaxy communicating their positions effectively in real time through the electric force of gravity it means there is a universal time. There can be no time distortion or time travel, something that common sense always told us.

And again, the powerful influence of common sense is clear here. Why science cannot make more use of it is a continuing mystery.

What about black holes? They are a mathematical fiction, a near-infinite concentration of mass required to explain concentrated sources of energy seen at galactic centres by employing the weakest force in nature, gravity. It is the high-school howler of dividing by zero.

Yet this howler continues to be made by deluded astronomers affected by the madness of crowds. It is high time they used their expensive telescopes to become less deluded. They might also consider taking some maths courses.

Plasma Cosmology shows that where electrical energy is concentrated at the centre of a galaxy, gravity can be ignored in favour of far more powerful electromagnetic forces. The collimated jets of matter coming from that focus are also replicated to scale in plasma labs. The jets are inexplicable if a black hole is supposed to be a cosmic sink for matter.

When you think about it, it never seemed likely that a place where matter disappears could simultaneously give rise to jets of matter. Science appears to have been on a wild goose chase here.

The implications for biological systems in this electrical model of matter are profound. A method of near-instantaneous signalling between resonant molecular structures within cells and on cell walls seems plausible and may provide a way of looking at the mind-body connection and other communications external to the body.

It may provide a link between classical Physics and the pioneering work of the Biologist, Rupert Sheldrake, in biological morphogenesis and telepathy.

This all seems reasonable.

Also, the work of the outstanding French Biologist, Louis Kervran, may gain a working physical model to explain how biological enzymes are capable of transmuting chemical elements at

body temperatures. It seems that by exquisite tuning, one resonant system of nuclear charges may be transformed into another. And like the decay of the neutron, ubiquitous neutrinos are implicated as a catalyst.

There is plenty of material here for an ambitious PhD candidate to get his or her teeth into.

It may be that the answer to our future power needs will be answered when we understand how to extract nuclear energy resonantly instead of by using brute force as we do now. The New Jersey based company, Black Light Power, seems to have stumbled upon a similar process using a resonance between hydrogen and the iron atom. It is interesting that biological systems also use heavy elements like iron and magnesium to perform their minor miracles of transmutation of elements.

It may be wise to invest in the Black Light Power company.

The electrical relationship between matter and mass allows us to understand how quasars can be newborn objects that have low mass and brightness and high intrinsic redshifts. With time, their mass increases and their intrinsic redshift decreases in quantum jumps. This shows that quantum effects also occur on a galactic scale. It is another powerful argument for the near infinite speed of the Electric force.

The quantum decrease in intrinsic redshift seems to have escaped conventional Astronomers. Perhaps they think their research grants will be stopped if they dare to question cosmological orthodoxy. They should be bolder.

The electrical nature of the universe reveals the currently accepted life story of stars as an elaborate fiction. Stars do not self-immolate. Plasma Cosmologists identify cosmic electrical power lines of unknown origin that shape galaxies and light the stars in our small corner of the universe.

These findings about intrinsic redshift and electric stars explode the big bang myth: The universe we can see is not expanding; it is only a small part of a universe that is of unknown extent and unknown age.

The Big Bang is only a theory. The fact that plasma cosmologists have identified cosmic electrical power lines should be enough to call it into question.

This outline may seem like the basis for a theory of everything but in truth the greatest mysteries remain. We cannot talk sensibly about a beginning of the Universe since the mystery of the origin of Electric charge and the nature of the Electric force remain.

When all is said and done this really is the greatest challenge. No scientist today can give a convincing explanation for the origin of electric charge or the nature of the electric force. It is to their shame that many of them are not even trying.

Are we to pretend that Skunkworks does not have antigravity? I just taped two hours of the History Channel series on the Universe and I have to say it feels like CIA disinformation propaganda, certainly feels like brainwashing after learning what I have learned, church of cosmology, religious dogma of a very high degree, with scientists speaking as if they were god himself, or priests and popes that speak for him. The whole issue is a can of worms and at some point it's gonna be opened.

One would expect that the History Channel of all channels would have done their research. Sadly, that seems not to be the case.

Just like I opened a can of worms with my statement that the magnifying transmitter was a theta pinch. Now that the cat is out of the bag where do you draw the line? How long NASA can talk about new discoveries and not mention Birkeland seems almost daily. At what point do you stop telling people to pay attention to history and the present?

This seems unanswerable.

I am mainly interested in what is gravity? I used it as a tool in my clinical practice for over 12 years and Ida Rolf always talked about gravity so it kills me to not know what it is as no one has found the charge carrier. Therefore is it not fair to ask the question is it electro-gravity?

A reasonable question. And one which science should address.

What technology would or could or has been achieved by industrial/military programs such as Skunkworks with a unified field theory based on public quotes and sightings of an electric universe and electro-gravity if one understood the first two questions? That seems logical enough.

Indeed.

I posted Wal's view of gravity and it is an electro-gravity model. Since all indications are that Wal knows what he is talking about and all indications are that Boyd Bushman knows what he is talking

about when he says that he is working on something that all you do is charge it and it loses weight.

There is no reason to doubt Boyd Bushman and Wal certainly seems to know what he is talking about.

So I have a real simple question. Do they have floating triangles? Yes or no. What did the Governor of Arizona, someone with military experience, see the night of the Phoenix lights that he finally disclosed on Larry King last fall, if it was not our technology, that means ET hypothesis allowed. He saw a massive flying triangle.

This is the Governor of Arizona on public record. He thinks it was ET. I disagree with him on that account. I would like to keep the conversation to it being from our planet. That would at some level keep it non-fringe? For some reason I still never get but I can swing that way.

I must humbly submit that we have flying triangles that use some manipulation of electrogravity and are built by Skunkworks. They have radar tapes from Belgium, video from Europe, sightings by very public officials here and abroad. It is not fringe nor is it imaginary. It is a fair and honest question. I see only one answer, yes we do.

Given this level of corroboration, yes, we have flying triangles.

Now why would it hurt so much to admit I believe that? I really have the ponderance of evidence on my side. You'd have to discredit public accounts by the Governor on the night of a massive sighting, military accounts from Belgium and Belgian police, radar technology and the electronic file of the evidence and video evidence.

I also have scientific statements by Bushman, Rich, and Thornhill to back up the rest of the evidence. In fact Thornhill and Bushman are my trump cards. They connect the entire picture with a wonderful frame. Would that not be a feather in the cap of the Electric Universe instead of a blemish, does it not support the view of Thornhill?

It is certainly most compelling.

The second statement follows my belief that no single and no-one's viewpoint is entirely objective or factual. I say that it takes a village of us, sorry for the cliché, to understand and reveal the underlying objective truth. Anyone who tells me that they have the correct understanding of _____ , or quotes the scientific consensus to support an argument, will get an eye-raise from me.

A fully justified eye-raise.

A kind reader sent me a link to an article in National Geographic and Nature this month July 2010 that reports particle Physics were shocked to discover the proton was 4% smaller than previously thought. This was discovered by firing a laser at a muon orbiting a proton, bumping it from one orbit to another. We are told that the standard model has no explanation for this discrepancy but that it may have something to do with the Rydberg Constant.

I have an answer for these particle physicists and it has nothing to do with the Rydberg Constant. It has to do with the unified field. Remember that all these experiments are taking place on the Earth, which not only has its own gravitational field but is in the larger field of the Sun. Quantum physicists always ignore this when they run their numbers but they shouldn't do that.

This is unfortunate. Perhaps a simple email to quantum physicists explaining that they should include the larger field of the Sun when they run their numbers might be the answer.

The masses and energies they are finding and are plugging into the equations are determined by the unified field, so gravity cannot be ignored at the quantum level. In my paper on the Saturn anomaly and before that in my paper on the perihelion of Mercury I showed that Einstein's field equations are wrong by 4% in the field of the Sun. That is to say gravity is generally miscalculated by that amount in our solar system at all points.

Since Einstein's field equations are really mass equations he does a mass transform to find curvature. All the masses in the solar system are wrong by 4%, including of course atomic and subatomic masses. This must be important because the equations that give us a proton size are dependent on proton and muon masses. Although these scientists aren't directly using Einstein's field equations in this problem they are using masses that are determined by the field. The 4% error in the field equations is directly causing the 4% error here.

4% is not much so perhaps that is why scientists didn't noticed it.

But there is more. This is not the only mathematical or field error in the quantum equations. It is the most obvious error, since it shows itself in simple experiments, but it is not the largest error by a wide margin. Even larger errors in the math are hidden errors since they offset.

In other words two or more mathematical errors, errors in different directions, nullify one another so that this simple experiment cannot show them. QED is full of offsetting pushes like this since it has not only been renormalized in multiple ways it has been jerry-rigged decade after decade. But I have shown that the proton size is off not by 4% but by around 170%. Yes, the proton is actually over 100 times larger than we think, not 4% smaller.

Physicists should, as a matter of urgency, re-examine their calculations. A 4% error can be forgiven. A 170% error cannot. How did Physicists miss this?

These scientists haven't even begun to uncover all the errors. To understand why this is so you have to read several of my papers including all my papers on Bohr. There has been a fundamental misunderstanding from the beginning concerning the assignment of the angular momentum of the electron orbit. Not only are the angular momentum equations faulty, the momentum itself has been misassigned.

When we add to this the fact that Coulomb's equation was always misunderstood we have a very large mess with many points of spillage. When we correct all these mistakes we find gravity at the quantum level at a size 1022 larger than we thought. We find a larger Bohr radius, we find a different Fine Structure Constant and we find no point particles. We find a photon with real size and spin and mass and so on.

If it is true that scientists haven't read all the papers on Bohr then they should do so. It is pointless science sticking to faulty equations when more up to date calculations are to be found in Internet discussion groups.

Yes, this is now the state of the art in Physics. This is what we see published in major journals. Nothing is now too ridiculous to propose. I have a question: Do you seriously imagine that physicists who would propose such things, or sit by and allow such things to be proposed, could be right about anything?

If any of the foundations were in order it would be impossible to propose or publish the things we now see proposed and published. As you can judge the seeds from the fruit, you can judge the theories by the articles. The popular articles are rubbish because the scholarly articles are rubbish and the scholarly articles are rubbish because the theories and math are rubbish. Physics is no more than photons escaping gravity.

I have rarely heard a more succinct summing up of physics.

First off I would like to say I am a huge fan of the Electric Universe models/ Plasma cosmology. I do have a couple of questions though if some of you would not mind answering. It is not my intention to start an argument if this is what these questions might sound like.

First off, do any of you have a Masters or PhD in Physics or cosmology/astronomy? If not how can you criticize the mainstream models without having a wide degree of expertise in it first? So that is my first question - your credentials.

Here we see a person entering the debate from the standpoint of mainstream physics. I shall let the proponents of the Electric Universe answer his queries for themselves.

There are plenty of credentials within the writings of Electric Universe, PhD's, Professors, Nobel prizes, etc. But as anyone who investigates scientific history knows some of the greatest contributions have come from self-educated Michael Faraday.

It can be a common misconception that Plasma Cosmology is non-mainstream science; this could not be anything further from the truth. The 'appeal to authority' 'non-mainstream fear agenda' that can often be pushed by any establishment to curious laymen I find very distasteful, and have often experienced it.

Once you meet this fear campaign with names like Hannes Alfven, Anthony Peratt, Halton Arp, Tom van Flandern, Kristian Birkeland, you realize the fear campaign is a pointless house of cards, credentials aplenty on both sides, it then becomes time to put the 'shiny badges' down and address the science.

I am just trying to see how all the mainstream scientists can be against the Electric Universe and Plasma models. It just seems like Plasma cosmology does not stand up to the computer simulations and other things mainstream science has done.

This is another misconception: I have seen a 1980's Plasma computer simulation of galaxy formation, from NASA JPL mainstream scientist Professor Anthony Peratt: I have not seen a stable galaxy formation simulation from the gravity cosmology.

No offense, but there are many computer simulations and real life results like Eric Dollard has created in vacuum tubes with Plasma, which are clearly spiral galaxy shapes.

I would ask you first to read the original works of Tesla, Maxwell, Faraday, the originals. Then read the work of modern Plasma Physics, from Birkeland onwards. Then connect the two, then and only then, start to look at the mainstream, for the mainstream leaves out the original, all the while teaching you that you are learning the original.

That in and of itself is a lie and a huge misconception. For instance, Maxwell's theorems are not Maxwell's at all; they are Heaviside's which is purely vectoral while the work of Maxwell is in quaternions and is a complex plane. So how can Heaviside's reduction be Maxwell's theorem? Please tell me...

When it comes to Tesla I will let Eric Dollard to the talking. So much to learn, so much not being taught, so much crap floating around as truth it makes me sick. So much money changing hands with those with credentials that they have no way to be unbiased to the corporate stink, rather follow the money, that's where the corruption lies. The assumption that a scientist will not fudge data to assist the corporate plan is pure lack of judgement of human nature on your part.

Maddogkull, your query is justice, and yes one could truly wonder why they don't see the full Plasma scenario. On the other hand one needs to take into account that the majority of astrophysicists are not necessarily putting too much work into the cosmology part. i.e. they are studying one specific aspect of the Universe, the sun or our atmosphere.

Also, much of the work astrophysicists/astronomers do, are very specific and narrow funded assignments from e.g. NASA or ESA; studies of the ionosphere, the surface of Mars, the corona etc. Just look at the areas of focus page for the Norwegian Space Centre.

As another illustration, a friend of mine works for a similar space centre and when I asked him about his view on cosmology related topics he replied that he frankly hadn't looked so much into that since high school as the stuff he studied and worked on since was more related to specific technical aspects about equipment and the higher atmosphere. i.e. they are specialists and the job they receive salary for is related to that.

I know what you are saying but it has to be more than just money and power. Lawrence Krauss is an astronomer and I honestly think he believes the Universe is expanding. How can someone who studies the Universe so much not notice the Plasma effects? That is what does not make sense. If you study the Universe everyday of your life noticing the Plasma effects would be 100% noticeable. This is what is confusing.

Hi Maddog, Lawrence Krauss is a mathematical theoretical physicist. Not an astronomer. Lawrence still rests on the assumptions that redshift equals distance known as the Hubble relationship or Hubble law. Edwin Hubble was the astronomer who discovered a slight similarity in his early results that as the redshift of a faint smudge galaxy increased the apparent distance increased.

It is now known that Hubble's law does not apply even remotely to the local group Virgo which he was studying. Later in his career Hubble began to retract his view of expansion but the inertia behind Einstein relativity had taken off and his cautions were ignored by the excitement that the Universe could be described by man's imagination and graph paper.

Halton Arp, Edwin Hubble's assistant, went on to study interacting galaxies. Arp noticed that high redshift objects apparently far away, if one takes Hubble law as gospel, were interacting with low redshift objects. This is impossible if Hubble law applies. Now remember very clearly Hubble and Arp are astronomers. Krauss is a theoretical physicist.

Hubble and Arp take measurements and spend their time looking at the sky, they have an appreciation for error factors, correction factors of the equipment used and of which objects they are studying. Krauss, on the other hand, takes mathematical formulas and try to build a Universe on graph paper focusing on a small set of parameters that 'bend and fit' the equations.

I thoroughly recommend reading Halton Arp's Seeing Red book. I also recommend reading any of Krauss's books side by side with it. You will notice that theoretical physicists pick and choose and ignore specific measurements taken of the night sky. Time after time they put their head in the sand when it comes to confronting data.

There is a clearly different head-space with these gentlemen. Once you notice the difference between imaginary ideas on graph paper and that large body of real measurements which they flat out ignore it becomes the most sorry realization. Krauss is clearly a

bright man but as long as real measurements are flat out ignored I just can't take him seriously.

Now a another voice joins in the discussion, seeking to theoretically underpin plasma cosmology with an improved theory of force.

You want my opinion? Fantastic. This is what Plasma cosmology needs. A theory that not only explains why the big bang theory, general relativity and special relativity are wrong but also aether physics and quantum mechanics. It's a theory that explains what no other theory has explained rationally as of yet which is pull.

Today's mainstream scientists have managed so far to admit that aether physics is wrong but they still insist that big bang theory, general relativity, special relativity and quantum mechanics are correct. Whist insisting that these theories are correct they have yet to come up with a convincing theory of pull.

I love it. I am completely sold on thread theory. I started a topic called Modern Physics where I summarize the theory in as few words as possible. I have some questions about the theory and some exciting ideas but I'm waiting to hear everyone else's reaction. You know my reaction, I'm sold. The threads would be filaments or plasma as far as I can tell so far.

This clearly links threads with plasma and so constitutes a viable synthesis of physical concepts.

I contacted Bill today and he isn't familiar with Electric Universe, Plasma Cosmology. His epistemology and metaphysics greatly resembles objectivism. He was aware of this but he is not hopeful for reason to affect mankind because he suspects we will all be extinct very soon. I haven't evaluated this area enough to comment on why he thinks this and if its plausible. I mentioned plasma filaments etc. He said he is doing some new videos soon.

The fact that Bill is not familiar with the way his ideas are used in the new synthesis and suspects we will all be extinct very soon in no way invalidates its importance.

One of my questions regards the nucleus of the atom under thread theory (TT). Bill Gaede describes the nucleus as a dandelion or a porcupine. The electric threads all terminate at the centre of the atom while the magnetic thread wraps around like a yarn ball, except it appears more like a shell because of the void separating the electron shell and the proton dandelion. I am interested in the nature of the nucleus.

This is important because modern physics does not yet have much idea about the nature of the nucleus. The dandelion/porcupine debate still rages.

Do the electric threads enter/exit in any kind of regular pattern that may be observable? If atom man were standing at the nucleus would he see V-shapes at the centre where the threads enter and exit by a nearly identical path or are the threads of the nucleus straight, entering on one side and exiting the other? Do all the threads intersect precisely or are they merely touching, forming a loose aggregate? If they look like v shapes do tips of the v's all intersect precisely or do they form a spherical shell of v-tips?

Science has not addressed any of this as yet.

These are important questions because answering them may lead to elucidation of the width and height of the thread. A loose aggregate explains nuclear instability and radioactivity while a precise, defined intersection may explain the relative impermeability of the nucleus.

Either way it seems that new physics will be uncovered. These are exciting times.

V-tips versus straight electric threads may be observable if the proton were sufficiently disturbed to shake the threads loose but not so disturbed that the v-tips were straightened too quickly. Lastly, there may be no regular structure at all, the threads may enter the nucleus and tangle with each other like Maniac Magee's ball of string. This could explain why radioactive decay appears unpredictable at the atomic level.

A fascinating proposition. The Large Hadron Collider could be gainfully employed disturbing the proton enough to settle this matter.

A question I'm having is what happens with the threads as far as entanglement is concerned. Would these threads allow other threads to pass through them? I think Bill mentions them colliding in his rod and axle example.

Bill should be brought on board and offered tenure by a leading University. Rod and Axle theory is ripe for development.

Look at nuclear decay. The geometry is self-evident in posts I have made. The nucleus forms in hexagons of neutrons and protons in complex nuclei. But let's take a step back. Start with the hydrogen atom. The hydrogen atom is special and will reveal all. The model of the hydrogen atom has never been balanced. At some point the electron either flies away or plummets into the proton nucleus.

Physicists are reluctant to mention this for obvious reasons.

I got started on this because of the hydrogen atom and its geometry that leads to hydrogen bonding. Of course life is dependent on hydrogen bonding. The relationship of these bonds and the fundamental geometry will help us create a structure and functional model that is based on angular momentum and three force models in most cases.

This could constitute a radical new insight.

In fact the point of reference seems to make fundamental relationships between the Universal Constants. We have seen that, given the principles of relativity and due to the gravitational conditions of the atom, a small variation in the velocity of the electron unbalances the forces such that when the point of reference is the proton the Coulomb force is greater than the centripetal force.

This difference is the gravitational force. When the point of reference is the electron the Coulomb force is equal to the centripetal force ensuring the equilibrium of the system.

This passage is highly technical but physicists who have studied to at least doctoral level should begin to understand it. That changes in velocity unbalance force is straightforward enough but nowhere do physicists change their point of reference from the proton to the electron.

Hydrogen bonds and the relationship to the special condition of a hydrogen atom are critical to understanding all forces.

The importance of this has largely escaped physicists up until now.

The strings are in constant vortical motion. The spinning charged electron creates an electric field between it and the nucleus with electric field strings that are in a vortex configuration due to the orbital motion. The electron of a hydrogen atom is travelling at 3 x 10^6 miles an hour, it would reach the moon in three minutes if travelling straight. It creates a magnetic dipole moment and a magnetic field. The hydrogen electron creates 1 microamp due to its spin. Each field is at 90 degrees.

The vortex configuration of electric field strings has yet to be confirmed but when it is, the creation of 1 microamp by the hydrogen electron due to its spin should automatically follow. Scaling up, we would appear to have a solution to the energy crisis.

That is an important key to the geometry of the vortex. The logarithmic ratio of phi will unlock the vortex.

Unlocking the vortex has long been the holy grail of Physics.

The phase shift of 120 degrees is universal as a constant. The double helix is two thirds of a triple helix.

A fact which seems to have escaped modern Biologists.

Aether is visible as the missing third helix. All is three. Three is fractal. 120 phase shifted, triple helix ropes, strings, threads, what have you.

The fact that three is fractal has been previously mooted by mathematicians. However, the fact that all is three is new and fundamental.

The geometric form is the star. It also looks like a poppy. Of course there is a very elemental reason the hydrogen atom and its electron form a star shape or a poppy shape depending on your relative perspective.

Electron microscopes are beginning to discern the shape of the hydrogen atom and its electron.

This is revealed in the relationship of Coulomb's Constant and the Compton wavelength and the path of the distributed charge unit as it expands and contracts towards and away from the nucleus in its orbital motion.

The path of the distributed charge unit has been measured but has so far been considered unimportant. Physicists may like to change their mind at this point.

Take a look at vortex math and harmony math.

This is undoubtedly a good idea. Especially harmony math, which seems rich in promise.

That is, in the hydrogen atom aka neutron, where the proton torus is squeezed within the electron lissajous-volume, do the frequency, wavelength and radii relationships [octaves] between the electron and the proton, do these 'notes' preserve binary integer relationship values?

Much more research is needed here.

Or, if we were to assume that the proton outer superluminal ring spin frequency was an 'octave' higher than the nuclear sub-ground-state electron frequency would that frequency be exactly a factor of two higher [or of four?] Or would it be 2.5 times two higher, or 2.5 times four higher, considering that the spin velocity is not c, but is 2.5 c? Or would it fall in some intermediary [binary integer?] Transitionary value?

These are vital questions.

Ditto the inner superluminal ring electromagnetic-wave frequency and wavelength of the proton's 'twist' smaller-ring topology?

The layman may be somewhat bemused at this point.

The twist velocity is 13 times c -- thus we ask is the twist frequency exactly a factor of two or four greater than the proton spin frequency ... Or 12.9/2.5 times two or four... Or some other intermediate value? In the 13 years since I published the electron, proton and proof papers I have never taken the few minutes required to sit down and calculate the answers.

Modern Physicists owe it to themselves to take the few minutes to sit down and calculate the answers. Afterwards, however, they must be careful not to claim the credit for themselves.

Entanglement is a difficult issue under TT (thread theory) as well as every other theory that I have encountered. Quantum cannot explain why a photon ball does not collide with another photon ball but rather passes through.

Physicists will argue that it doesn't struggle at all but they are looking increasingly self-deluded at this point.

The wave theory of light requires a medium just as sound waves do, which implies a cosmic aether, which has its own problems. Not the least of which is the fact that disturbances in this aether should logically interact with each other just like waves on an ocean.

This problem with the wave theory of light is such that some Physicists have abandoned it.

In his book Bill Gaede gets to the crux of the problem which, as expected, is a matter of clear definition. In particular we must define the word touch. What does it mean for two objects to touch one another? If they merely come extremely close they cannot influence each other.

This is axiomatic.

But if there is no space separating two objects then they no longer have separate surfaces and then they become a single object.

Bill Gaede concludes that touch does not happen at the thread level, it is an observed phenomenon of the macro world. Touch is a phenomenon arising at the surface level. An object must own a surface in order to touch another object.

Bill has hit the nail on the head here, (or if not, has come extremely close!)

The matter of owning a surface or not isn't clear, it is purely a matter of velocity.

The modern theory of touch hasn't yet addressed the velocity dependence of surface ownership. Again, more research is necessary.

At high velocities the electron shells of two atoms will pass through each other until only the nuclei collide. At low velocities the shells collide. Electric threads in magnetism collide because of speed.

Nuclei collision has been observed in the Large Hadron Collider. Low velocities are not yet achievable so shell collision has yet to be observed.

Under thread theory we must also clarify which type of touch we are referring to, push touch or pull touch. Two atoms may not be in contact yet they can affect each other by pumping torsion waves.

The clarification between push touch and pull touch is one of the hot topics in Physics today. The jury is still out and the final explanation may have to wait until the construction of a Superconducting Supercollider expected towards the end of this decade. Pumping torsion waves have not been detected with certainty in the LHC but tantalising hints of their presence have been seen.

This electromagnetic touch is an alternating pressing and tugging through the torsion mechanism. This explains why quantum mechanics detects light as a particle upon impact but not as it propagates.

The exact torsion mechanism is a matter of debate. However, the explanation of why quantum mechanics detects light as a particle upon impact but not as it propagates is of profound importance. Pumping torsion waves offer an intriguing explanation.

Thus the interface between the two objects is not a surface but a signal. Atoms do not actually intervene in the process, they are simply endpoints. The push touch of our everyday experience is mediated by matter itself whereas gravitational and electromagnetic touch is mediated by the rope.

The extension of thread theory into rope theory is reminiscent of Einstein's extension of special relativity into general relativity, but is clearly more wide-ranging.

Interatomic Touch: Two electron shells colliding. Gravitational/electromagnetic touch: The alternating expansion and contraction of an electron shell that torques the electromagnetic rope inducing other atoms to expand/contract at the same frequency and retransmit the signal.

A neat summing up of a complex theory.

Bill Gaede ultimately says that the thread, although three dimensional, is nothing like the thread of our everyday experience. Indeed it can't be, it is the fundamental constituent. It is conceptually made of a single piece. There are certain cases where the Universe appears to be composed of surfaces and when it appears not. He proposes the 2-d rule and the 3-d rule.

The 3-d rule: Two surfaces cannot cross each other. At the subatomic level there are three ways to generate a surface.

1. An electron shell comprised of magnetic threads.

2. A spherical region within 0.8e-15 m of the proton or neutron centre comprised of electric threads.

3. A bundle of curved threads swinging around a series of atoms as seen with axle and rod in the magnetism video.

Modern Physicists accept the third rule because it is made clear in the video. They do not accept fully the first two, though opinion is swinging in their favour.

The 2-d rule: A rope can cut through a thread, surface or another rope. It has properties we typically associate with planes. The thread is so much thinner than the radius of the smallest chunk of matter (H-atom) that for all practical purposes it behaves as if it were 2-d.

The 2-d rule is now largely accepted. It is so thin the exact measurement is yet to be made with accuracy. However, it does indeed behave as if it were 2-d.

The 2-d rule may seem supernatural. Bill proposes it is so only in terms of ordinary matter and everyday things. The thread is the fundamental constituent. It's is fundamentally different than the matter we observe. We observe matter by detecting the light emitted by it or by simply colliding with the matter.

We cannot observe threads or ropes in the former sense because they are light. The torsion that propagates along the rope simply serves to excite the atoms in our eye to pumping faster.

The next generation of electron microscopes are expected to allow Opticians to see eye atoms pumping faster.

Therefore the thread is a defined entity: It exists by definition, not because we observe it. We can only observe threads in the latter sense in magnetism because threads are colliding with other threads. There is no matter that can collide with a thread because the thread

is so thin compared to the smallest chunk of matter (atom) that the thread does not present a surface.

This is highly intuitive.

I disagree with Bill in this discussion. However, he proposes an alternate explanation but ultimately doesn't make it a part of his theory. I prefer it because it does not require us to make specific cases for when the Universe appears to allow objects to pass through each other or not. The theory is that the cosmic rope is static.

For example, when you shine your flashlight at a wall and then move it you may be tempted to think that the ropes connecting the bulb to the wall have moved with the flashlight and crossed countless threads as you sweep the wall. In fact this does not have to be the case. The flashlight is stimulating threads that are already there. All the atoms comprising the bulb slide along existing ropes like beads on an abacus.

Modern Cosmologists are bidding for time on the world's largest telescopes in order to observe the cosmic rope more closely.

Matter is essentially a bundle of signals sliding along a permanent, static web. The atom moves and leaves some of the rope it was composed of behind and takes up some new rope in the direction of travel. It slides along somewhat like a bead on an abacus. At its new location it continues to pump and stimulate torsion waves but in a new section of threads. In this Universe it is perfectly reasonable for the thread to always possess a surface.

Modern electron microscopes have come close to observing atoms moving and leaving some of the ropes behind. It is only a matter of time before resolution is such that they will be able to image atoms pumping and stimulating torsion waves. When they do, the pictures promise to be stunning.

The rules are reformulated more simply: 3-d rule: Two surfaces cannot cross. Objects can only touch. Touch is defined as 0 distance between the surfaces of two objects i.e. the surfaces remain distinct though there is no space between them. Touch comprises two frames of the Universal Movie by definition. In frame one two objects are touching, in frame two they are separate. If they are not separate in frame two then there were not two objects in frame one but rather a single object.

If you see a dumbbell-shaped object and it does not separate into two spheres you conclude by definition that it is a dumbbell-shaped object. If instead the dumbbell separates into two spheres you conclude by definition that the object was simply two spheres touching. Touch is inherently dynamic whereas most people think of it as static, i.e. they try to visualize touch in a photograph when it is really a movie. This is why the concept of touch has been so difficult.

And this contrast between photograph and movie fully brings out the depth of the theory. It can seem surprising that the Universe works in a way which we can easily visualise using commonplace analogies but perhaps we are wrong to be startled. Everything in Quantum Mechanics can be modelled using everyday instances with which we are all familiar. Schrödinger's Cat is one well-known example.

In the Static Web Dynamic Touch Universe there is no 2-d rule. Electric threads responsible for magnetism do collide with atoms but they do not disturb the atoms appreciably. The thread slides around the outside of the atoms and continues. The thread is so small and moving slowly enough that it does not knock the atoms appreciably out of position. In this formulation, a combination of Bill's Static Web Universe and Dynamic Touch, I believe the issue of entanglement is solved.

An elegant solution to a puzzle which is currently baffling Physicists, who are making no headway by sticking inflexibly to their Standard Model.

There is one last issue with the fundamental nature of the thread that Bill Gaede does not solve to my satisfaction. That is the issue of the thread's rigidity. The thread is conceptually made of a single continuous piece so how can it bend? A fundamental constituent certainly cannot break by definition. Bending also seems impossible by definition since this would require the opening up of spaces within the thread and the compression of spaces elsewhere.

Bill claims that continuity does not have to imply rigidity. He correctly points out that we have no examples of a continuous structure in our experience. As a result he concludes that we cannot ascertain if continuity is an indicator of rigidity or fluidity at the rope level and that we must simply make an assumption that continuity does not imply rigidity based on what we observe. I disagree. Conceptually, that which is made of a single piece simply cannot bend.

It is fascinating to see science carried out 'in the raw'. Here, the issue of the thread's rigidity is under examination. We see agreement ('We have no examples of a continuous structure in our experience.') followed by a crucial point of divergence ('I disagree. Conceptually that which is made of a single piece simply cannot bend.'). This is the essence of science.

Exchanges such as this also demonstrate what the Internet was always intended to be, a forum in which informed minds could come together and through knowledgeable discussion bring rationality to bear upon the scientific problems of the day. The Internet has many detractors and when one views some of its content one despairs. But there are also vast swathes of what the Internet can be, and discussions such as this reinforce that positive view.

There is an alternative and that is a jointed structure. Essentially the thread is composed of individual pieces called chinks each with a male end and a female end. The thread behaves like a chain at this level but looks like a thread at the rope level. The joints are closed, the thread cannot be broken. The architecture proposed, individual chinks of thread, explains why the thread does not have to remain straight. Additionally it is interesting to note that with this architecture the rope has a characteristic bend radius below which it cannot bend further.

Radius is not really the correct term. What I am referring to is, if you were to take a bicycle chain and wrap it around in a circular shape, really polygonal, and pull on one end to decrease the size of the n-gon, eventually you would end up with three chinks forming a triangle. Pull further and there will no longer be a shape inscribed. The size of this triangle is what I'm referring to.

And here we see an example of a theory being extended. Real science done in real time is highly exciting. Even more exciting is the certainty that by the time you read this the discussion will have moved on to further fruitful areas.

The continuous yet flexible thread Bill Gaede proposes, when subjected to this treatment, has a smallest shape of arbitrary size. This size could only be described as depending on the flexibility of the thread which has little meaning in the context of a continuous discrete object.

Why should the fundamental constituent have a characteristic flexibility of 1, 10, 100, or whatever? The jointed thread possesses a fundamental smallest producible shape that is characteristic of the length of a single chink. Under these constructs I believe the problem of thread entanglement and flexibility are solved.

Whilst persuasive, this extension of thread theory will require careful consideration to find the fundamental constituent's characteristic flexibility. As noted, this could be 1, 10, 100, or whatever.

Has science taken a wrong turn? If so, what corrections are needed? Chronicles of scientific misbehaviour. The role of heretic-pioneers and forbidden questions in the sciences. Is peer review working? The perverse consensus of leading scientists. Good public relations versus good science.

Students of the philosophy of science will find much to consider here.

The functional aspects of the geometry of the hexagon are revealed in this in-depth book. Watch how it then relates to the icosahedrons. This paper on the hexagon of graphine will explain some fundamentals of structure and function. Notice how it relates to the spiral.

The spiral is overdue for reassessment.

On the basis of these new experimental findings on needle morphologies I propose a new growth model for the tubular needles. That is individual tubes themselves can have spiral growth steps at the tube ends. It is worth mentioning that the spiral growth steps, which are determined by individual hexagon sheets, will have a handedness.

The growth mechanism seems to follow a screw dislocation model analogous to that developed for conventional crystals but the helical structure is entirely different from the screw dislocation in the sense that the present crystals have a cylindrical lattice.

Needle morphologies are at the cutting edge of analytical geometry, a flourishing topic in mathematics. Here we see it allied to thread theory. The fact that the growth mechanism seems to follow a screw dislocation model is authoritative evidence in its favour.

Therefore the threads do cross and in fact join to every third string from the vortical centre. That is the fundamental triple helix. The aether is invisible and therefore we see a double helix but the space for the aether helix is right there, each one 120 degrees out of phase. Look at the Birkeland current or the DNA, notice it is missing one thread or rope.

Form work done in the 1950's Crick and Watson developed the 'Double Helix' theory of DNA but times move on. Since then there has been the feeling that a helix could be missing from their initial analysis and now it seems that third helix has been identified. The fact that the aether is invisible has held up progress but the space for the aether helix has always been recognised to be there.

I had a huge paradigm today. This forum and the members and their links work wonders for me. I try to explain these ideas to others as I learn and integrate them as they evolve and relate. I found myself today thinking the primary angular momentum is the rotor and the atomic geometry is the stator.

Consider my recent work on nanotechnology. Consider your pencil lead. Imagine telling your teacher in 1970 that this material is a superconductor and stronger then steel. They would look at you funny and send you to the special class. Yet this is in fact true.

Teachers in the 1970's were not noted for their empathy.

This shows why the graphite of your pencil as cubes is soft and writes on paper. Yet take that same material, go nano and line them up one at a time and they make hexagons. Then all of a sudden this graphine is now a superconductor and stronger then steel. Structure truly is never separate from function. The hexagon stator is much more integrated geometrically with the angular momentum rotor which I believe forms a six sided star.

The rotor/stator argument has been made before but the concept of a six sided star is a real advance. Also, mathematicians may like to consider, aside their engineering colleagues, just how much more a hexagon stator is integrated geometrically with the angular momentum rotor. Such discussions could quickly lead to surprising new insights.

This changes the properties of the same material by changing the geometry because we now have a different relationship between the rotor and stator that is totally different than the cube. Remember my finding last week of the hexagon nucleus geometry? Think about carbon hexagons. Is the body a supercomputer, superconductor?

Biologists are divided on the question.

Realize the cells make tetrahedrons as they grow. Realize the diamond is tetrahedrons while coal is the cube. Remember we are a liquid crystal? So of course we grow as tetrahedrons, not cubes. We have hexagon atomic geometry and tetrahedron cellular division with icosahedrons cellular geodesic structure. We are diamonds, not coal; we are truly liquid crystal superconductors.

It is undeniable that we do not grow as cubes. From this, much follows.

When we work with nanoscales or molecular scales we are **building chassis; the motor is supplied by the Universe. But truly we are building stators for the ever present rotor.**

Engineers will relate to this analogy.

Since the G-force of APM is a push and a pull it stands to reason the orbital motion of the hydrogen electron does make a six sided star or daisy flower type configuration. When they show the electron making perfect circles that's never right.

'It Stands To Reason' (Latin: Stantes Rationem) is a formal method of proof in logic theory. An example from popular culture: Everton Football Club have never won the Champions League. *Stantes Rationem* Everton are rubbish.

The more you can conceptualize the electron distributed charge configuration the easier the knowledge to build a proper stator because it must relate to the rotor.

This is self-evident.

When I ask about the geometry of the nucleus I am referring to architecture, not Euclidean geometry. I misspoke when I said geometry.

Here we see someone acknowledge in public they may have mislead their peers. This is refreshing and is something science would do well to emulate. How often do we see a scientist make such an admission?

Architecture refers to concrete shapes and objects; geometry refers to conceptual shapes and objects, idealized figures if you will. A 1-d line is impossible whereas a 2-d plane is only a conceptual object. A 2-d plane does not exist on its own but only as a geometric figure indicating the surface of a 3-d concrete object.

A solid produced by translating or rotating or whatever a 2-d plane is not a solid object, it is a movie of a conceptual 2-d object moving from one location to the next. Mathematicians claim a sphere is hollow and has an infinitely thin exterior. Again this is a conceptual object. Anything infinitely thin is non-existent.

Modern mathematicians continue to make the same error.

Any good theory of everything is just different views of the same thing. There is more than one theory of everything, just check your feet.

This is good advice under any circumstances.

The reason there is more than one good theory of everything is due to the fractal nature of the whole thing. The geometry I mention is architecture. Study tensegrity, a structural engineering principle,

to see that there is no difference. Structure and function cannot be separated, that is architecture and 3-d geometry. All the geometry I speak of is atomic, or molecular, therefore its 3-d.

This is the standard view.

It's all relative. There is no clutter. It's the same picture. Thread theory, String theory, Tensor theory, Tensegrity theory, Electric Universe, APM, Scalar theory, Tempic theory, etc. A rose by any other name. All Theories Of Everything that are the same item from different viewpoints. But its only one view of many, all equal. Each leaves the standard view on the floor or in the garbage.

The standard view has been in trouble for some time and this concatenation of many seemingly different theories merely strengthens the argument against it.

Thread theory seems very similar to my punctual field of vector pressures. Ropes are capable of passing through surfaces like gravitation of a book on a shelf. While the shelf prevents the effect [acceleration due to] of gravity from being observed, gravity is oblivious to the shelf's presence.

Opaque objects are like this, allowing some of the vectors of light to intersect unhindered, yet preventing their full effect from being observed. Radio intersects the walls of a building unhindered, yet visible light effects are not observed. Heat similarly reveals the blockage of light effects behind an opaque surface yet is itself a consequence of the pressure field which signals the light. 2-d characterizes basic vector geometry.

Tug one end of a rope and the opposite end of the rope instantaneously responds. The tug need not travel at any speed, let alone the c-rate. A rope is of course made of material, so this makes immediate common sense, but a vector connecting two points across a distance of space is a little harder to visualize.

It is surprising how many Physicists have not tugged one end of a rope and observed the opposite end respond instantaneously. The experiment is not difficult to perform but the ramifications are great.

Like voltage, an energy flux at the ground point or the dropping of an electron to a lower energy state, changes the field potential throughout the entire field instantly. This field change need not take any amount of time, other than the split second time required for the centroidal flux to occur.

This could easily be shown to be the case. Modern atomic clocks are more accurate than a split second.

The direction of this pressure tug is more difficult to ascertain in some cases because all that is observed is a change at the peripheral point of detection.

It remains the case, sadly, that pressure tugs in general are difficult to work with.

With gravity of course the centrally directed [centropic] nature of the field is plain. With light effects the presumptions comprising our view of light's nature determine our conclusions about what happens.

If light is emitted as a wave or particle at a constant and limiting speed then all of the mental contortions of Einsteinian relativity must be considered along with its obvious contradictions and baseless hypotheses of dark matter and energy, black holes and the like as well as all of the consequences of Young's disprovable interference/diffraction paradigm.

The contradictions of Einstein's relativity are becoming more apparent as time goes by. Young's interference/diffraction paradigm is likewise widely thought to be on its last legs.

If as I claim light is the detection of a field compression through the vectoral connection between the peripheral point and the light source/sink then having that vector directed centropically [toward the source/sink] is a sensible conclusion.

The vectoral connection stands to be confirmed in the near future. When it is, light can, at last, be properly defined as the detection of a field compression.

You are right about a good theory of everything. However, you've already stated that APM does not include gravity. Thread theory does. Therefore thread theory is the true theory of everything and we will speak in the language of thread theory when discussing on this board. Your theory is wholly unfamiliar to me and I would familiarize myself if it seemed a productive activity.

Defending competing theories in this way is the essence of the scientific method.

I believe Bill Gaede correctly explained mass in his book. This is how we conceptualize mass: In everyday life, if you swing a ball around you feel a centrifugal force pulling on your arm. However, imagine a Universe with only two hydrogen atoms and they are

rotating around each other. If the length of the rope between them does not change there will be no pull.

It is difficult to imagine how there can be.

If one atom does not tow the other toward itself then the system is simply in equilibrium.

This appears to be a clinching argument.

There is no reason for the resistance known as inertia unless these two atoms are, in fact, pulled by every other atom in the Universe. If every atom in the Universe is interconnected by electromagnetic ropes and possesses a detectable property known as mass the source of this parameter should be the aggregate pull of matter outside the H_2 system.

There is a quantity of matter sitting there, two H atoms, but if they remain stationary they feel no pull, no inertia, and have no detectable mass. As soon as they move they necessarily feel the tug of every thread connecting every atom bearing down on them.

It is slowly permeating throughout the Physics community, the idea that if atoms remain stationary they feel no pull. However, when they move it is a different story.

When you lift a bowling ball you feel resistance because it is attached to every atom in Earth by a rope. You are dragging the ball against this composite tug we call gravity. If the ball were not attached to anything it would simply float freely in space. Without strings attached the ball can affect nothing else in the Universe.

Again, a common sense view which has the advantage of being intuitive. This could be usefully contrasted with the non-intuitive theories of modern Physics which fail the common sense test. It should be a priority of science to indentify the ropes and demonstrate that by cutting them a bowling ball will simply float freely in space. A fraction of the money currently spent exploring the outer reaches of the solar-system should be enough to fund the experiment.

Electricity is best described as a drill bit turning in place.

I have been unable to discover a better description of electricity.

When magnetic threads sweep around an atom they collide with adjacent loose threads inducing them to spin in the same direction, which in turn induce the loose threads on the adjacent atom and so forth. Electricity is stimulated by light, a torquing of the electromagnetic rope. The Unified theory of light via the dual strand entwined rope is shown here.

Physicists have been searching for the unified theory of light for a long time and not once have they paused to consider the dual strand entwined rope.

Electric and Magnetic fields are a mathematical description of a physical phenomenon. When the electron shell of an atom expands and contracts it naturally torques the electromagnetic ropes to which it is attached. This produces a torsion along the rope that, when it arrives at the next atom, causes it to expand and contract.

Modern electron microscopes should be capable of displaying the expansion and contraction of the atom's electron shell showing clearly the torsion along the rope. This experiment is still waiting to be performed.

If there are relatively loose threads in the electron shell then when it expands the loose threads can become close enough to adjacent loose threads to collide with them. This induces them to sweep/spin in the same direction. This process propagates down the wire or whatever.

The sweeping threads are rotating perpendicular to the direction of propagation. If the wire is brought next to another wire the sweeping threads can collide. Depending on the orientation of the wires they will either smack into each other, repelling, or hook around each other, pulling.

The wire, or whatever, will be easily visible. Sweeping threads smacking or hooking may be only a matter of magnification.

However, you've already stated that APM does not include gravity.

Totally wrong, I never said that. APM is three forces, one is gravity.

This is Science in the raw again. Competing theories argued from firmly held theoretical positions.

Your definition of mass does not work for me. In fact each one of your descriptions is not a definition. Mass equals a dimension consisting of a perfect circle of two dimensional string. Mass never changes.

The mass of the atomic distributed charges is not equal to energy. Energy is a sum of five dimensions. Electricity equals atomic charge, E-field equals light threads, B-field equals aether threads.

Here we see three new equalities postulated, each of which is worthy of a separate field of physics in itself.

Perhaps you were referring to another theory, I apologize I don't remember exactly where it was said, it doesn't matter much.

In the intellectual rush of theory development this kind of thing is common.

Three forces? There is push and pull, how can there be any more than two forces? Either two objects collide and knock each other away or one object pulls on the other via a rope that connects the two. If there is a third force you will have to make a picture or movie of it because I cannot conceive of a third force. A very simple diagram will do. Just show me objects influencing each other through a mechanism besides push or pull.

It is still the case today that a third force is proposed yet it is also still the case that no convincing diagram or movie of it force has ever been demonstrated. Therefore the third force remains conjecture.

The definition of mass is perfect. An atom is connected to every other atom in the Universe via ropes. If this atom moves it necessarily feels a tug by every other atom in the Universe. This tug is a resistance to the atom's movement which is detected as mass weight. Did you even watch the video?

Referring to the video demonstrating pull and push.

Mass is a circular two dimensional string? A two dimensional object is an abstract, conceptual object. A two dimensional object cannot be said to exist because it does not have location. Something that is less than three dimensional is called nothing. Unless you can, in fact, point to a two dimensional object. If you cannot then your definition of mass is entirely nonphysical and has nothing to do with Physics.

This analysis of mass has never been seriously disputed.

Even at the tiniest of scales every concrete object, those objects that exist, must be fully three dimensional. It makes no sense to talk about a two dimensional concrete object. A two dimensional concrete object is a logical impossibility. If you turn it on its side it vanishes. There is nothing there.

What you saw originally was not an object but a geometric figure. A geometric figure symbolizing a two dimensional object is simply a three-dimensional object seen from a head-on perspective. A two dimensional object does not exist on its own independently. It is part of a three dimensional object.

Your theory of mass is a non-starter; it does not pass the first stage of the scientific method. If mass is defined in terms of anything

but a three dimensional object or combination thereof it has nothing to do with Physics.

You say that electricity is atomic charge. In order for someone else to understand this definition you must, at a minimum, show us charge. You must either point to the concrete object charge, show us a model of charge or show a movie of one or more concrete objects interacting and name the interaction charge.

In order to be Physical in the strong sense the topic under discussion must be shown. Charge is a word bandied about yet no-one has ever shown charge. Part of the problem with modern science is that it almost wilfully refuses to provide concrete objects.

Why is that so difficult? An urgent priority of science should be to build three dimensional models of the things they discuss. Other examples which spring to mind are quarks, neutrinos and photons. Models of these should be made and put on display. Einstein claimed a photon was a needle several miles long. Is this really the case?

When I say what is electricity and you say it is charge you have not given a definition. You have given a synonym. If you knew very little English and were to ask me what is a plant and I responded it is flora you would be annoyed unless you already knew the meaning of flora.

I don't know the word electricity or charge so the easiest way to resolve this is if you make a picture or movie. At the very least you must describe a picture or movie of three dimensional objects interacting and name it electricity or charge.

This is a small favour to ask. People will be forgiven for thinking scientists are being obstructive and seeking to make their subject mysterious for the sake of it. Making models and movies would demystify science overnight.

What is a light thread? What is an aether thread? Remember, answer these questions in terms of concrete objects, those objects that exist, else they have nothing to do with existence. If you wish to claim a two dimensional object exists just send me a model of one.

A challenge I do not expect to be answered soon!

What is a thread? Since you do not understand light threads and aether threads then what is your thread made of? So you believe in threads and ropes but not string? How do the ropes make changes in weight via gravity yet mass stays the same? Since you equate mass with connected threads well it's very quickly becoming a mind experiment. Is your thread mass? If not then what is mass?

Again you are giving a description not a definition. I am not asking what makes the effects, i.e. ropes in your theory, I am asking what it is. If I cut the ropes do you lose mass? Can you lose mass? Does mass change with velocity or gravity?

These are penetrating questions.

My mind experiment that the mass is a circular string is just as effective as your ropes. 2-d is perfectly accepted if mass is a fundamental dimension, you're thinking of it as a sum and limiting its definition by doing that, the trap of all modern attempts to understand mass properly.

It also provides a strict definition in four sentences. Please do the same with the ropes. You still only give explanations, not definitions. I can do both in four sentences:

It is a fundamental dimension, a building block of matter. A perfect 2-d circular string is mass, it has no length. It is always the same mass per atomic charge which never changes or varies. Nothing ever happens to mass. It does not change, via gravity or velocity.

These are actually five sentences but such is the originality of thought we will let that pass.

There is only one force, a push-pull that is showing itself in three ways, electrostatic charge, electromagnetic charge and gravity. Just like white light through a prism, the push-pull splits three ways.

Charge is explained by Coulomb's Constant. It is 4pi. All charge is distributed and here we can agree. You cannot conceive of a 2-d mass string, yet I cannot conceive of a 1-d charge. Yet Physics does it all the time.

An explanation of charge has been a long time coming. Finally it seems that the problem is open to solution. As usual in physics, simplicity provides the answer. Charge is 4pi is as simple as it gets.

The relationship to primary angular momentum and charge is inseparable from the conductance of the aether and the circular string that scans it. Structure and function cannot be separated at any level. The question what is the thread made of is an incorrect question.

And again, with apologies to the layman, an incorrect question is one which cannot be posed in physics. Therefore it is not allowed and discussion of it must cease. Only if a question is technically correct may discussion continue.

The thread is the fundamental constituent. It is conceptually made of a single continuous piece. There is no smaller constituent pursuant to the definition of fundamental constituent. The definition of fundamental constituent is simply that which has no smaller part. The thread is the thread is the thread by definition and cannot be cut, by definition.

A 'by definition' argument is extremely powerful. Such an argument has never been known to be proved incorrect. So when a theory leads to a statement such as 'the thread is the thread is the thread by definition,' it is a sign that the underlying theory is watertight.

You are free to hypothesize your own fundamental constituent that is not made of something. If we were disallowed from hypothesizing such a fundamental constituent there would be no consistent and valid theory of Physics.

Which is a restatement of the above.

How does the rope make changes in weight when location changes? You did not watch the video. It is explained right there. I have pointed you to the answer. Please do not waste time and board space asking questions I have already answered. If you watch the video and still have questions I will try to explain it differently but please tell me what in the video didn't make sense first.

It is often frustrating when newcomers to an established field fail to assimilate the evidence first.

The definition of mass is resistance to motion. Motion is defined as two or more locations of an object. An object with mass resists motion. An object without mass does not resist motion. Mass is not an object i.e. does not have physical presence. You cannot draw mass and point to it. Mass is inherently conceptual because it requires two or more objects. We cannot conceive of mass in a Universe with a single atom. Again, a concept is a relationship between two or more objects. A concept can only be described.

This is a very clear conception of mass. But to fully emphasise the point using imagery familiar to us all:

I cannot point to love or justice. I can show you a movie of two people having sex or a judge in a courtroom. These are descriptions of love and justice. As long as my concept involves only concrete objects then it has a bearing on Physics. Mass, energy and time are all concepts. They are illustrated by description in terms of concrete objects which are in turn defined as that which has shape and location.

We can use the physical concepts we have described to explain observations, such as gravitation. Let me break it down: The Universe is composed of a dual stranded rope that interconnects and comprises all atoms. The rope is taut. The rope is a 3-d concrete object and is the fundamental constituent. Therefore it is made of a single continuous piece.

This theory needs to be tested by science as soon as possible. It is highly likely that the Universe could be composed of a dual stranded rope. String theorists may consider entering the debate at this point as rope theory impinges on their field. The main point of difference could be that rope theorists identify their fundamental constituent as being taut whereas string theorists seem to be confused as to the exact state of their strings.

Humans observe that one body attracts another. When a ball is thrown away from a body it necessarily returns to that body. We explain this phenomenon as taut ropes connecting the ball to the body.

This is an empirical and observational point in favour of rope theory.

Your two dimensional string is completely invalid. A two dimensional object is an abstract object. It does not have location. There is no two dimensional object in the entire Universe. Again, if you were to try to show me a two dimensional object I would ask you to rotate it.

When you do it would either a: disappear or b: have width, in which case it is in fact three dimensional. An object that disappears spontaneously is not only illogical but would have not had the ability to interact with other objects.

Which seems reasonable.

Again this is by definition. I have defined touch as resulting from the law that two surfaces may not cross. A two dimensional abstract object does not possess a surface i.e. it does not have an exterior. Therefore it may cross anything in the Universe and can have zero effect on anything.

Are you proposing something which cannot physically affect anything else? Why would I care about such an inconsequential thing even if it does exist? If you would like to propose your own law of touch for the purposes of your theory you are of course entitled to do so but under the definitions laid out your theory is entirely nonphysical.

The search for the Law of Touch is a major topic in physics and much research effort is being expended on it. Tentative conclusions are being drawn but none seem as promising as the two surfaces conjecture as outlined above.

Geometry has no bearing on Physics at all. Physics is the study of shapes. Three dimensional shapes, concrete objects, are the primary study of Physics. They are not made of one dimensional lines or two dimensional planes, indeed nothing physical can be composed of such abstract things.

It has taken a while but finally Physics is coming around to the view that three dimensional shapes are its primary field of study.

The minimum prerequisite for a physical theory is that every object in the theory must be shown. If the actual object itself cannot be presented then a model will do. This is the minimum. If you cannot show me what you're talking about it exists only in your head and I can only believe your theory based on faith. I am not a man of faith, faeries, or energy.

This is the realist interpretation of Physics and it is gaining ground. .

Therefore when you propose that mass is a two-dimensional circular thread I have a number of problems with this. You must be clear if you are talking about two physical dimensions, length and width, or two mathematical dimensions meaning any point on the circular thread may be specified by two numbers.

If you claim it has two physical dimensions this is a logical impossibility, that which does not have height looks like nothing. You may argue that concrete objects don't have to be three dimensional and do not have to possess a surface. That's fine, simply show me your two dimensional object or a model of one and your theory is a success, you win, I concede.

If you propose it only has two mathematical dimensions then I walk out of the room because your theory has nothing to do with Physics, the objects of your theory do not have physical dimensions. It's that simple.

It is always the case that simplicity must rule in Physics. It may seem drastic but in the end it is reasonable to walk out of the room if the objects of a theory do not have physical dimensions. There is little point in staying.

In particular I would like to see a picture or movie of charge and a model of a 2-d string.

Physicists who routinely use these concepts might like to take the opportunity to rise to this challenge. The fact that they haven't makes one afraid they don't entirely understand what they claim to be talking about.

Starting all over again I think thread theory is a very good explanation of light. However, I question Gaede's ideas on gravity and magnetism. What I think thread theory explains so well: The Michelson-Morley experiment without the use of aether or space-time. His argument goes that space should always be regarded as a dimensionless medium.

This seems obvious enough.

What I wonder about: The diameter of a thread. How is the signal sent? How does a rope change its frequency? His version suggests that the speed of light is infinite.

Which could be the case. The jury is out.

What about atoms attracting each other? Like gold atoms? I can clearly see links between them on pictures and why would hydrogen atoms form H_2 and not gold?

This is a deep mystery.

I don't understand his version of quantum jump and can it explain why an electron needs a special wavelength to jump/inflate?

Quantum Jump theory is in its infancy and the inflation of electrons is one of the primary areas of research in this field.

The number of ropes per atom due to being directly connected to each other would be 10^80... How can that all fit in one electron shell?

A major field of research in particle physics concentrates on just this 'Packing Problem' as it is known. Results are being held up by science concentrating on the search for dark matter.

What I have problems with: If a neutron is a convergence of ropes, how can it have weight? And why does it always have the same weight?

Neutron Theory is at the boundary of modern particle physics. The neutron is a poorly understood particle at present. The solution to the invariant neutron weight conjecture may have to be postponed until the Large Hadron Collider receives its next scheduled upgrade in the year 2018.

He does not explain why the force (tension) of gravity is so much weaker than magnetism. In fact, it seems like in his versions of

magnetism and gravity, gravity would come out as a much stronger and important force/tension.

In some scenarios this is exactly the case but as always experiment will decide the issue. The LHC is going to be busy in the next few years if physicists manage to get their priorities right and use it to investigate these fundamental issues.

He also does not explain Mercury's elliptical orbit. Also, if Peratt's galaxy formation simulation can be regarded as evidence that one does not need Newton's gravity, Einstein's relativity or Ricci/Gaede's tensor but is purely a result of electromagnetism perhaps gravity is diamagnetism? Just a wild guess.

But physics is full of wild guesses that became true. There is no shame in guessing wildly.

One does not need a special explanation of gravity. As mentioned by Altonhare and others, in his version of magnetism E-M ropes suddenly become obstructive.

It would be fascinating to know why.

A side question: He claims that since gravity is a tension it is instantaneous. He shows a rubber band and says wouldn't that mean that sending messages faster than light is theoretically possible?

Rubber Band Theory promises to deliver faster than light messaging as a by-product of the tension hypothesis of gravity. One of science's next steps should be to confirm the gravity/tension premise and thus open the way to circumventing what looks increasingly like a mistaken speed limit for the Universe.

Another side question: I'm still confused about the guys tugging a rope over long distances, like from one galaxy to another. I've come up with the following thought experiment. Take Alice and Bob. Alice is in space at the edge of the Milky Way galaxy closest to Andromeda. Bob is at the edge of the Andromeda galaxy closest the Milky Way. Between them is a humongous ring.

My question is what will happen when Alice tries to spin the ring clockwise while Bob tries to spins the ring anticlockwise simultaneously?

This is a question of importance both in Physics and Astronomy. It is still undecided.

The diameter of the thread is theoretically estimable if the number of atoms in the Universe were known. This number could be used in combination with the Rutherford radius 0.8×10^{-15} to calculate the diameter.

Physics is unfortunately nowhere near a reliable estimate of the number of atoms in the Universe. Opinions vary widely. The Rutherford radius was a breakthrough in the last century when it was first calculated. The diameter of the thread promises to provide a similar breakthrough and may fairly be described as the discovery of this century, when it occurs.

If you read the book carefully you should be able to answer the second question. The electron shell expands, pulling in some rope. If the length of the rope before the expansion was l and it had n links its frequency was n/l. If the amount of rope pulled in is m then the new frequency is: n-m/l. His mechanism does seem to suggest that the signal should propagate instantly since the rope is a continuous entity. If one part of a continuous entity moves the rest of the entity must also move simultaneously, by definition.

However Bill Gaede claims that a continuous entity does not have to be perfectly rigid. If it can deform then it has a characteristic flexibility which would limit the speed of light. The alternative explanation, and the one that I prefer, is that the rope is made of smaller chinks. It may be somewhat like a bicycle chain or simply a series of pieces with male/female ends that fit together. This way there isn't some arbitrary flexibility granted to what should be a perfectly rigid entity. A chain-like structure has an inherent radius of curvature even though its constituents are perfectly rigid. To form a spherical shell this structure would be forced to wrap around in an integral numbers of chinks.

Chink Theory is on the cutting edge of physics and more investigation is needed, though the pointer towards male/female ends that fit together is a promising development exciting research labs throughout the world.

The diffraction patterns observed for a single electron are an important phenomenon for thread theory. The first question we have to ask is what, exactly, is the beam of electrons? When a filament is heated does it excite the atoms to spinning their ropes faster and faster perhaps c/w, increasing their angular momentum so they spin further and further?

It had previously been assumed that the atoms were spinning their ropes ac/w (anti-clockwise) so suggesting they spin them the opposite way is a dramatic proposal and could lead to new insights.

If there is an extractor biased at some voltage then its atoms will be spinning their ropes also c/w so that they pull on the extended

ropes in the filament. This could pull the loose ropes even looser and looser down the various lensing and deflecting elements until it arrives at the slit.

Now, the question is, does this long, loose rope with all its inertia actually move through the slits? The likely explanation is that it does not. More likely it collides with threads comprising the material and sets them spinning also. Where there are slits the electron rope will only induce atoms of air to spinning faster and possibly knock them around some but not much else because the air atoms possess their own relatively high velocity/inertia.

The atoms in the slit material are relatively stationary and thus all their loose threads are stimulated to spinning largely in a similar direction so the material becomes magnetized or, alternately, has a charge build-up.

This is a far-reaching suggestion yet is not as unlikely as it sounds. A material becoming magnetised or becoming charged has not yet been confirmed experimentally but there is nothing in quantum mechanics which expressly forbids this. Air atoms being knocked around some should be measurable and science should make the detection of this a priority.

I'm sorry this was a very difficult phenomenon to explain in words, I hope it makes sense. If not I'll think about it carefully and try again. I think the purpose of this thread is to pose hypotheses for others to evaluate, letting many minds work on the same problem. So, onward.

As usual we see the humility of the true scientist here. This is to be applauded. It is a shame that mainstream scientists cannot follow the example instead of appearing arrogant when in the public eye. That sort of attitude does no-one any favours.

The problem is that modern physicists do not actually understand what light is. If they do not know what light actually is how can they count it?

Yet physicists continue to insist they can.

How can they be sure that the internal processes of the material itself do not depend on its relative velocity rather than some kind of vague and undefined flow of time? If the internal processes of caesium are not perfectly regular in all instances than it is simply a bad clock.

Physicists are loath to consider that they may be using a bad clock presumably because so many of their published results rest upon it.

This cannot be ruled out since physicists understand neither the light they observe nor the behaviour of the fundamental constituents of matter.

Though physicists will not admit it in public, in private they have admitted they don't understand light and are struggling with the behaviour of the fundamental constituents of matter as well.

In fact this is far away the most likely explanation since time travel and time dilation lead to paradox, duality and self-contradiction.

Einstein never once considered that he was using a bad clock. Physicists should make it an urgent priority to go over his calculations but this time include the idea of a bad clock. If relativity survives such an examination all well and good but if contradictions arise then the theory is open to doubt.

Finally, I alluded to the reference standard. If the two caesium clocks are being compared against each other they are both in the experimental group of the experiment. What are they themselves being compared to? What's in the control group? If we keep one clock here, clock A, and send one into orbit, clock B, we must start counting radiation from each one at the same time and stop counting at the same time.

What clock are they using to ensure that clock A emits radiation for precisely the same duration as clock B? If the caesium clocks themselves are the reference standard, the control, then the experiment is moot because the only way to ensure both A and B emit for the same interval is to stop counting at precisely the same number of light quanta. This is by definition.

So, again, I ask: What is their external reference standard? What exactly are they actually doing?

This question has never been satisfactorily answered. It is to physicists shame that they continually avoid the question and refuse to provide a straight answer. Perhaps the question could be put once and for all at a future funding committee meeting and if a straight answer is not received then funding should be cut immediately subject to a straight answer being received. It really is not good enough for physicists to pussyfoot around not providing straight answers to straight questions.

I don't understand what Harvard Tower experiment is all about. Can anyone explain me that and what it has to do with thread theory?

The fundamental flaw in the interpretation of this experiment is that they attempted to measure an ill/vaguely defined term, energy, using the poorly understood Planck Constant and a poorly understood effect, the Mossbauer Effect.

Energy is indeed ill-defined. The Planck constant is notoriously badly understood and the Mossbauer Effect shouldn't really be a part of physics, so poorly understood is it. An immediate aim for should be for physics to withdraw discussion of these items until they are understood more.

Relativity actually predicts a shift in frequency with proximity to objects with mass, whereas the researchers measured an energy difference via the MB effect. While they may have the equations to get the numbers right they have no physical interpretation.

Why Physicists just use equations to get the numbers right while having no physical interpretation of their results will be beyond the comprehension of most people. And if they measure an energy difference when their theory predicts a frequency shift surely common sense would indicate that something is wrong.

Perhaps one of sciences next steps should be to ensure that physicists pause and think about the disparity between measurement and theory before they leap into print.

The all-consuming pursuit of a Nobel Prize may be to blame. If so perhaps the prize should be made less valuable. Then there wouldn't be such a rat-race to try to win one.

What we have here is an observation, what we don't have is why? Explaining this experiment in terms of thread theory is beyond me right now.

Which is not surprising. This theorist is working alone and is not in receipt of the millions of dollars which seem easily available to more pushy people.

I think there are even more fundamental things at the thread/atomic level to work out for thread theory such as the nature of the electron serpentine. Is the thread composed of individual chinks with some relationship to the Planck Constant/length and the nature of fusion. This comes before highly complex phenomena involving massive quantities of matter we don't understand on the most basic levels yet.

The nature of the electron serpentine isn't even high on the list of topics of research in particle physics laboratories according to a recent survey. This should be rectified.

What about atoms attracting each other? Like gold atoms? The answer is simply that electron yarn-balls become enmeshed in each other.

On the other hand, Yarn-Balls theory is today as firmly established as Yang-Mills theory.

The hypothesis that the thread itself is composed of individual chinks adds a potential new layer to the quantum jump phenomenon. The number of ropes per atom due to being directly connected to each other would be 10^80... How can that all fit in one electron shell?

There's no lower limit on the diameter of the thread, so I don't really see the problem. If we assume that the Rutherford radius is a result of the convergence of every thread into the smallest possible spherical region then the volume of the thread in the nucleus is roughly 4/3*pi*.8x10^-15^3/10^80~2.145x10^-125 and its diameter is 2.145x10^-125/pi*1.6x10^-15 ~ 4.27x10^-111 meters. I don't really know exactly how the threads arrange themselves in the nucleus but this provides a lower limit on the thread's diameter. In any event, it's incredibly thin.

No mathematical physicist disputes these calculations so they are good evidence of the theoretical upper bound for thread dimension.

He only shows hydrogen atoms, what about atoms with multiple electron shells?

The fact is thread theory is a brand new theory and there are zero experiments actually attempting to investigate the nature of the elements in terms of thread theory. There is no reason to think, however, that thread theory should have any problem with the other elements.

Science should hang its head in shame that there are zero experiments attempting to investigate the nature of the elements in terms of thread theory. There is a good reason for this and the reason is that science itself feels threatened by the emergence of new theories.

A paradigm shift in science is resisted by the old guard. A true revolution can only occur when enough have disappeared. In the case of thread theory let us hope that hidebound theorists make themselves scarce as soon as possible. Thread theory won't be the first theory to uncover a whole pack of dogs who refuse to learn new tricks.

One reason the neutron appears to have a consistent inertial mass is that the distribution of threads in the Universe is more or less uniform. Another reason is that there is more going on than just a convergence of ropes. One could imagine the ropes coming together, converging, and then some winding around each other.

Thus a neutron would be a tangling of a convergence of ropes. A tangled convergence is similar to a proton except the proton is only thread whereas the convergence is rope. A hydrogen atom's mass is a contribution from both threads of the rope. This explains why the neutron's mass is similar to the hydrogen atoms. Such a tangled convergence would be pulled in all directions just as a H-atom would.

This seems plausible.

One does not need a special explanation of gravity. Magnetism is not perfectly understood under thread theory and not at all otherwise.

The mystery of magnetism is one which continues to confound physicists to this day.

However, a good reason for the strength of magnetism relative to gravity is that it involves collision between threads with a high angular momentum.

No physicist I have spoken to has even considered this.

A good reason for the weakness of gravity is that it is a pulling competition between every atom in the Universe. If we imagine a Universe with only the Sun and Earth there is no competition, there is tension between only two bodies. The more bodies we introduce the more competition and the overall tension between any two bodies is dampened. Note I am not saying that the Earth is gravitationally pulled away from the Sun by rest of the stars in our galaxy and that causes us to observe smaller weights in our vicinity.

An analogy is the case of two top predators that meet each other. There is great tension between them; they run together, one kills the other, done. Alternatively if there is a large group of top predators the total tension increases logarithmically so the tension per predator goes down. No individual predator wants to get in a fight because that will bring the other preds over to take advantage of his vulnerable state.

This proof by analogy provides good evidence in favour of thread theory.

Note that the total ropes in the Universe increases faster as we add each atom. 10 atom Universe has 45 ropes, 11 atom Universe has 55, 12 atom Universe has 66 etc. Quantitatively we can say that the total tension in the Universe is independent of the number of atoms so the tension per rope decreases as the number of atoms increases. It may not be immediately apparent why this is so but it is intuitive.

Intuition is not always given the credit it deserves in science. Too often physicists find themselves stuck on a problem and when calculations fail they just give up. Here, however, we see a quantitative analysis backed up with intuition. This can lead to new truths without the theorist becoming bogged down in detail.

Imagine two bodies connected by ropes under tension. Now introduce a single atom at a third location and attach it via taut ropes to every atom constituting the two bodies. This single atom now pulls on all those other atoms cancelling out some of their pull on each other by a balanced amount.

If this were not true it would be equivalent to saying the new atom had no influence on the rest of the Universe. Gravity is observed to be weak because its effect between any two bodies is inversely proportional to the total number of atoms in the Universe.

In an even simpler example imagine two atoms connected by a taut rope. They pull hard on each other but they also make an angle of 0 degrees with each other, call their common axis the x axis. Now introduce a third atom equidistant but not directly between the first two it makes an angle of 45 degrees with the x axis and so is directly on the y axis. Attach it with taut ropes.

It will now pull on the original two atoms equally. The component of the third atom's pull in the x direction will cancel out some of the original atoms' pull in the x direction. The total tension remains constant and the tension/rope weakens. As mentioned by Altonhare and others, in his version of magnetism, e-m ropes suddenly become obstructive.

Given this sophisticated analysis the result follows inexorably.

What is at the very centre of the galaxy? Absolutely nothing. There is not a speck of matter at the core. That's what I predict. There is however the vacuum spark, an event that is not fully understood, emits no visible light, and does not jump from point to point.

Astronomers continue to insist there is a black hole at the centre of our galaxy but they have been unable to provide completely convincing evidence. The vacuum spark is therefore a valid candidate. Like a black hole, it emits no visible light and does not jump from point to point. It is therefore in line with observations.

This dark sun exists at the centre of all electron flows that have a gap in them due to tension and is the source of energy at the centre of most spherical celestial objects including the Earth.

This provides further evidence for vacuum spark theory. Electricians will confirm that electron flows which have a gap in them could be described as having a dark sun at their centre. This could be seen on a much smaller scale if electricians brought microscopes with them when they came around to rewire a house, for instance.

The centre of the galaxy has a really powerful vacuum spark. From the vacuum spark are built up all the electromagnetic structures that lead to the formation of our material Universe.

It is likely to be so powerful that detection should be relatively straightforward and one can only wonder why the observation hasn't taken place sooner. Perhaps astronomers are too preoccupied looking for comets so they can name them after themselves.

Yes, a vacuum that works with the same Physics as a sink draining. When you compare this to any vortex in a fluid you will see that there is nothing at their cores. The eye of the hurricane is calm and void of any material. What is so hard to believe about an empty core at the centre of the galaxy?

Proofs, logically, divide into proofs by deduction and proofs by induction. The proof by induction is well represented here. Logically, if one thing looks like another thing then they can be related to each other formally, by induction. In this case a spinning galaxy looks like a sink draining. Therefore there is nothing at the core of a galaxy.

Space is fluid. Everything is in a fluid state; the only thing that changes is the viscosity. The vortex is at the heart of anything that requires energy and what is at the heart of the vortex? Nothing. I rest my case.

It is a succinct yet compelling argument. The case may well be rested at this point.

There are some Electric Universe theorists who seem to forget gravity and that is unfortunate for them and the mainstream scientists they are trying to convince. You see, the mainstream scientific community sees all the same electromagnetic structures in

space that the Electric Universe theorist does. **The mainstream scientist just can't accept that charge separation can occur on such size scales and therefore must rationalize away the observations.**

The history of science is unfortunately full of such rationalizations. It is a shame that modern science continues to fall into the same trap.

Without gravity the Electric Universe theorist is unable to explain why there should be charge separation on cosmic scales. As we have seen in posts on the global electrostatic charge of stars, on any scale where we can have gravitational structure, we not only can, but must have charge separation on the same scale.

Astronomers and Physicists must have their eyes closed not to notice this self-evident fact.

In short, what an Electric Universe theorist should be telling the mainstream scientist is this: If gravity can act on cosmic scales and protons are more massive than electrons, then Electric fields and currents must exist on cosmic scales too. Simple as that.

Protons are known to be more massive than electrons and gravity obviously acts on cosmic scales. These observations give strength to the argument.

In the Electric Universe all stars are variable stars whether they have companions or not. At least in the sense that their output is dependent on an external power source, galactic Birkeland currents, and in the course of its travel through the galaxy the star would encounter differing electrical environments.

Astronomers are slowly coming round to the realisation that a stars output is dependent on an external power source. Previously they have assumed, without first-hand evidence, that stars are their own energy source.

Most stars systems are multiples, single stars such as the Sun are less common. One of the reasons for this is that stars often fission under electrical stress. The red giant Betelgeuse may turn out to have a companion, or not.

The Fission Theory of stars has gained much ground in recent years. Our own sun may soon fission into two stars under electrical stress, or not.

Its large size is somewhat illusory in that it has an enlarged corona. Most of what is perceived as a giant star is actually a huge but tenuous plasma in glow mode.

Occasionally one will find an Astronomer who disputes this but they are becoming fewer. Giant stars being a plasma in glow mode is entering the scientific mainstream.

The European Space Agency's (ESA) X-ray observatory has imaged a spectacular set of rings which appear to expand with a speed a thousand times faster than that of light. I thought nothing can travel faster than the speed of light. So says modern Physics. Interestingly this article just sluffs it off like it's nothing new.

But since this is the case then we are dead wrong about the constant speed of light so relativity just got thrown down the crapper with one smooth observation. In my opinion it is the same function of the universal magnetism in which we live. Cheers.

All scientific theories eventually run their course and there isn't one which survives from the dawn of physics up to the present day. Relativity seemed like a good idea a hundred years ago but times move on. It is a shame to throw it down the crapper but physicists will get over it in the usual way that scientists get over their pet theory being proved to be wrong.

A clue is given at the end of this illuminating debate. Universal magnetism promises much in the way of new theories. The overall message is that science should stop being afraid of change and instead embrace it.

This should certainly be one of science's next steps and we can all raise a glass to that. Cheers!

Chapter 21: Products

Demand switches: Protection from electrical radiation. When an electrical circuit is not in use demand switches cut off the electricity at the fuse box providing valuable protection from electrical radiation while you sleep and the body repairs itself. Electricity has become indispensable for everyday life. The use of electricity inevitably causes alternating electrical and magnetic fields.

The effect of these electrical and magnetic fields on the human organism has been the subject-matter of many International Studies with all in all alarming results. Best to be on the safe side with a precautionary reduction of at least the home-made exposure to alternating electrical and magnetic fields.

This is sound advice. Since we cannot shut down our electricity supply lines or the National Grid it is best to be on the safe side.

A reliable demand switch optimised according to building can substantially contribute to a reduction of your daily exposure without you having to forfeit any comfort. It is for this reason that experts recommend the decoupling from the mains, often called isolation, as the first and most important technical step to be implemented with respect to alternating electrical and magnetic fields.

Science should really if non-alternating electrical and magnetic fields also have the same potential detrimental effect. Until this is done it seems best to decouple yourself from all sources of alternating electric and magnetic fields whenever possible.

Our demand switch is installed in modern DIN distribution boards or next to older fuse boxes and it is wired inline to the circuit to be protected. When there is no load on the circuit, that is when everything is switched off, the demand switch automatically disconnects the line at the fuse box. When any switch or appliance in the circuit is turned on, it instantly restores power. Price: £165.78

Inline wiring is state of the art. However, installation next to older fuse boxes is built in to the specification for use in older fuse boxes and this means that everyone can be protected without having to go through the expensive rewiring of a domestic supply which could run into thousands of pounds. This makes the demand switch price extremely competitive. More research would drive this price down and such work should be undertaken as a matter of urgency.

Organic Andrographic Paniculata: Andrographis is a small annual plant that is easy to grow and its use in no way threatens any natural wild populations. Known as 'the king of bitters' it is a truly disgusting tasting herb that we have packed into a taste-free vegetarian capsule. Herbal remedies have been used safely and effectively in all traditional societies for thousands of years.

This herb is organic and traceable back to the field from where it was grown. Certified organic by the soil association. Dosage: Take 2 capsules 3 times a day. Use short term for infections. Don't use if you are pregnant. The body may utilise this herb to boost the immune system, support the liver, prevent blood clots, support irregular cell behaviour and prevent or reduce cold and flu symptoms. Price: £14.95

This seems reasonable. Especially when one considers the cost of making the 'king of bitters' taste-free. Science should move quickly to determine exactly how the Andrographis supports irregular cell behaviour as this could have wide-ranging ramifications in the future.

Genuine real active allicin: All the benefits of garlic but without the smell. This is a genuine allicin product and guarantees active and real allicin unlike many other allicin products available today. The active part of garlic but without the smell. Genuine allicin products are also capable of destroying a wide range of bacterial and fungal infections.

Published work shows excellent activity against staphylococcus aureus, candida albicans, streptococcus species, Escherichia coli, salmonella species and helicobacter pylori. Methicillin resistant staphylococcus aureus bacteria MRSA treated with allicin powder, liquid and crème, shows a large zone of dead bacteria usually left untouched by pharmaceutical treatments. No zone of inhibition. Price: £24.99.

Again, this is very reasonable and again, it is good to have the smell removed. Science should concentrate more of its efforts on removing the smell from other products. One thinks of poo. It would be a major step forward if science could remove the smell of poo and a good starting point might be licencing the smell removal technology demonstrated above.

Hospital doctors should be in the forefront of using active allicin in their battle against MRSA. This antibiotic resistant bacteria is gaining a greater foothold in hospitals and science shows no signs of coming to the rescue with a remedy.

The Cisca Saltpipe is a simple and natural way to make a real difference to your health. It uses the benefit of ancient salt-cave therapies to maintain an optimum respitory function and to help you to breathe easier, alleviate sneezing, coughing, shortness of breath and leaves you feeling revitalized. Salt crystals have been used for decades in the treatment of respiratory disorders and to help strengthen the immune system by cleansing the lungs allowing them to absorb more oxygen. Price: £29.99

This seems very reasonable for a product which cleanses the lungs allowing them to absorb more oxygen. There is also the bonus that it uses ancient salt-cave therapies and there can be few still unaware of the wisdom of ancient people in these matters.

Go-Nitrix with l-arginine for better cardiovascular health. Can help to keep arteries younger, prevent heart attacks and strokes. Helps to increase circulation providing more nutrients and oxygen to the cells and can therefore help with overall energy and stamina levels. Price: £39.95

Science should pay this inexpensive price and research its effects in keeping arteries younger. This is a major problem in today's Western Industrialised society and medical bills are soaring as a result.

Resveratrol, found in the skins of red grapes, has long been associated with maintaining and supporting cardiovascular health. It has more recently been found to have many other associated health benefits as well, such as anti-aging, and with illnesses like Alzheimer's and even cancer. Price: £29.95

There is an increasingly large amount of money being wasted in the attempt to find a cure for cancer. Yet the benefits of Resveratrol have never been fully explored by science presumably because it is easily found and it does not require expensive chemistry to prepare a dose. Modern pharmaceutical companies don't like it for those reasons and have therefore dissuaded scientists from working on it in favour of developing complicated drugs which they can then licence on exorbitant terms.

Between 85/95% of people are thought to be deficient in omega-3 fatty acids. Poor diets and food lacking in good nutrients make us susceptible to free radical scavengers and poor antioxidant qualities which in turn leads to increases in our degenerative state of health. Unfortunately, eating more fish, the traditional source of omega 3,6 and 9, is not necessarily the answer. Fish found in most of our oceans today are contaminated with industrial pollutants and toxins like mercury, PCB's, heavy metals and radioactive poisons.

Few biologists will deny that the fish found in most of our oceans today are contaminated. So eating more fish is obviously not a good idea. But the same biologists will also agree it is vital that we do not leave ourselves susceptible to free radical scavengers.

Cardiu krill addresses this shortfall. Cardiu krill is farmed from the unpolluted southern oceans. Krill is a small crustation, the staple diet of whales, and combines a unique potent antioxidant with essential omega 3, 6 & 9 oils in a 100% natural source. A good krill oil is 300 times more potent than vitamins A & E and 48 times greater than omega 3 found in standard fish oils.

Cardiu krill represents a renewable and sustainable nutrition source. In fact a good quality krill oil can have 47 times the antioxidant power of fish oil and 34 times the antioxidant power of coenzyme q-10. Price: £29.95

This is very reasonable for a unique potent antioxidant.

Natural combination of ionic sea minerals, macro minerals, trace minerals, vitamins, Sea Veg, algae, powerful antioxidants, Bioflavanoids, Quercetin, enzymes, co-q6,7,8,9,10, amino acids, Pau Darco, aloe vera, astaxanthin and much more. Price: £39.49

This is even more so, especially when one considers the inclusion of co-q6,7,8,9 and 10 as well as Pau Darco.

Frequensea with marine phytoplankton: The most complete wholefood created naturally by Mother Nature's oceans. All the nutrients the body needs for cellular repair and healthy cellular regeneration, increased oxygen levels and much more. Price: £35.00

This seems highly economical. Particularly impressive is the fact that all the nutrients the body needs for cellular repair are included.

SerraPlus capsules: Serrapeptase enteric coated capsules giving 40,000iu activity, MSM 350mg, 73 trace minerals 50mg. Contains 60 tablets x 400mg. Price: £19.95

This seems a reasonable price to pay especially when one considers the 40,000iu activity. Science has rarely been able to go beyond an activity level of 20,000iu under controlled conditions so SerraPlus capsules represent a real advance.

900gms Shalkur apricot kernels. Price: £32.20
This does not seem such a reasonable price to pay and it could be that science needs to step in here and initiate research with the aim of producing Apricot kernels at a lower price.

Liquid ionic minerals in glass bottle: 100ml of ionic minerals from the Great Inland Sea. Price: £14.99
This can be little argument with this price, especially when one considers the source of the ionic minerals is the Great Inland Sea itself.

The easiest way to better health by Dr. Lamar Diltz N.D., M.T. Have you ever wondered what the essence of life is? The answer is oxygen. Your body is composed of 59% oxygen. The average person consumes 6 to 8 lbs. of oxygen, 4 lbs. of food and 2 lbs. of water per day. Twice as much oxygen is needed as food or water. Oxygen is your most prevalent nutrient yet it is the most overlooked and neglected.
Dr. Lamar appears to have the nail on the head here.
Oxygen is a natural antibiotic which dissolves plaque, stones and cholesterol in the arteries, joints, capillaries and works as an antibiotic against anaerobic bacteria, spurs cell immunity, helps against septicaemia, carbon monoxide poisoning, peritonitis and intestinal blockage. Many of our problems are in direct relationship to a lack of oxygen which allows anaerobic bacteria in candida, cancers and diabetes. Ever wonder why you are so tired after a big meal?
Science has researched tiredness after a big meal but has come to no definite conclusions.
Let's take a look at some of your oxygen robbers. Sugar, shallow breathing, the destruction of our green forests, lack of exercise and stress. In some cities the oxygen level is as low as 7 to 10 percent, unoxygenated tap water with contaminants and cooking depletes oxygen.
Both smoking and overeating will cause severe oxygen deficiencies. Not only is more oxygen required by the oxygen

reduction system to complete the digestive process but because of the excess of toxic metabolic waste products the body has to deal with.

The warning signs have been there for many years now.

Research shows that oxygen deficiency can be the single greatest cause of disease. According to the National Institute of Health's latest research 80% of us spend 90% of our time in a controlled environment breathing the same air over and over. Oxygen deficiency is behind a lot of disorders and health problems.

Immune system, energy production, detoxification and overall health depend on a high level of oxygen in your body. You can starve your body of oxygen. Just like you take vitamins to supplement what you are not getting in your food you need to take liquid stabilized oxygen to supply your blood with extra oxygen.

This advice seems scientifically sound. If the body is not receiving as much oxygen as it should then you need to take liquid stabilized oxygen.

Remember, you need twice as much oxygen as you need food. I tell my friends in the health food industry that everyone who enters their stores needs oxygen. Liquid stabilized oxygen improves the integrity of all inter-cellular fluids. Everybody wins with liquid stabilized oxygen; the elderly, those with chronic fatigue and weakened immune systems, active children, teenagers, athletes, parents and busy professionals. Because medical science and treatments are so complicated we think the answer has to be. This is not true. Life begins with oxygen and ends with the last breath.

Drug companies are on the point of introducing complicated medicines aimed at improving the integrity of all inter-cellular fluids. They will not take kindly to uncomplicated competition.

As I have lectured throughout the world the most asked question has been: If there was only one thing I could do for my health what would you recommend Doc? My answer has always been the same. Before there was food, there was oxygen. Without oxygen nothing lives. I consider liquid stabilized oxygen one of the biggest breakthroughs in medical history.

That oxygen came before food is undeniable. This is good evidence for liquid stabilized oxygen.

One of the most valuable benefits to be derived from adding liquid stabilized oxygen regularly to your health regime is the fact that it destroys the putrefying bacteria in a person's digestive tract. A large percent of the food we eat literally rots before it is completely

digested and absorbed. This is evidenced by the fact that most bathrooms have an air freshener of some kind.

It is an insight not granted to many to discover one of the biggest breakthroughs in medical history. The evidence has been staring us in the face, or rather the nose, for a long time, in that most bathrooms do indeed have an air freshener of some kind. But science has been slow to wonder why.

Liquid stabilized oxygen is an aqueous solution of non-toxic sodium chlorite NaClO$_2$ and pure deionised water. This unique formula stabilizes thousands of oxygen molecules in a solution with electrons galore. With these negatively charged oxidizers viruses, bacteria, and protozoa are kept from causing disease in the body. It is the only non-toxic, virtually tasteless product I know of that will add extra oxygen to your bloodstream.

Chemists should hang their heads in metaphorical shame at not discovering the fact that thousands of oxygen molecules can be successfully stabilized in a solution with electrons galore. There have been reports that chemists have tried to stabilize oxygen molecules by using a few electrons but somehow they never made the final, crucial step.

This highlights a major problem with modern scientific research. If an experiment does not yield immediate results, scientists seem comfortable with writing up a null-result paper and presenting it to journals for publication. They seem to lack the drive and curiosity to take things further. One wonders how many advances have been delayed by the lackadaisical approach of scientists when it comes to completing an investigation.

Because of the importance of oxygen there are many products on the market today that claim to be oxygen supplements. However, most of them are designed to stimulate oxygen in the body, they do not add extra oxygen to the body. A lot of them are no more than salt water. Sodium chloride is table salt. Most of these products, when tested in the lab, do not have enough oxygen to measure.

The stabilized oxygen in Dr. Lamar's liquid stabilized oxygen has been used and proven to increase the oxygen level of the blood since 1971 and is the one that others are trying to copy. Why settle for less than the best? For a few cents a day liquid stabilized oxygen is the best insurance I know for a healthy life full of energy and vitality.

As always, copyists abound. But Dr. Lamar has evidential proof going back decades and such proof cannot be ignored.

Stabilized oxygen: 132ml of Dr. Lamar's stabilized oxygen, over 25 years of scientific study behind this product and the one others try to emulate. Price: £32.99

This seems a remarkable price to pay.

Science should move immediately to see if there are any other substances which can be successfully stabilized and therefore be of use medically. It is unlikely that they will discover one as important as oxygen but until science tries it will never know.

Taking its lead and inspiration for Dr. Lamar and his team of researchers should be one of sciences next steps and science should take that next step in the very near future.

Chapter 22: Radionics

Radionics: Rooting out problems: First cut a strand of hair from your head. Next, fill in a questionnaire about your state of health and send it, with your hair, to an address on the other side of the country. Then sit back and, while not exactly by return of post, you will in due course receive relief from whatever ailment is troubling you.

It could come in the form of a pill or a potion but it's just as likely to come in the form of healing vibrations transmitted from the person to whom you've sent your hair. What is it? Magic? Witchcraft? A load of twaddle? No, it's radionics, the largely unexplained art of healing someone you've never met who is hundreds, even thousands of miles away.

This is definitely something which science should investigate.

There are only 80 or so practitioners of radionics in Britain and Rebecka Blenntoft is one of them. She's also the secretary of the UK radionic association and, like her colleagues, she gets to the root of her patients problems by holding a pendulum over their hair sample or witness, as it's called, and seeing what happens.

It certainly seems to follow the scientific method of inquiry. Seeing what happens is the basis of all experimental science.

We get the information by interrogating the witness, she claims. I will ask question after question, some looking for a yes or no answer, some looking for an answer that will quantify the health or otherwise of the patients various physiological systems aural, visual, skeletal. So, as well as rotating in a clockwise direction for yes, and anticlockwise for no, the pendulum also gives scores out of 100 when placed over a sort of healthometer chart.

The variation of question type is also typical of science. Sometimes scientists are interested in a yes/no response but at other times quantification of response is more appropriate. Also, note, we have a clear methodology at work. A definite and unambiguous rotation direction is indicative of the binary (yes/no) response while a further graded numerical scale is applied in the case of the sort of healthometer chart.

It's quite a time-consuming process because you have to go through every part of the body, says Blenntoft. It's also quite tiring,

because you have to stay very tuned in and focused on the person you are treating.

It is not unexpected the process is time-consuming, so is an MRI scan and for much the same reasons. That practitioners become tired is hardly surprising. General Practitioners in a busy surgery also report fatigue and again it is for exactly the same reason. A GP also has to stay tuned in to his or her patient and remain focussed at all times.

Once she's identified the problem area she enters an eight-digit numerical code into a black-box-like radionics machine (they prefer the word instrument), either via a digital keyboard or a set of dials. Followed by the relevant treatment instruction 'restore', 'rejuvenate', 'elasticise', for example. Almost simultaneously, it is claimed, the patient will experience some form of improvement in their condition.

The parallels with modern scientific practice continue. Specialists in hospitals often enter numerical data into their diagnostic devices and, to the layman, these machines could well appear to be 'black boxes' in that their inner workings are known only to experts. Doctors do not claim to be engineers and Radionics consultants are no exception.

You don't believe it? Neither did Blenntoft until she saw the effect a radionic diagnosis had on a dog in her local village. The treatment can be used not just on humans but on animals and even crops and soil. This dog was in a terrible state, itching and scratching its skin red raw she recalls. A radionics practitioner discovered it was allergic to everything that came out of cows. And within a few days the dog was fine and running around.

This is good scientific evidence. The fact that the dog was fine within a few days and running around is proof that the radionics practitioner was correct.

For the Radionic Associations Chairman, Geoffrey Bourne, the proof came in two-footed form. A local farmer had a very bad recurring kidney infection but had become allergic to penicillin, he recalls. There was nothing the doctor could do but the radionics practitioner traced the problem back to a tetanus injection the farmer had been given at the age of 10. It took a year of treatment but that farmer, who was in his seventies, went on to live till the age of 96.

This is impressive. In cases where there is nothing the doctor can do the significance of the radionics treatment approaches 100%. A year of treatment is hardly excessive when one considers the extended periods required for modern medicine to even begin to alleviate long-standing symptoms, but the definitive proof comes in the increase in longevity due to the radionics treatment.

So how exactly does it work again? Best guess is that we all plug into some kind of universal energy grid and radionics constitutes a kind of battery recharging rescue service. From afar.

Given the facts, science should now determine the exact nature of the universal energy grid. It will not surprise many that science has not identified the grid so far, after all this is the same science which has failed to identify the nature of dark matter and dark energy. It seems that science is falling behind in determining the nature of the Universe and this is one of the reasons why many observers are losing confidence in science as practiced by today's scientists.

With the exception of Professor Brian Cox it is difficult to think of a scientist who is making the fundamental discoveries which were so common in the past. It could well be that science today does not attract the most intelligent people, as was formerly the case, and so standards have dropped over the years. It certainly does seem that science is running out of new discoveries.

If asked to name the last big discovery a few people might mention the Higgs Boson but after that the list of achievements becomes remarkably short. Charitably, it could be that scientists are simply involved in the wrong areas of research. Do we really need to know how spherical an electron is? Proving beyond doubt the properties of the universal energy grid into which we are all plugged would be a good start.

Those physicists who complain that this would take them away from their own pet projects might like to remind themselves that plenty of new physics will come out of investigating the universal energy grid and Nobel Prizes may be at stake. It is a shame that Physicists have to be motivated in this way but if that is what it takes…

You can get a list of practitioners by contacting the UK Radionic Association. An analysis costs from £45.

This seems reasonable.

Five facts about radionics: Radionics was invented by an American doctor, Albert Abrams 1863-1924. Hair is the commonest witness; photos, blood spots and signatures can apparently work too. The witness provides up-to-date information on your health months later. The eight-number codes are called rates and are listed in a

large directory. The radionics instruments are not plugged in to electric power.

The fact that radionics was invented by a doctor automatically makes it worthy of investigation by scientists, a number of whom are doctors themselves. It seems obvious that hair and blood spots can be a witness but it is also intuitive that photos and signatures can work too. This makes the radionics process available to photographers and graphologists to name but two.

Sometimes a missing persons inquiry will lack hair and/or blood spots but will have available a photo. The radionics protocol will determine the health of the subject months later and indicate whether the missing person is still alive. If the former, a search of hospital records could yield results. If the latter, then family and friends can be reassured that although a loved one is absent at least he or she is in good health.

There is also good news for physicists. For a long time they have been bound by something called the Principle of Conservation of Energy and it has been something of a roadblock in the development of the subject. Now that radionics instruments have been shown to work without being plugged into electric power, physics can finally abandon the outmoded principle.

All our instruments are hand made. Simply the best quality. Evolution-MXP: US$1,750.00.

It is a homoeopathic remedy simulator which can make simple and complex homoeopathic remedies from rates provided with the instrument using lactose pills or alcohol/water liquid. It can produce radionic frequencies which can be broadcast through the instrument to any patient anywhere in the world using a blood or hair sample as witness. Can diagnose, analyze and prescribe remedies. Comes supplied with thousands of frequencies for organs, tissues, muscles, bones, remedies, homeopathic remedies and psychological and mental conditions.

It goes without saying that hand made instruments are superior quality. And $1,750.00 is a remarkable price to pay for such an instrument.

Science should investigate the expertise behind a device producing radionic frequencies which can be broadcast to any patient anywhere in the world using only a blood or hair sample as a witness. Physics has been completely unable to replicate this result.

Rumours of individual scientists combining their results to give some frequencies for organs, tissues, muscles and bones have surfaced but none of them have claimed thousands of frequencies and certainly none have worked out the frequencies for psychological and mental conditions. This shows just how far science is behind the burgeoning field of radionics.

The 15 regulators can be tuned to one or more remedies simultaneously to target specific areas and conditions of the body. There are thousands of homoeopathic remedies and you can create your own complex remedies. There are 12 levels of operation ranging from the physical through mental, emotional, all the way up to cosmic energy levels. Any of these levels may be tested for best function by means of dials and set to balance energy as required.

Features of the instrument: The instrument is capable of making radionic and homoeopathic remedies in any potency required from 1x to 999,999mm. It can copy any remedy by use of the energy rate supplied with the instrument. The copy and imprint buttons will produce remedies in a matter of seconds, a real boon for a busy practitioner, from either the original remedy or from its frequency in the manual supplied. It offers diagnostic, analytic and radionics broadcast facilities. Simple and complex remedies can be quickly made, and, if necessary, stored either on the ECS card or in the memory of the instrument.

The copy and imprint buttons ability to produce remedies in a matter of seconds will impress research teams throughout the world who have so far only managed to implement button protocols which take a day or two to achieve the same outcome.

The manual contains thousands of frequencies including human anatomy, tumours, parasites, fungi, bacteria, toxins, allergies, nosodes, vitamins, energy, chakras, aura, minerals, colour therapies, Bach flowers and California flowers.

Even the best equipped research labs have been unable to deal with California flowers though results have been reported sporadically for nosodes.

It is possible to carry out biophysical tests with the instrument measuring the variation between the current state of the patient and the healthy state. It can be used with equal success on humans, animals and plants. The energy level of the patient can be measured on the instrument before there is any physical manifestation enabling energy treatments to be given before the condition becomes apparent and before it can take hold of the patients system. The instrument is light; it comes with a strong and excellent quality case.

Science has long dreamed of such a device, one which can measure the variation between the current state of the patient and the healthy state. MRI scanners were supposed to do this but have failed and CAT scans also promised much but their worth has been overestimated. In the event, neither device turned out to be able to allow energy treatments to be given before a condition becomes apparent and that has been the big disappointment as far as science is concerned.

Amplifying radionics antennas: Our radionics antennas have the purpose of amplifying the energetic information generated by our equipment. This means that a radionic rate or a group of rates can be transmitted directly from our equipment to the antennas and the antennas amplify the energy of the rate or rates that creates an oscillation power field between both antennas.

Amplifiers have been known to science for a long time and have been successfully used in a variety of devices. But science has long been on the trail of an amplifier which can successfully create an oscillation power field between two antennas. Though considered theoretically possible no working device has yet been demonstrated so this is a step forward. Science should investigate the proof of concept demonstrated here and, if patents are involved, invite the leaders of the research teams involved to share their knowledge whilst maintaining prior art for the discoverers in order to protect their intellectual property rights.

This field of energy which arises between the antennas can penetrate all type of living tissue and reactivates the cell structure allowing the fastest assimilation of the information generated by the selected rates.

Biologists will also be interested in the cell structure reactivation properties of the antenna energy field and so should accord the inventors the same privileges.

It is possible to treat all the body as well as specific parts of it. Also, with the help of the broadcaster, the antennas create a strong treatment at a distance. The antennas have 4 timers for treatment 15, 25, 35, 45 minutes and these go off automatically past the time. Also it has the option of continuous operation. Specification measurement: 26 x 10 x 26 cm. US$7,500.00.

This is a very reasonable price to pay for strong treatment at a distance.

Our computerized system has a friendly environment and it is easy to work with, therefore any therapist with basic knowledge of

computers is able to use this program without troubles. The patient information data base is well done because it is based on the Mexican official norm 168 of the clinical file; all these data are stored in the program and are available for view.

As anyone with a background in the field will recognise, Mexican official norm 168 is the industry standard.

Scanning the sample of the patient the system obtains automatically the personal general frequency. With this program you have the possibility of broadcasting; it means that you are able to broadcast any frequency or therapy to any patient to any place of the world. And not only to one patient. You can broadcast all the patients you need in different schedules.

The personal general frequency had been thought so difficult to obtain that the patient's physical presence was automatically assumed to be mandatory. However, this now appears not to be the case.

Satellite companies and other worldwide broadcasters may also be interested in investigating the means of achieving global coverage of any frequency to any place in the world.

A problem could arise if this new technology threatens to displace or depose existing means of disseminating broadcast information. The very satellite companies who may be interested could well attempt to buy the technology and refuse to let it be licenced as it would disturb their carefully built cartel and lead to diminished profits. As technologically aware readers will know, this would not be the first time an advance which could benefit all has been delayed by the shareholders in a technology which seems ripe to be superseded.

The frequencies database is distributed in categories and this is not all, also you can create your own categories and subcategories without limit. You can add more new frequencies to any categories subcategory. Also a search engine is available to find out any rate quickly. You are able to store infinity of remedies and formulae, also you are able to store them with the name and order that you want.

The important word here is 'infinity'. For the first time infinity has moved beyond the purely theoretical into the firmly practical. For all of recorded history, and this is not an exaggeration, human inventions have been merely finite.

Owners of a home computer will be familiar with the limited storage capacity of their machine and will no doubt have upgraded it as technology advances and finance allows. Yet their storage capacity is still finite. Now, for the first time, we have an infinite storage device.

That an infinity of information can be stored could mean that several of the 'laws' of Physics may need to be looked at again.

Testimonies: First and foremost let me commend you on your excellent pre and post purchase service. I will admit that I held a few reservations as to authenticity and safety when purchasing this machine via the internet but all of which were quickly dissolved by your service and assistance.

Plenty of prospective purchasers will no doubt have heard scare stories concerning items ordered from the internet and this testimony should certainly put their minds at rest.

As for the machine I have yet to find a more comprehensive machine at the price which I purchased it for. The machine itself is beyond compare to any standard radionics machine and still impresses me with its vast range of frequencies available. Its uses are amazing and the results impeccable.

I've been researching its efficacy before consistent use in my practise and one of the experiments was an MRI and IVP comparison of a patient with an analysis using the MXD. Amazingly enough, when the results were compared, they were in complete agreement with each other. Thus the benefits for the patient using this machine is tantamount in terms of costs as healthcare costs and special investigation costs in South Africa are shockingly expensive. Thank you again for your helpful advice and dependable service. Kind regards Dr. Vernisha Moodley, South Africa.

Dr. Moodley is not the first medical practitioner to be impressed by the vast range of frequencies available which are, as he states, beyond compare. That the MXD shows complete agreement with expensive Magnetic Resonance Imaging and IntraVenous Pyelogram scans is good scientific evidence and should provide a suitable starting point for validation of the MXD by the scientific community.

I have been using one of Homoeonics radionics machines for a year and I can recommend his business without any reservation. The machine is fantastic and in addition to potentising my homeopathic remedies, in any potency I want, thus saving hundreds of dollars in buying my remedies from a laboratory, it can diagnose and treat any ailment by broadcasting to anywhere in the world. This is of exceptional value, as many of my clients live at a great distance.

Diagnosis and treatment of any ailment anywhere in the world is an obvious step forward in medical intervention and science should take on board the evidence presented here and initiate a research programme aimed at extending the range of the radionics machine so that it can also be used off-world, for example in the International Space Station.

Much effort and expense is required at present to provide a fully equipped clinical environment aboard the ISS but if diagnosis and treatment could be provided remotely via the MXD then this would constitute a significant diminution of such resources which could then be reallocated to scientific experiments. It would also give astronauts peace of mind knowing that any treatment required will be transmitted directly from Earth so there would be no need to await the arrival of a supply vessel in the case of an unforeseen medical procedure becoming necessary. Given the limited volume capacity aboard the ISS, not every eventuality can be catered for.

The broadcasting facility has saved me from having to send a bottle of remedy through the post with the danger that it will not be treated correctly in transit, the possibility that the postal service will lose the parcel, and from the added expense of buying the remedy, the bottle, the packaging material and cost of postage. Mary Edwards DIHom British Institute of Homeopathy, Australia.

To which one might add how much more so this applies to a remedy needing to be sent to the ISS. With the added danger that the rocket used might blow up.

I wanted to present a few words about the owner of this company. I have just completed my second sale with Homoeonic, purchasing two of his higher end models of radionic computers.

They represent sheer excellence of hand production and manufacturing. I am obviously very pleased with his product, both with quality of components and pristine assembly. He is a business man with huge integrity and is interested in his customers total satisfaction. I must admit I was sceptical about ordering over the internet and from a foreign country. I could not have been certain of what obstacles I might encounter particularly on a first time ordering. He put me totally at ease, not just once but twice. My MXD was delivered with great expedience, impeccable of workmanship. Homoeonic company keep up the great work providing a great product that is well needed. Gordon H. Brown, Alamo, California. USA.

This speaks for itself.

Voice Programmed Remedy Maker: Frequently we have been asked if it might be possible to make some sort of Remedy Maker that might be small enough to fit in the pocket and yet have all the

functions necessary to originate Homeopathic Potencies. After searching in vain for suitable components that might be small enough to make such an item we had pretty much given up until we hit on the idea of making it 'Voice Programmed'.

The advent of the Voice Programmed Remedy Maker constitutes a major advance in vocal control and one which science should investigate as a matter of priority.

The Voice Programmed Remedy Maker that we have designed gets over this difficulty. Measuring only 130mm x 65mm x 25mm and weighing 106 grams, the device is fitted with a special kind of microphone which converts speech into subtle vibrations. These are then stored in a temporary memory inside the device and then amplified and fed to a tiny 'Well' that is fitted in the instrument in which the bottle of blank unpotentised tablets are placed.

Science has been seeking a special microphone which can convert speech into subtle vibrations but so far without success. Now that this device has been invented the way is open to numerous applications.

In use the device is held close to the mouth. A button on the side of the Remedy Maker is pressed in and you speak the name of a Remedy you want immediately followed by the Potency (if any). The button is then released and a 'beep' sound is heard confirming that a remedy has been recorded and stored. Then the device is placed on a table or flat surface and a small bottle of tablets or even just one or two tablets can be placed in the small 30.3mm diameter Stainless Steel Well that is fitted in the device. A switch next to the Well is operated and held down for about 3 seconds and then released and the device again beeps to confirm that your remedy has been made.

As with all breakthroughs, once made it seems obvious. Nevertheless, an enormous amount of research effort is doubtless hidden behind the simplicity of its operation and the research team is to be commended not only for their demanding work but also for making the interface so user friendly. Science could learn a lot from this approach.

How many times has an invention been presented to the public without the bugs fully worked out? One thinks of the various incarnations of 'Windows' for example. And how often has a new device been offered for sale with an almost incomprehensible user manual? Older readers will remember trying to program an early Video Cassette Recorder (VCR) to tape a TV programme a day or so in advance. Younger readers will no doubt have their own horror stories to tell.

If required more bottles of tablets can be placed in the well and potentised simply by again operating this switch, or if required a lead can be plugged into an output socket on the side of the device to feed remedies to a larger receptacle (such as a metal baking dish) so that you can potentise a large amount of a substance all in one go. Then the temporary memory can be cleared by operating another small switch on the side of the device so that another remedy can be inputted.

Clearly this device has not been released without much thought going into how it will be realistically used.

The Remedy also incorporates an 'Automatic Lock' facility. Each time a remedy is made the vibrations are immediately automatically locked into the carrier material so that they cannot fade even if the tablets are then touched by the hand, dropped on the floor, or stored next to strong smelling substances. This technique of locking the vibrations into the carrier is a result of work pioneered by White Mountain and to our knowledge no other company offers devices that incorporate this facility.

And to my knowledge the 'Automatic Lock' is also new to science. Locking vibrations into a carrier material has been theoretically proposed by a few scientists working in this specialist field but none have been able to put forward a hypothetical model let alone produce an implementation which works on the scale described. Science should invite the White Mountain team to lead a conference aimed at wider dissemination of their work and invite leading material scientists to make up the audience.

Remedies can even be made and given to patients, or oneself, even if no blank tablets are available simply by placing a finger in the metal well on the device, operating the Make switch and transferring the vibrations directly into the body. This makes the device ideal to use as an emergency first aid device, for example it could be used to make Apis Mellifica, which is a powerful Bee Sting Treatment, or one could give a dose of Malaria Officinalis, which is considered by many to be useful in Malaria prevention.

The health benefits of this go without saying. The United Nations under the guise of the World Health Organisation (WHO) could find their tasks worldwide made less complicated by implementing the use of a device which can deliver remedies direct to patients via their finger.

The special beauty of this device is that within seconds you can be making vibrational remedies from literally anything you can

think of, even the illness itself. 'My Throat Problem', for example, or 'The Pain in my Leg' etc.

And there is no reason why this cannot be extended to situations where surgery is contraindicated. 'The Tumour in the Middle of My Brain', for example.

The Voice Programmed Remedy Maker comes in a matt black case and is available for just U.S $395.00 (plus U.S $10.00 Shipping by Insured Priority Mail).

This is a very reasonable price to pay for the advanced functionality available. Science would do well to divert a fraction of the funding aimed at training astronauts to go to Mars into purchasing a number of VPRM's and distributing them amongst the world's advanced research laboratories so that the technology can be studied intensively.

Note: This device cannot make Medical Drugs. It will make any Homeopathic Remedy, in fact any kind of vibrational remedy. No claims are made that the device can cure any specific illness but in the hands of an experienced practitioner the device will prove invaluable. Warning: Medical Drugs can be damaging to the health and cause serious side effects, even death.

Science should make it perfectly clear that medical drugs are not the panacea they were once held to be. What is needed now is a concerted campaign aimed at educating the public, informing them of the dangers and guiding them towards the kind of therapy offered by the Voice Programmed Remedy Maker, for example.

Special Version: Voice Programmed Remedy Maker Mk2: This special version of the Voice Programmed Remedy Maker looks identical to the standard version but it is fitted with a different output switch and modified circuitry that enables remedies to be remotely and continuously transmitted to patients in one position of the switch, or in the other position remedies can be made exactly the same as in the standard version.

Scientists specialising in modified circuits should be redirected from their present research goals so they can study and understand the advance which allows remedies to be remotely transmitted.

The remote transmission is achieved by placing a hair sample from the patient in the little well of the device once the remedy vibrations have been created and then the output switch is set on continuous.

Given that hair has been identified as the active factor in the process the addition of a trichologist to the investigation team would seem advantageous.

The Voice Programmed Remedy Maker Mk2 is available for U.S $425.00 (plus U.S $11.00 Shipping by Insured Priority Mail).

This is a most reasonable price for the additional functionality. Scientists have estimated that even if they could faithfully reproduce the available operational parameters to within their individual tolerances their prototype cost would probably run into the thousands if not millions of dollars. This cost-analysis alone tells us how far science is behind in investigating one of the most important technologies to become widely available in recent years.

Question: How does the device manage to make a remedy from words spoken by the user?

Answer: It must be understood that this is not just some clever new electronic gadget that performs basically conventional functions, this is a one of a kind device employing radically new concepts. Using Psychotronic Technology sound waves are used to modulate free undifferentiated energy within the microphone resonator and these weak subtle vibrations are then fed to a second resonator for amplification before being passed through circuitry to make the remedy.

Physics has been resting on its laurels for a while now. The last radically new concepts it invented were relativity and quantum mechanics, both of which are nearly 100 years old. This is not the record of a progressive science and physics is in danger of becoming a waning science, satisfied with the old triumphs of the past and unwilling to entertain new and exciting ideas.

A look at modern day physicists tends to confirm this. Where are the young, keen scientists with new ideas that lie outside accepted views? Unfortunately they seem few and far between. Physics appears to have entered a stage of 'steady as we go' with no revolutionary theories on the horizon. Practitioners in the field will claim that finding the Higgs Boson represents progress but Peter Higgs first thought of his Boson in 1965 and physicists have only recently got around to discovering it. In the meantime dark matter remains a mystery and dark energy is just a name for something that may not even exist.

Perhaps physicists simply aren't as clever as they used to be, or perhaps they are guilty of not trying hard enough to think up new ideas and are satisfied with investigating around the edges of existing knowledge without having the confidence to strike out on their own and produce something truly original. The development of psychotronic technology shames them.

Question: Can the device recognise different languages or accents and will the device still recognise what I want if I have a cold?

Answer: Words are used to represent a thing or situation. Many different words (even different in language) can be used to represent the same thing. Therefore it makes no difference what language a person speaks in, as far as this device is concerned they will get a remedy from whatever they ask for. Regional accents also make no difference but it is important that the user speaks clearly into the device.

If this problem has been solved it represents a real step forward. Voice recognition technology has been developed by scientists working on AI implementation but a stumbling block has been the parsing of regional accents. A particular instance of this has been trying to get systems to understand what someone from Glasgow is saying, though this is also a challenge for non-AI systems, human beings for example.

Question: Can I use the device if I am in a noisy environment, such as a railway station or amongst a crowd of people?

Answer: It must be understood that this is a Psychotronic device and as such it is capable of performing functions that a solely electronic device could not do. The microphone input is designed to only take information from the voice of the user, e.g. the person who presses the record button, therefore the voice of anyone one else who speaks in a crowd or any other sound will not be inputted.

The invention of the mono-user microphone has also been pursued by science but with limited success. Psychotronic technology is therefore ahead of the cutting edge and represents an advance on purely electronic based devices which must now be seen as an almost redundant technology.

Question: Will the device work if I press the record button down and someone else speaks?

Answer: NO, for the same reason as given above.

Just as quantum mechanics, in its day, was seen as something strange so psychotronic technology appears now. Yet there is every reason to suppose that it will become just as familiar and for exactly the same reason. A hundred years from now people will be wondering what all the fuss was about.

Question: How do I know if the gadget really works?

Answer: Most people who work with vibrational medicine find at some time it helps to be able to dowse, usually with a pendulum, or to use muscle testing to find out if a remedy helps. If you can dowse with a pendulum ask it, 'Does the Voice Potentiser work'? If you cannot please ask someone to dowse the question for you.

People who have dowsed or used muscle testing appear in little doubt when answering the question.

Question: I noticed that the device has stopped beeping, what should I do?

Answer: It is possible that the batteries have developed a little bit of corrosion or in some other way they have lost contact. The solution is to take the back off the case and move the batteries around a little bit, being careful not to damage them or to touch any other components. If the batteries are still good this will solve the problem. If this does not work you may have to put in another set of batteries, the device uses 2 CR2016 batteries. After removing the old one make sure you load the new batteries positive side up.

Manufacturers of other types of devices supplied to the public should take note of the level of detail supplied in these instructions and follow suit in their own literature. For example, in the above it could easily have said 'move the batteries around'. But here we have the added detail, 'move the batteries around a little bit'.

New Combo Remedy Maker: The Combo Remedy Maker was designed for the user who wants to 'Have it all' - that is the combination of a Voice Programmed Remedy Maker Mk2, a Voice Programmed Healer and a Radionic Potentiser for P.C, all in one small hand held device, measuring 5½ inches x 3¼ inches x approx 1½ inches deep.

Science is as yet unable to explain how all of the functionality of the Combo Remedy Maker can be made to fit in these dimensions. Experiments have led to the development of a prototype Combo but its volume was three times that stated above and it could hardly be described as 'hand held'. The situation is as it was with the development of the first mobile phones in the 1980's. It seems that science is at least 30 years behind in the technology of Combo miniaturisation which is quite a blow.

Get a Rose Quartz Dowsing Pendulum FREE when you buy a Combo Remedy Maker (whilst stocks last).

This is a very generous offer.

The built in Potentiser for P.C. also includes an input jack and we supply a special Pickup Sensor which can be connected to this jack which is able to 'pickup' vibrations from pictures and photographs it is placed on, these vibrations then being converted into remedies by the Combo Remedy Maker or transmitted directly to patients for distant healing. This pickup can also be placed on bottles of remedy so that the vibrations can be copied to make more of the same remedy or the potency can be altered as the user wishes. You can change background colour, potency, power, timer, add multiple remedy combinations, auto save, manual save, jump to colour remedy, programme remedy or lock remedy pages!

Science has yet to invent a device which can 'pickup' vibrations from pictures and photographs let alone proceed to then convert these vibrations into remedies. A small investment in the special Pickup Sensor would therefore seem to be wise. This would represent money well spent, especially when one considers the billions wasted by American particle physicists on preliminary work intended to establish the Superconducting Supercollider in Texas, a misadventure which was wisely cancelled by Congress when it became clear that costs were spiralling out of control on a project which had no clear scientific objectives but just seemed to be part of a very expensive wish-list aimed at outstripping CERN's development of the Large Hadron Collider.

The device comes with instructions, Computer Software on CD Rom, Output lead and clip, Transmitting Pickup to place on your computer Monitor or attach to the side of your laptop using a small piece of easily removable Velcro and a Pickup Sensor to provide the Copy Function. Also included as a free gift is a selection of White Mountain's favourite music MP3's, many of which are impossible to obtain anywhere else.

It is one of the faults of mainstream science that it is perceived to lack the human touch. This is a prime example, not only do we have a device which is highly functional and effective in its own right we also have a free gift of the manufacturers favourite music provided at no extra cost. Rarity value is always welcome but a start could be made just by making the gift free and from the heart. If individual scientists were to provide a list of their favourite songs to go along with the devices they supply then they could begin to make a connection with their clients possibly leading to a bond being formed strong enough to encourage repeat business.

If you were to purchase the Voice Programmed Remedy Maker Mk2, the Voice Programmed Healer, and the Potentiser for P.C. as separate devices the price would be U.S $1,244.00, but now you can have the functions of all these devices in one instrument at a cost of only U.S $760.00 - plus Insured U.S Priority Mail $19.25.

This is a very generous price when one considers the innovative technology involved and the wide range of functionality available.

Interested? Please email and let us know you would like to purchase and please supply some address information - at least your town and country, so we can calculate shipping costs. We prefer payment by PayPal but there are other options. Devices are built to order so please allow at least 1 week for this device to be constructed after you have paid for it.

This is another area where science can learn an important lesson. Having a device built to order means that the latest developments can be incorporated in real-time. It is another advance on the traditional scientific method of having a device sitting on the shelf for months before someone actually buys it. The idea is that your device is always state of the art and the customer is therefore receiving the maximum value for his money.

Did you lose your Transmitting Pickup? We have developed a Universal Transmitting Pickup (UTXP) that works with any Combo Remedy Maker that is within a range of 10 feet (304.8cm). This is to save customers having to return their equipment to us to have a new input circuit installed. The price for a UTXP is just $30.00 plus shipping.

They think of everything.

People should keep a close eye on science to see how it deals with radionics. All too often, science, and individual scientists, lose themselves in a fit of pique and refuse to follow research leads which did not come from other scientists to the detriment of us all. Indeed, the history of science can well be read as the history of important ideas and breakthroughs not given prominence because a *scientist* did not make the initial discovery. This freemasonry simply has to stop. It is impeding progress on so many levels.

The development of an open mind for all scientists should be one of science's next steps.

Chapter 23: Reiki

Physicists tell us that life comes from the space inside an atom and even from the space within the space within an atom. Everything comes from within not from without. Life itself is created and sustained from that which resides within.

No physicist has ever gone on record to dispute this.

Physicists will also tell us that matter, the human body for example, is inert. Matter can only 'come alive' or become animated when it is infused with the ether, space/energy.

Matter is indeed inert and so it seems logical that something else is required to make it come alive.

For those of you who have read some of my other articles you may remember the mention of an atom being 99.999% space and as we and everything else are made up of atoms then we consist of 99.999% space, only it doesn't appear that way. Let us take this a step further. An atom is also, therefore, made up of 0.0001% matter; therefore we are made up of 0.0001% matter too. The human body consists of 0.0001% matter and 99.999% space/ ether/light/energy - spirit.

These figures are not doubted by any competent scientist.

We all have this wonderful space inside us, a space consisting of light and truth and a great expansive realm of possibilities to explore and enjoy.

Modern day scientific instruments are capable of confirming this. Photometers have been developed for the former and galvanometers ('lie detectors' in popular parlance) for the latter. Where these register zero that is proof that truth is present.

Everything is energy and everything is made up of a variety of vibrations, either a single vibration for the very basic form of existence or as in the human body a multitude of vibrations.

As noted elsewhere, modern theory due to Einstein proposes that everything is energy. Everything is also vibration, hence modern string theory which postulates that everything is made up of tiny strings in various states of oscillation. Obviously the more complicated the system the greater the number of vibrations which will be present.

Ahnu's are comprised of vortices of energy which spin both ways within themselves. These energies are connected at the outside at the top, three dimensionally and at the centre and the bottom three dimensionally where they form a perfect heart. **Each and every one of our chakras spins both ways and like the Ahnu is comprised of twin vortices of energy. All chakras spin both ways at the same time creating balance.**

The modern day principle of Moments, in which two forces try to spin in opposite directions creating balance, echoes this essential truth.

Science has discovered that a proton and a neutron, for example, contain energies which spin both ways. They have twin vortices of energy. Scientists have also demonstrated that imparting a one way spin into a twin vortex of energy, i.e. energies spinning both ways at the same time, causes an imbalance within these twin-spinning vortices of energy.

Particle physicists spend a proportion of their time doing just this.

We are a part of everything yet we are also everything at the same time. We are merely a holographic image of the whole of existence. When a hologram is broken each piece contains a picture of the whole and just as each cell of our body contains the blueprint of the whole body so too do we contain the blueprint of the whole Universe. So how can we not be attuned, aligned and connected with everything that is when we hold its entire blueprint within us?

This is highly persuasive.

When we focus our attention outside of us we resonate with energies that exist outside of us, mostly the ego consciousness, feeding ourselves everybody else's thoughts and energies and everybody else's rubbish. Many of the symbols used in various healing modalities are also designed to attract lower astral vibrations into our energy fields.

Why would we want to connect with, invoke and channel these man-made lower vibrational energies and the potential problems they bring with them? After all, this is the pea-soup creation of the ego which contains the thought-form, predatory type energies, which serve only to weaken our energy fields.

Astral vibrations are yet to be discovered by science but it is surely a matter of time. Thought-form predatory type energies and their role in weakening our energy fields are other examples of energies which should be investigated by physicists. What magnitude are these energies for example? Can a barrier against them be developed and if so how quickly?

Each of our cells has its own electromagnetic field. This electromagnetic field is the cells' energy field, auric field, or more correctly it is the 'consciousness' of the cell. Consciousness is the precursor for our DNA structure and it is this information, stored within the cells' energy field, that determines whether the cell reproduces itself in a healthy or unhealthy way.

If chaotic and disharmonious information is stored in the electromagnetic field of the cell, the original information held there is altered, and the consciousness of the cell, the energy that feeds and instructs the cell how to behave, becomes imbalanced.

The investigation of the cell's energy field is likely to become a multi-disciplinary effort for science involving as a minimum biologists, physicists and cell energy experts. Such a team should be assembled immediately.

Scientists have recently discovered that the 'memory' of a cell holds information which enables the cell to reproduce in its own likeness. In the case of a tumour, this is in the same diseased way. The new tumour is replicated and reproduced from the memory of the cells of the previous tumour, quite simply because the consciousness which feeds the cell and pre-determines its structure has become altered or unbalanced.

The information, which the tumour cells' consciousness now holds, has become compromised by chaotic information energy which has affected the ordered and harmonious information of its original blueprint. Hence the tumour reproduces itself in an unhealthy way and the tumour continues to exist.

Cell Memory theorists may also take a close interest as their field is likely to be expanded quickly and irrevocably by such a study.

Russian Scientists have discovered that a healthy cell has a uniform, balanced and harmonious energy field around it, whereas an unhealthy cell has an aggressive, irregular and unbalanced energy field. Further, they discovered that because of the aggressive nature of unbalanced cells, their energy fields or consciousness, these cells had the ability to affect balanced and healthy cells, and, over time, influence them into becoming like-minded unbalanced cells through what is known as resonance. Rather like a tuning fork setting off other tuning forks around it.

Not for the first time we see Russian Scientists leading the way.

As biophysicists have discovered, it is only when the defence mechanism electromagnetic field of the body becomes weakened that we become ill. When our energy field is balanced and strong physically we are in good health, but when our energy field is over-exposed to chaotic, disturbed and unbalanced energies we have ill health.

The importance of the work of biophysicists cannot be over-emphasised. Medical doctors have been slow to realise that we only become ill when the electromagnetic field of the body becomes weakened. They have concentrated too much on bacteria and viruses and have ignored our energy field.

Our existence is intimately connected with the quality of water available to us but for many people water is no more than a liquid for quenching thirst, for bathing in and for putting out fires. However, water has some very profound properties. Wolfgang Ludwig, physicist and advisor to the World Research Foundation states: The water molecule is a polar molecule - it has positive and negative charges separated by a dipole length.

In the same way that a magnet has a north and south pole, so the water molecule has a positive and negative pole and is therefore, an electric dipole. Because of its dipole structure, water has the property of storing information. The dipole structure of the water allows information to be stored as in the case of magnetic tapes or magnetic disks.

The World Research Foundation is the most respected research foundation in the world. Otherwise it would not be allowed to name itself so. Other, more minor foundations, must call themselves by lesser names. An example would be the National Research Foundation of South Africa but there are many others. Wolfgang Ludwig is one of the most respected contributors to the World Foundation and it would be fair to say that when he expresses an opinion physicists all over the world are primed to listen.

Water stores information and therefore, has memory. This is exactly how homeopathy works although the traditional stance adopted by the medical profession still claims that homeopathy cannot work in such infinitesimal doses. The science of biochemistry claims that anything diluted beyond d23 retains nothing of the original substance and is, therefore, returned to pure water or saline depending on the original solution.

For example if belladonna were introduced into pure water and diluted to d23 dilution 23, that is $1:10^{23}$, or 1:100000000000000000000000, biochemists would claim that nothing

of the introduced substance belladonna would remain at this dilution. On a molecular level this would be correct but it is only one part of the story.

Biochemists have made important advances in science but they can be blinkered when it comes to new ideas. Biochemical colleges have historically concentrated on the molecule as being of fundamental importance but new paradigms have arisen and those trained decades or even just a few years ago may not be aware that the molecular level is only one part of the story and not the most important part at that.

The science of biophysics, on the other hand, has shown that when carrying out this same experiment they can introduce a substance to pure water and dilute the solution to d200, that is $1:10^{200}$, or 1:100000000000000000000000000000000000000000000000000000000000 0000000000000000000000000000000000000000000000000000000000000 0000000000000000000000000000000000000000000000000000000000000 00000000000000000000000000000 and they can not only measure the electromagnetic field of the substance introduced but also photograph the energy fields of the original substance introduced at this dilution.

This result is highly significant. Not only does it contain a verified way of measuring the field , it also gives a practical method of establishing that the energy field is present. No-one can argue with a photograph.

Both testing methods of biophysicists clearly show that the presence of the original substance introduced i.e. Belladonna is still in the solution at d200 in an energy form. In other words the energy imprint, the consciousness of the original substance introduced, has been recorded and stored in the solution. This is absolute concrete, unequivocal proof that water stores information, that water has memory.

No reputable physicist has come forward to contradict the finding.

Water also has other properties. According to tests made by Engler and Kokoschinegg in 1988 water has a structural memory and a structural variability, and because of this it can store acquired information over a long period of time and pass it over to the body. Thus water can share its memory with our bodies.

This finding has never been challenged. This means that Water-Memory theory has the same status as the theory of Gravity. Until science finds an apple that falls upwards the theory of Gravity is secure.

Likewise, until science finds that water does not store information over a long period of time and pass it over to the body Water-Memory theory is secure. This is how science works.

Furthermore, of what is the Earth mostly comprised? Water. The surface of the Earth is covered by approximately 80% water. The Earth's atmosphere is full of water molecules. Thus it becomes obvious that the Earth is literally one big memory bank. A huge library of stored information.

Geologists occasionally dispute that the Earth is mostly comprised of water. They have not presented completely convincing proof. Atmospheric scientists will confirm the Earth's atmosphere is full of water molecules. Therefore it does indeed look as if the Earth is one big memory bank. Logicians, probably the most pedantic of all scientists, have never been able to point to a logical flaw in this argument.

Does the concept of the consciousness of water explain more about what happens at a cellular level? We know that the immune system sends killer t-cells to fight infection or invasion by intruding aggressive organisms but how does it know where to send them?

This unsolved question has come to be known as the 'Hard Problem' of Biology.

Biophysicists have been able to determine that when a cell association dies it dies as a complete entity, not cell by cell. Physicists can also measure the light emitted by the body after it dies. Clearly there is some form of collective communication going on here.

Physicists have been intrigued by measurements of the light emitted by the body after it dies but have been unable to come up with an adequate explanation.

Thanks to advances in science we now have it confirmed that we are indeed made up of light - we are in effect beings of light, or light beings.

This may come as a surprise to some but it must be remembered that the theory of relativity came as a surprise to many when Einstein first announced it. Einstein won a Nobel Prize for his contributions to physics and it is almost certain the discovery that we are made up of light will receive the same accolade in the future.

The Earth's Assemblage Point runs through its axis, between the north and the south pole. Is it a coincidence that these areas are inhospitable to man? Is this why icebergs are pure water representing pure consciousness and not salt water? Could this be

why we have the occasional 'ice-age'? Could this be the Earth nature attempting to purify her own consciousness water?

Science is usually behind the curve as far as new paradigms are concerned and here we have another example.

In the book The Secret Life of Plants, Peter Tompkins and Christopher Bird chart some very interesting scientific developments relating to the consciousness of plants and the energy interaction and influence between all living things. It is an exciting discovery of how energies interplay with each other and which can directly or indirectly influence other living systems; discoveries which were so profound by their nature, they naturally aroused the interest of the intelligence agencies and the military in the 1950's and 60's in what may well have been the springboard for their respective present-day non-lethal weaponry - a unique way of influencing and controlling populations of people through radio and microwave radiation without causing a fuss or drawing untoward attention to their actions. This weaponry was reportedly first used in the UK against the protestors at the Greenham Common cruise missile base.

Where, it is worth noting, plenty of plants were growing at the time.

Tompkins and Bird start by looking at the work of Clee Backster, one of America's foremost lie detector examiners. Backster soon discovered by playing with his galvanometer and attaching electrodes to the leaves of plants he could witness a response in plants to his own thoughts.

Science has been slow to publicise these findings and it is probable that such publication has been blocked by the same intelligence agencies. Prince Charles, whose conversations with his own plants have been widely reported, has allegedly been advised against commenting further on grounds of National Security and it is no coincidence that he has become quiet on the subject.

Galvanised and excited by his findings Backster started an intensive programme of research into the extraordinary perception of plants. He discovered that plants reacted not only to threats from humans but also from other animals. Backster's investigations continued to establish if a plant had memory. He set up an experiment where six of his students, blindfolded, drew lots from a hat.

On one piece of paper only were the instructions to actually damage the plant. Two plants were used; one was duly damaged by the student acting on the instructions on the piece of paper and the other plant acted as a witness to the event. Backster showed that the surviving plant gave no reaction to five of the six students but whenever the actual culprit approached it, it reacted and caused the meter to go wild. Backster had effectively demonstrated that plants have memory.

Backster went on to extend his field of investigation into mould cultures, scrapings from the human mouth, blood cells and even sperm. The results were just as interesting as those he discovered with plants and sperm cells were quite uncanny in the way they seemed capable of identifying their own donor.

After further experiments his findings were published in 1968 in the International Journal of Parapsychology. Backster's work was providing a foundation of belief that there was some form of cellular consciousness common to all living things.

The International Journal of Parapsychology is, of course, one of the major journals of Parapsychological research and is not noted for publishing articles without foundation.

Other scientists followed up on Backster's work with their own experiments and one, Sauvin, found he could control his model train set or even start his car engine by transferring his thoughts to a plant while the plant was connected to the device in question. Sauvin went on to show that this phenomenon could be repeated by controlling the flight of model aeroplanes. By transmitting thoughts to a plant Sauvin could start, stop, or alter the speed of a model plane in flight.

Rumours persist that certain intelligence agencies depend on daffodils for their most covert operations.

Tompkins and Bird go on to describe an eight week experiment carried out on summer squashes by broadcasting music to them in their respective chambers from two Denver radio stations. One station specialised in heavy rock and the other in classical music. Those squashes exposed to Haydn grew towards the transistor radio, one of them twisting itself lovingly around it.

The other squashes, subjected to rock music, all grew away from the radio and even tried to climb the walls of their glass cage. Similar tests were carried out on other plants and vegetables with the same results. Rock music proved to be detrimental to the wellbeing of living matter, whilst classical music proved to be beneficial.

Of all the scientific experiments carried out over the past decades this will surely cause the least surprise. Rock music is now confirmed to be detrimental to the well-being of living matter, something I am sure everybody has felt all along to be the case. To further the claim for plant behaviour mirroring human behaviour one only has to imagine oneself being subjected to rock music, say Ritchie Blackmore's Rainbow, from a radio. Is it so hard to visualise yourself, like the plant, moving away from the radio and even trying to climb the walls in order to escape it?

Other tests, report Tompkins and Bird, show that mimosa pudica, which naturally fold their leaves as darkness falls, showed that even if kept in a darkened room the leaves would still 'close' as the sun was setting. Which, Jean-Jacques Dertous de Mairan concluded, meant plants must be able to sense the energy of the Sun without being able to see it.

This is highly intriguing.

This knowledge of the interaction of energies between plants, vegetables, humans, sound and vibration led to further experiments whereby the yield of crops, vegetable or fruit could be greatly enhanced when stimulated with energy frequencies of the right nature. Joseph Molitovisz invented an 'electrical flower pot' which could keep flowers much longer than is normally possible. Over time this knowledge led to the use of radio broadcasting for removing pests from agricultural land, a totally pesticide and chemically free form of eradication.

To this day it is notable that vast swathes of land surrounding the repeater aerials for commercial radio are devoid of sentient life.

A company was set up called U.K.A.C.O. inc. to relieve farmers of unwanted pests and to increase their crop yields. This process was so successful that when insecticide salesmen visited the farms under treatment they were informed their products were no longer required. This upset the American insecticide industry to the extent they ridiculed U.K.A.C.O. inc. by calling them fraudsters and charlatans.

In the early 1950's pressure was put on the government to stamp out this new method of pest control as it severely threatened the profits and therefore the position of the insecticide industry as a whole. Sounds like a familiar story. All the resulting pressure and smear campaigns from government and industrial big shots finally forced U.K.A.C.O. inc. to close its doors. Ego, fear and greed had won the day once again.

A depressingly familiar tale.

Also in the 1950's the de la Warrs were achieving remarkably similar results in improving crop yields with their 'black boxes' and

similar episodes were occurring on both sides of the Atlantic throughout the 1960's and 70's.

It is wearingly commonplace how powerful industry moguls can force the abandonment of new ideas when they feel their profits are endangered. One is reminded of the reaction of the major oil companies to the rumoured discovery that cars can run on water with the addition of a pill containing a few household chemicals. We have heard no more of that and the question must be asked over and over again until an answer is forthcoming. Why?

Of course, the approach of the intelligence agencies was quite different to the fear approach of the insecticide industry. Realising that if insects could be affected or even killed by 'radiating' poison at them then it followed that the same technique could be used by the military against concentrations of troops or even whole populations of towns and cities.

The US army took an interest in developing ways of measuring emotional responses in people via plants. Funds were provided for research into this field and even the US Navy showed an interest. The operation centre, at the Naval Ordnance Laboratory in Silver Springs, Maryland, were repeating Backster's experiments. And it was not long before we had the development of non-lethal weapons which today we recognise as microwave and radio frequency mind control machines.

Mind control machines have also been supressed and removed from the public view. But there is little doubt that they exist. One scientist who is not willing to go on public record has joked about the microwave in your kitchen being an object of suspicion in this regard. He later claimed that he was half-joking but refused to say which half was the joke.

All this work and research eventually led to the first 'energy machines' being developed. Georges Lakhovsky developed his multi-wave oscillator to successfully treat arthritis, chronic bronchitis, and other ailments in the 1940's. Lakhovsky had laid the foundation for what was to become known as radiobiology. However effective this multi-wave oscillator proved to be, the medical profession refused to pursue Lakhovsky's findings and the use of the multi-wave oscillator for medical treatment became officially illegal throughout US health authorities. Nothing appears to have changed today.

It is instructive to compare the situation with the prevalence of large gas-guzzling cars and trucks on American roads and large electricity consuming devices present in American kitchens. Smaller, energy efficient models are available but are not replacing this dinosaur equipment. It is not difficult to guess the reason.

Dr Abrams developed what he called his 'oscilloclast'. The oscilloclast or wave breaker could emit specific waves capable of curing human afflictions by apparently altering or cancelling out radiations emitted by various diseases. Abram's 'box' represented a threat to the egos and profits of the medical profession and Abrams was vilified even well after his death in 1924.

The same technology is used to cancel engine noise in the passenger compartments of modern aircraft. The engine noise is recorded, inverted, then rebroadcast into the cabin. Noise cancellation is achieved. That this technology can be used in the aviation industry but not in the medical profession is one of the scandals of our times.

Dr. Ruth Drown made refinements to Abram's machine and although she received a British patent for her design and apparatus, her claims were relegated by the FDA Authorities to the realm of science fiction and her equipment was confiscated in the early 1940's. She, like many before her, was branded a charlatan and a cheat and allegedly died of grief, a broken woman.

It is deplorable that a British patent can be awarded for design and apparatus yet American authorities feel themselves able and competent to reject it. Perhaps this another case of jealousy. In the past, the American government was so disillusioned that its scientists had failed to invent commercial supersonic flight that it banned Concorde from flying over the continental United States. This fit of pique eventually led to the demise of Concorde.

These advances in energy machines eventually led to the development of Radionic and Radiesthesia 'boxes' to treat humans and in the 1960's we witnessed the development of Kirlian photography in Russia - the ability to capture energy fields of plants and humans on photographic plates.

Science has been slow to confirm the importance of Kirlian photography. Perhaps this is a hangover from the problems Soviet science experienced during the Stalinist era. But it is clear progress was never completely stifled.

Fortunately, the fear-mindedness of the ego-consciousness in its desire to stop these beneficial developments helping humanity - and what may otherwise seriously deprive its multi-nationals of profits - has merely delayed the evolvement of these energy machines. The

ego-consciousness may have berated and condemned many people whose only intention was to benefit humanity, and may have willingly and intentionally destroyed them and their inventions in the process, using them instead for their own manipulative purposes.

But thanks to these wonderful people, their determination through good intent, and the dedication of others who have followed up on their work, truth and balanced light consciousness have managed to survive.

The ego-consciousness of individual scientists has always been a problem but science is, or should be, about the search for truth.

These remarkable advances in medical treatments may have been delayed but fortunately for us all we are now seeing in Germany and Russia energy machines that can identify illness and disease before it becomes apparent in physical form.

Unfortunately the US and UK continue to lag behind. A whole generation of patients will not thank the healthcare systems of these countries for failing to deploy energy machines which can identify illness and disease before it starts.

There are four stages to illness and disease. Stage four is the physical manifestation of disease, a condition whereby medicine can determine abnormal changes in the physical and biochemical structure of tissues. In Germany, due to the science of biophysics, machines can now detect illness and disease at stage two, while it is still in an energy form and before it has manifested into a physical problem such as a growth. This naturally makes the treatment far simpler and far more effective and less demanding on the individual.

Unbelievably, many medical practitioners still insist on waiting until stage four prior to beginning diagnosis and treatment. Even intervention at stage three would be a tremendous advance but like the ostrich with its head stuck in the sand, they refuse to listen.

Due to the research into bio-resonance therapy by German and other physicists we can now witness and experience machines which can take energy from the human body and split it into its endogenous naturally occurring form, and its pathogenic disease causing form. These machines can then amplify the endogenous fields, which strengthens the immune system, then 'invert' the pathogenic frequencies to effectively cancel them out. These energies are then fed back into the body.

Why science fails to offer this treatment can only be because it would seriously upset the traditional medical applecart. No more expensive drugs need be designed and manufactured. No more easy money for scientists who take decades to design such drugs which even then don't work for everybody. Science should hang its head in shame. Energy machines should be part of the fight against disease starting immediately.

One can also identify hormonal deficiencies, mineral deficiencies and so much more. Science now has the frequencies for almost all disease causing agents so these too can be identified and treated accordingly. Conversely, and perhaps more worryingly, so do the military and intelligence agencies. Compare this treatment with what is available in America and Britain today - the insecticide and pharmaceutically produced chemical and toxic equivalent.

Science seems to be keeping these frequencies to itself.

What becomes clear from the science presented by Tompkins and Bird is that all living things are constantly interacting with each other and sharing information on an open basis. From my own research into the science of biophysics I found that the Earth's Schumann waves resonate at the same frequency as our hippocampus found in the brain, and that the hippocampus of every mammal resonates at this same frequency too.

Interestingly, the hippocampus is part of the limbic system, which functions in the emotional aspects of behaviour related to survival and is further related to functions in memory.

It is an important result that the Earth's Schumann waves resonate at the same frequency as the hippocampus. Science has been slow to recognise just how important this is and has therefore found itself left behind in the study of the limbic system.

Here we have a direct link with nature and our interaction with it, related to survival and memory. That any interference with nature and our interaction with it is important to our well-being is evidenced by the fact NASA built Schumann wave generators into their space craft - in space the astronauts are away from the Earth's magnetic fields, i.e. Schumann waves - to keep their astronauts physically and psychologically sound and to stop them becoming ill.

And of course this worked. The astronauts in Apollo 13 did not panic when the explosion occurred which would stop their moon-landing mission. Also, the astronauts who regularly broadcast from the International Space Station are noted for their equanimity.

Maybe some indigenous people are truly knowledgeable and it us so-called civilised Westerners who are the barbarians. Maybe we could actually learn from some of these people rather than allow ego to portray them all as being stone age Indians. It becomes obvious from Narby's work in studying certain indigenous peoples that their knowledge does come from the 'inner journey' by taking one's consciousness down and even beyond the molecular level, through DNA, and into the light which DNA is composed of.

It follows, therefore, that some of our ancestors, shamans of the past, also used this same path to gain knowledge, spirituality and truth and this 'way' has been passed on through generation after generation in some indigenous tribes today.

This work assumes even more importance now that, using nanotechnology, we can go beyond the molecular level ourselves. We could retrace the same route taken by shamen of the past. The wisdom of ancient people is often discounted by science, almost always to its detriment. One of sciences next steps should be to realise just how much ancient people knew and respect them more for it.

As an aside, it may be interesting to point out here that Andre Simoneton was able to establish empirically that he could measure specific wavelengths from food that indicated both vitality and freshness, or not. Regarding meats, the only one Simoneton included in his top list of edible foods was freshly smoked ham.

He found that freshly killed pork radiates at 6.5 thousand Angstroms but once it has been soaked in salt and hung over a wood fire, its radiance rises to 9.5 or 10 thousand Angstroms. Other meats, he discovered, were almost valueless other than as a digestive exercise. Simoneton established that the human body in health resonates at 6.5 thousand Angstroms.

Specific wavelengths from food have largely been ignored by science. These ideas however may be overdue for re-evaluation. It will come as no surprise to most that other meats apart from freshly smoked ham are pretty valueless. So, scientifically, it begins to appear that Vegetarians have it almost right.

Science holds the belief that 97% of our DNA is junk. Narby preferred to call it 'mystery' DNA. My contention is that human DNA may well contain the blueprint for the whole of creation which could explain similarities of DNA coding across many varying species of plants and animals and of course humans.

We know from the science of biophysics that our DNA is composed of light and our DNA emits electromagnetic frequencies and also produces sound. In the 1980's, German physicists demonstrated that cells of all living beings emit photons and showed that DNA was the source of this photon emission.

Science hold many beliefs which will be shown in the future to be completely wrong. One of the main problems with science in general and with individual scientists in particular is their stubborn insistence that they are right and only they are right. They should look carefully at the history of science and when they do they will find that science is littered with 'laws' which proved to be wrong and theories which, in the light of later discoveries, simply don't make sense.

Biophysics on the other hand has not set itself above any other evidence but merely reports the facts that our DNA is composed of light and produces sound. It leaves it up to others to formulate theories as to why this should be so.

So DNA is a crystal which amplifies the light preceding it and which it is comprised of and amplifies its DNA structure or coding for life into this outside world. Has the frequency of our 'complete' human DNA merely been broadcast into the Earth's consciousness, its water, where this information, this coding for life has then been stored and over time led to physical existence as we know it?

In the beginning was the word. Well DNA is the word, the coding, the sound, the instruction or blueprint for physical life to evolve. Is it possible that 97% of our so-called 'junk' DNA is not junk after all and may be, along with the other 3%, the precursor for all living organisms? Does our DNA, therefore, hold the answers to the Universe and the whole of existence?

These are important questions. The Human Genome Project has been working along the same lines and attempts have been made to show that the frequency of our DNA could have been broadcast into the Earth's water. Now that the project is complete, results are expected soon.

In a recent terrestrial TV program on organ transplants it was shown that the recipient of the donor organ, in many cases, had undergone a change in personality. One young girl was describing how she kept seeing and feeling the death of this young child, being able to describe quite graphically how the child had been bludgeoned to death. When these unusual circumstances were checked out the

young girl's experience described exactly how the 'donor child' had died.

It is unfortunate that science generally refuses to admit as evidence people's experiences as documented on TV shows.

Another 'organ donor recipient' at a gathering went up to a lady and said 'Hi Mom, it's good to see you.' The lady turned out to be the mother of the organ donor and the recipient had immediately recognised her. Another was where the personality of a middle-aged man completely changed. A man who used to suffer from stress and become angry now found he carried out housework when he was stressed, something he never did before he received the heart of a female donor.

And there are so many similar cases of this type of occurrence. This happens simply because the 'consciousness' of the donor is retained in the donor organ and its memory, or identity, has been placed within the body of the recipient and is able to express itself by sharing its 'consciousness' with that of the recipient – even after the death of the donor.

Reports of this kind are far too common for science to continue to ignore them. It is intuitively likely that the consciousness of the donor is retained in the donor organ. The fact that it can be transferred into a sharing of consciousness with the recipient is what should interest scientists.

Research recently published in the Journal of Near-Death Studies, conducted by Professor Gary Shwartz, Professor Paul Pearsall and Dr. Linda Russek of the Universities of Arizona and Hawaii confirmed that people who have organ transplants could inherit the habits of their donors. Their findings are very similar to those experiences detailed above.

One male who received a woman's heart had a new passion for the colour pink, a colour he had previously hated. He also had a newly acquired taste for women's perfume. Similarly, a woman who received a heart and lung transplant developed cravings for chicken nuggets, a food she never ate. Uneaten chicken nuggets were found in the jacket of the young male donor when he was killed. It goes on and on in the same vein, with alterations in people's sense of smell, food preferences and emotions to changes in tastes of music.

This is very telling evidence. Probably every transplant patient has a similar story to tell. When evidence mounts up in this way science has a duty to investigate.

The scientists involved in this study are embarking on a second, bigger study involving more than 300 transplant patients. Will they be allowed to complete it?

And if not, why not? Who has something to hide by stopping such a study?

It begs the question: What chance has human consciousness to evolve into a more enlightened state now that we have genetically modified pigs specifically created to act as organ donors for humans? Maybe as recipients of these organs we can all look forward to the time when we have flashbacks of happy family get-togethers, many of us mingling together with our noses deep in swill?

And all displaying a deep aversion to sausages.

Could the Sun be a cold dark place? Could the Earth and every planet in our solar system be hollow? Is our food being negatively energised? Has science found alien DNA in our human make-up? You may think I've lost the plot, lost my marbles, but I assure you I am as sane as the next person is. I will fully explain each of these statements later in this article and you will hopefully realise we do live in an illusion and realise just how effective and dangerous this illusion really is.

The explanations to these questions and statements will seriously challenge some peoples conventional belief systems. That said, nature does not lie, only man is capable of telling lies and believe me some large porkpies have been told for thousands of years and are still being told even now.

There can be little doubt that large porkpies have been told for a long time and only the most optimistic would claim the practice has ceased.

How do we know whether we are being deceived or not? How do we know if we are living in a possible illusion? Well I would now like to put forward some feasible explanations.

This is not only important, it is refreshing. How many times do we hear of views and opinions put forward which are so obviously not feasible? Science has as its starting point the feasibility of explanations so those detailed here deserve to be taken seriously and investigated accordingly. Scientists, above all others, know that cranks do not offer feasible explanations.

If we look at nature we understand and observe that in fact everything is in motion; it ebbs, it flows and it spirals and spins.

No serious scientist doubts these findings.

I remember as a child playing on the roundabouts when we used to play a game where we spun the roundabout as fast as we could and the last one to remain on, or the last one to be thrown off, was the winner and I wondered why we were flung off? I also wondered why the nearer the edge of the roundabout we were, the more unbalanced we became and the easier we were flung off it, yet the nearer the centre we were, the less we were affected by the spin, and the more control and balance we had.

Take a look at all those wonderful photographs of spiralling galaxies and look closely at what is in the centre. Space is in the centre, though generally depicted and photographed as light.

A photograph of any spiral galaxy will confirm this. Likewise, scientists involved in the study of spin states in general will confirm the observations obtained from roundabout studies.

Remember all those nature programmes about the destructive force of our weather, hurricanes in particular? What is at the centre of each and every hurricane, whirlwind, tornado, cyclone or storm? At the centre of each and every one of these wonderfully destructive natural forces is the eye of the storm. What is the eye of the storm?

The reader should consider this carefully. From it, much follows.

Look at the aerial view of a storm and compare it to the photographs of the galaxies. They are very similar to look at, especially when we know what is at the centre of the storm, space. Everything that spins is covered by the basic laws of nature and therefore can only have space as its centre. When something spins then everything is naturally flung away from the centre to the periphery; which is exactly what happened when we were children playing on the roundabout.

Look at the way water flows down the plug hole. If the plughole is big enough we can even place our finger in the centre of the plughole while the water is draining away without our finger becoming wet. Why? Because everything that spins has space at its centre.

A few maverick scientists claim that not everything that spins has space at its centre. They may be right but they are not proceeding according to the rules of science. The scientific method relies above all else on evidence. Where the evidence is lacking a theory is interesting but it remains no more than a theory.

Now what does the Earth and every other planet in our solar system do? They all spin. Everything in nature that spins in a physical sense has a hollow centre if you like. Ridiculous I hear you say. If it is so ridiculous then why does the Earth resonate like a bell when seismic charges are set off? Only hollow objects resonate this clearly, solid objects do not resonate so readily. Does this notion really seem so ridiculous after all?

The combination of two separate observations which at first glance have little in common yet together provide an explanation for a third is one of the cornerstones of genuine scientific progress. It is known as the inductive method. Here we have an observation concerning spin, which is not disputed, allied to an experimental result from seismology, which is also well-founded. The synthesis of the two leads to a new truth, and one which was not intuitive. This is science at its best.

For a number of years I have been uncomfortable with the notion that our Sun is a huge fiery yellow ball with extremely high temperatures which radiate heat and light to our solar system. Why? Well, I could not understand why as you climbed further up a mountain the colder it became, yet you were nearer the Sun.

To this day no convincing argument has been advanced to counter this observation. As anyone knows who has actually moved themselves nearer to the Sun the temperature drops. It really does become colder the further up a mountain you go.

But this is counter-intuitive. When moving closer to any source of heat you should become warmer. Thousands of people through the ages have noticed this and have looked to science to provide the answer but what is science's reply? A deafening silence. There appears to be no universally agreed upon scientific response to the conundrum.

I was also a big fan of the space program and used to follow its every course and action, and I gradually became more dissatisfied with our conventional conception of what the Sun was. We were being informed that space was black, the deep blackness of space, space was cold, the icy cold of space, and we were informed that space was indeed a vacuum.

Now, were we not told at school that heat could not pass through a vacuum by conduction, convection or radiation? Isn't this exactly how a thermos flask works? No I was not happy with this conception; the Sun could not radiate heat and light to the Earth even if it was a huge ball of fire. There had to be something else responsible for creating heat and light as we know it on Earth and I wanted to know what that was.

Given this recollection from schooldays it is no wonder that the writer went in search of an explanation.

I found my answer to this enigma while I was reading Living Energies by Callum Coats, explaining Viktor Schauberger's brilliant work with Natural Energy. Here, in this marvellous book was a full chapter just on the Sun, explaining the Sun's energetic interaction with Earth, and most importantly just what does create heat and light on our planet.

There is not the scope within this article to give a full scientific extrapolation here so I will condense the important parts, as I understand them. The Sun's energy is kinetic not thermal and interacts with the Earth's energies and in particular the Earth's atmosphere where these energies are converted into heat and light.

The consequences of the Sun's kinetic energy is a poorly understood topic in Solar physics at present. Everyone is encouraged to read 'Living Energies' for a fuller explanation, especially those scientists who have given up trying to understand how the Sun's kinetic energy interacts with the Earth's atmosphere.

To be more specific it is the water molecules in our atmosphere that contain the energy of heat and light. Light is a function of the atmosphere as is heat. This process is known as the hydrological cycle and involves our oceans and rainforests in a process similar to precipitation. This is where the water molecules from the oceans combine with the water molecules of the rainforests to provide a perfect balance in our atmosphere which can then utilise the energy from the Sun to provide heat and light.

It is our atmosphere that contains and provides us with heat and light, the very spark of life; in fact our atmosphere is responsible for sustaining all life itself. Therefore, as we climb higher up a mountain our atmosphere becomes thinner, so there is less energy converted to heat and therefore the colder it becomes. Interstellar space has a thermal temperature of $-273.15°C$ (0°Kelvin) so how can we possibly obtain heat directly from the Sun?

This argument has never been refuted. Also, the measurement of exactly zero Kelvin for the temperature of interstellar space is extremely telling evidence in favour of the cold sun hypothesis, though it must be added that certain stubborn space scientists still cling to the belief that space is warmer than this.

Even some of our ancestors thought the same way. Within Living Energies, published by Gateway Books, there is a reference to a

translation of the Naacal tablets, page 81, from James Churchward's book, the Lost Continent of Mu. I quote 'The light was contained in the atmosphere. And the shafts of the Sun met the shafts of the light in the atmosphere and gave birth to light. Then there was light upon the face of the Earth. The heat was also contained in the atmosphere. And the shafts of the Sun met the shafts of the heat in the atmosphere and gave it life. Then there was heat to warm the face of the Earth.'

Even in translation, the explanation of how the interaction between energy from the Sun produces heat and light via our own atmosphere could hardly be clearer. Scientists could learn a lot about solar radiation from the inhabitants of the lost continent of Mu.

A second quotation taken from 'The Life and Teaching of the Masters of the Far East' by Baird T. Spalding, which records his three-year visit to Tibet at the invitation of High Lama states, 'If we take the science of things, we know there is a legend told here that all the heat and light and many other natural forces are contained right within the Earth itself. The Sun, of itself, has no heat or light. It has potentialities that draw the heat and light from the Earth. After the Sun has drawn the heat and light rays from the Earth, the heat rays are reflected back to the Earth by the atmosphere that floats on the ether.'

It seems the Masters of the Far East were fully conversant with global warming.

What do you believe? Vital though the Sun's energy is to life on this planet, the Sun is probably a cold dark place.

Until science can come up with a more persuasive theory this is the conclusion that must be drawn. Startling it might be at first sight but when logic is followed the inference is inescapable.

The basic law of physics tells us that when we have energy fields in contact with each other there will be an interaction between the different fields. But is the interaction harmful or beneficial to our biophysical body? It is harmful as I have scientifically demonstrated in my document The Dangers of Electromagnetic Fields (EMF's).

The effects can also be demonstrated to be harmful by kinesiology testing, or dowsing, or by using a Lecher Antenna which is a scientific apparatus developed in France used to measure subtle energy fields.

There can be little doubting the basic law of physics.

A computer screen at 20,000 Hz is hitting our energy fields at 20,000 times a second. This is 72 million times an hour. This is literally blowing a huge hole in our energy fields. Some mobile phones operate at gigahertz. (GHz) 0.8/0.9 GHz and 1.8/1.9 GHz, the latter being nearly two million or could that be 2 billion cycles per second or 1.14 billion trillion times in just ten minutes.

Imagine what this does to our energy fields. Especially energy fields around the head and brain. Protection from these fields is essential to both prevent and reverse illness and disease.

Science has been relaxed in the face of this evidence but numbers do not lie. Individual scientists are slowly becoming aware of the danger but their voices are being drowned out by those in charge of mobile phone companies for whom profit is the only motive. They know exactly how often our energy fields are being hit by their products yet they choose to remain silent. A step forward would be for mobile phones to be reconfigured so that they hit our energy fields only a few times per second, say somewhere between one and ten. In that way our energy fields would have a chance to recover.

EMF's produce free radicals. EMF's cause electromagnetic stress which reduces immune response. EMF's cause movement of electrons in our cells. Our cells then have to work hard to maintain their own electromagnetic fields expending unnecessary energy.

Movement of electrons in our cells is certainly something we could do without.

EMF's also move ions into and out of the cells, again expending unnecessary energy by making the body work hard to correct these moves and restore balance. This reduces the energy potential of the cell because the cell is expending unnecessary energy to do this, thus weakening its defence mechanism, or immune system.

Biologists will immediately see the problem here.

An example of this weakening effect would be a virus, which is not mobile and is like a little crystal and can only attach to a cell membrane when the energy of the cell is down. Viruses do not affect us unless the cell's energy potential is weakened. EMF's promote the effect of known carcinogens and increase the rate of tumour growth by up to 24x.

Crystals are noted for their immobility, even little ones, and it is surprising that science continues to insist that viruses are mobile in the face of the evidence.

Magnetic fields travel right through the body and through every cell, affecting every cell. Electrical fields are absorbed by the skin and affect the peripheral energies of the body. Studies show that this type of radiation can cause all types of cancer, headaches, fatigue, M.E. heart disease, eye problems, memory loss, Parkinson's, Alzheimer's, cot death, central nervous system problems, tumours, especially brain tumours, miscarriages, deformities in embryos, breast cancer, childhood cancer and leukaemia to name but a few.

These studies are extremely persuasive and the reader is invited to review the original research papers.

The majority of pathogens, agents capable of producing disease, are anaerobic and proliferate in the absence of oxygen. Reduced oxygen levels in the body will lead to illness and disease. That is fact, there is no question and no doubt about this. In our attempt to prevent and reverse illness we must start by increasing the oxygen levels in our bodies.

Science has been lax in determining the amount of oxygen which should be present in our bodies and despite numerous searches I have been unable to find an agreed upon figure.

As part of my research, I naturally researched the validity and efficacy of various products available in the market place. From my findings there is only one make of stabilised oxygen, 25 years scientific study behind this one product alone, and one make of ionic, sub-colloidal vitamin / mineral solutions that I use in my treatments and that I would recommend.

It is important to establish this expert has no connection with the companies he mentions. He has conducted independent research as detailed above and is reporting his findings. Science proceeds in the same way.

From the only company who has the technology to produce the minerals and vitamins in this form, a form in which natural vegetation presents them to us - hence the high absorption factor – 98%. Other oxygen products, especially those that contain hydrogen peroxide or magnesium peroxide I would not recommend at all after reviewing the latest scientific studies into the effects these have upon, and within the body.

With regard to the antioxidants, there is again only one product I would recommend because it contains grapeseed extract, maritime pine bark, circuminoids, vitamins C, E, alpha- and beta-carotene plus++ and is the most powerful antioxidant on the market today.

Because of the cost of this one product alone, and the fact that a/o's are not as critical in the whole treatment protocol as the other elements involved, then other cheaper products which contain only Grapeseed extract or maritime pine bark, for example, may be utilised to do the same job if cost becomes an important factor in choice.

It is science's fault that money should become a factor. If science had realised the value of the constituents mentioned it may have been possible to synthesise them for a fraction of the cost. Cheaper products are available (again no thanks to science) but these circuminoids.

My mum is a walking, living example of the above outlined treatment. My mum was scheduled to have an operation to remove a lump from her leg having had some seven or so years earlier a malignant melanoma removed. This operation was then postponed so she could have a huge lump removed from her breast. The consultants felt that this represented a greater threat to my mum's health than the lump in her leg did.

To cut a long story short my mum's operation was scheduled for 29th June 1998. Only minutes before my mum went down to theatre she complained that she had not been marked and was naturally concerned that the correct breast was operated on. The surgeon came to the ward, apologised to my mum who was naturally quite distressed, and proceeded to palpate the breast with the intention of marking it there and then.

The surgeon ended up sitting on the end of the bed shaking his head and apologising to my mum because he didn't know what was happening, he could barely locate the lump, it had virtually disappeared, she didn't need the operation and he was going to send her home. My mother at first broke into tears and then eventually started laughing and said it's not often you get to go into hospital for bed and breakfast; she was admitted the night before.

His mum was lucky. How many patients with similar prognosis have either been operated on needlessly or, worse, not had access to a treatment which science has simply ignored. Too often, and it is worth a study as to why this is the case, surgeons operate purely for the sake of operating. They are 'knife-happy' to use the industry term.

On August 4th my mum got the all clear, no trace of the lump in her breast whatsoever. On August the 11th, surprise, surprise, there was also no trace of the lump on her leg either. Coincidence?

Miracle? You'll have to decide for yourself what you think. But in three months and one week after undergoing the above mentioned natural treatment something amazing happened. My mum's body had somehow repaired itself, restored its own balance.

Restoration of balance is the key here. Natural treatment promotes this but to read any medical in-house journal you would think there is no such thing. Instead a regime of drug taking and operation after operation is the standard response to almost every illness. When was the last time you heard of a consultant taking a fresh approach and recommending that tumours be treated with maritime pine bark?

Distance healing in the beginning, protection from all electrical equipment, stabilised oxygen, nutrients and antioxidants as previously discussed were all that my mother used to help her body achieve balance, thus restoring health. She refused to have any medical treatment after she saw how my father suffered from chemotherapy before he died of cancer at Christmas. Proof? If this were the only case of its kind then I'd be sceptical but hopeful, but it's not, and therefore I am not sceptical but very excited instead.

Science should also be excited but stubbornly refuses to. One day it will seem obvious that cancer patients should be offered protection from electrical equipment but that day is not yet.

To date my Mum continues with a maintenance programme using Dr. Lamar's stabilised oxygen, nutrients and antioxidants. Plus she carries with her at all times Equilibra's personal energy enhancer and also places this under her pillow every night. This helps to strengthen her energy fields and helps her energy fields ground and disperse chaotic energies emitted by electrical and electronic equipment, other people, unhealthy environments and aids the body with its natural repair and detoxification processes.

This is a regime which has been fully documented. The introduction of a maintenance program will be familiar practice to oncologists and here we have a proven scheme. Stabilised oxygen is a key element but the use of a personal energy enhancer should also become part of clinical response. Placement under the pillow is a primary technique but science should lead the way in determining if this is precisely the usage indicated in all forms of cancer or whether placement in other positions could be more effective given the nature of the oncological presentation.

There is no greater force than that of equilibrium. Everything, every energy, every vibration ultimately strives to reach its own perfect state of balance.

This is on the verge of becoming a Law of Physics. After hundreds of years of research no scientist has found any evidence that there is any greater force than that of equilibrium. From Newton to Einstein the story has been the same.

Almost all illness and disease originates from an imbalance in our energy fields. This is science, not myth or Eastern esoteric philosophy, but hard scientific evidence on how and why we exist and can and do become ill.

Stubborn to a fault, scientists still attempt to deny this.

Would you like to: Help protect yourself from the effects of electromagnetic radiation, microwave radiation, geopathic stress, mobile phones, people's negative dispositions and other negative energies? Improve the health and well-being of your physical and spiritual body? Help to eliminate the effects of and transform the chemical and toxic ingredients in your water, drinks and food today?

Energise your water, food, supplements, Bach flower remedies, essential oils, stabilized oxygen and nutritional products by increasing their biophoton levels life-force energy, which, when consumed enhances the ATP levels in your body leading to better health and longevity?

Help your body to align and improve its electrical communications? Strengthen and balance your natural energy fields? Align and correct the flow of the assemblage point, front and back? Improve the well-being of your animals, plants, living and working environment?

Nobody would wish otherwise.

How our energy products work: Everything that exists in the Universe is made up of many energies and different vibrations of light.

This finding is now commonplace and probably does not need repeating but there may still be some people who are unaware.

We humans are no different either in the way we are composed. Just as we put together letters to form words to express ourselves, we can also put together various vibrations of energy to produce specific and synergistically balanced energies, which, through resonance, can help us recover and maintain sound physical and spiritual health in an otherwise chaotic and materialistic driven world.

In understanding how the products work we first need to examine and comprehend some of the properties and abilities of water and how the quality of water can determine our health and physical well-being.

One of the important properties of water is its ability to store information, or energy, and in particular electromagnetic waves in the form of photons. Photons are waves/particles of light which are also found in the living cells of plants, animals or humans and because they are found in what is referred to as biological systems, they are also known as biophotons.

Physicists have been slow to acknowledge the storage property of water but since this is how the products work they will have to change their tune. In order to depose the theory science must first come up with a more convincing explanation of how the products work. Science has failed to do this up to the time of writing.

As our cells hold between 75% - 90% water they naturally store and emit light or photons which control vital process. Our DNA is composed of light and emits photons or electromagnetic waves which carry instructions - information - to our cells on how to replicate and behave.

Only the most unadventurous biologists dispute this.

If this mechanism does not function correctly the organism becomes diseased. The light from our DNA, the photons emitted into intracellular fluid, determines the health of reproduced cells.

The discipline of intracellular fluid research has been invigorated by these findings and younger biological science specialists are moving into the rapidly growing field in greater numbers each year.

We can see from photographs of ordinary London tap water that there is little photon activity or light present. There is hardly any life-force energy in the water therefore it is 'life-less'.

It is true that London's tap water emits hardly any light and it is not only Londoners who have noticed. Visitors to the capital have remarked on the same thing and if confirmation were needed one need only look at the number of people in London who carry bottles of water with them wherever they go.

Equilibra's products have been scientifically proven to increase the photon absorption capacity in water through a process known as 'resonance'. Our energy products reproduce the natural frequency of coherent light-energy, which then resonates into food, drinks, the atmosphere and even directly into the energy fields of the body to be

utilised by our biophysical and biochemical processes to enhance spiritual and physical well-being.

Resonance is a well-known phenomenon and has been studied extensively.

Each product is designed to work as nature works - simply but effectively and totally naturally.

It is rare to find a biologist today who will dispute the simplicity of nature. Biologists have been guilty in the past of exaggerating and claiming that biological systems are complicated but recent research is against them. It is becoming so obvious that leading universities have started to downgrade degrees in the biological sciences precisely because nature is so simple and therefore does not require many years of study to understand.

The science of biophysics has established that only when the defence mechanism of the body becomes weakened, our electromagnetic field, do we then become ill. Our electromagnetic fields are where imbalances first start and if these imbalances are not corrected, they can manifest into a physical problem resulting in illness and disease. We are, after all, as the science of biophysics has demonstrated, primarily composed of light, photons and energy.

Biophysics has taken priority over biochemistry in recent years and its results are extremely interesting. It is often the case that one science will firstly compete with and later subsume another (survival of the fittest!) and we seem to be spectators at that process occurring today. Biochemistry claimed that the defence mechanisms of the body were chemically based and that entities such as antibodies, white blood cells and interferon were important. But biophysics has subsumed this and has provided a wider and more holistic paradigm of pathogen antigenesis which has left biochemistry not redundant exactly, but in an inevitably diminished role when it comes to the true explanation of illness.

Doctors should have instruments made available which will enable the detection of imbalance in our electromagnetic fields. Members of the public who feel their own health providers may be unaware of the latest findings should lobby to ensure that suitable products are available in hospitals and health centres. If resistance is encountered from elected representatives it should be sufficient to quote the finding from biophysics that we are primarily composed of light. It will be a brave public official who feels able to contest that.

The scientific testing of our products was carried out by D. Schweitzer in his London clinic. Our universal harmonisers have been further tested by Masaru Emoto in Japan.

It would be difficult to find more respected testers than D. Schweitzer and Masaru Emoto. That they are involved at all should give reassurance that the products have been fully validated.

Professional grounding mattress pad - universal worldwide grounding rod. Larger than the home/office mats to lay under or over the mattress for electromagnetic discharge, protection and full body grounding to Earth during sleep. Comes with external grounding rod only. This connection kit is universal and suitable for use worldwide and in any country providing you do not live more than two floors above ground. Price: £134.99.

This is a very reasonable price. Full body grounding has been researched by scientists for decades but they have been unable to produce a working model. Others have claimed to be able to produce a universal connection kit but their work has yet to be peer-reviewed and remains theoretical.

The fact that such a kit is now commercially available speaks volumes for the inertia of present scientific research despite the millions of dollars devoted to it. Scientists are often distracted from their primary research goals by the latest fad and this is increasingly becoming apparent as they are overtaken by specialists employed by commercial companies who offer contracts specifically targeted at success in one endeavour at a time. Science, with its generalisations, is losing out and losing out more often as their own specialists become side-tracked.

A recent example occurred when scientists supposedly investigating emf/emr discharge and purportedly developing their own external grounding rod became involved, of their own volition, in the search for the Higgs Boson. Work on the rod stopped and in that time commercial companies took the lead. This should never have occurred. The scientists at CERN should have solved their own problems and external grounding rods should have been developed earlier and made available to health practitioners.

To make up for this, some have suggested that these same scientists should investigate why the grounding mattress pad is not effective over two floors above ground. They should research a mattress which is effective at all floors and make their results freely available so manufacture and distribution can begin as soon as possible.

Elanra MKii - Professional Ioniser: Most ionisers will clean the air of pollutants but the negative ions they produce are too large to enter the bloodstream and so you miss out on the really important health benefits that breathing small negative ions can bring.

Large ion chemists had been well on the way to proving that they are too large to enter the bloodstream so they will be pleased that final proof is now available.

Only the Elanra ionisers have the patented technology capable of consistently reproducing the small, breathable negative ions of oxygen found naturally in healthy environments such as forests, mountains or by the sea. Price: £395.00.

This is an extremely reasonable price. Scientists have been able to produce small negative ions themselves but tests have found them to be unbreathable. Other scientists have produced breathable ions of oxygen but they have been positive, not negative. Still other teams have produced breathable negative ions of oxygen but they have not been small enough.

For energising water and protection from electromagnetic fields. For energising water and food, improving and enhancing the energy in personal care products, nutrients, flower remedies, aromatherapy oils etc. Can be carried about the person to enhance and strengthen one's energy fields which helps protect us from harmful EMF's, mobile phone radiation and other people's negative energies. Price: £40.00.

Science has never been able to provide a similar product which can be carried on the person. During simulations for the International Space Station, a device intended to provide protection from harmful EMF's was developed but the prototype had a volume of over several metres cubed and could hardly be described as portable. In the case of the Space Station this didn't matter because the device could float but its use on Earth was limited and the problem was never overcome. Science simply gave up.

Universal harmonisers without crystals: A set of two harmonisers for energising water, food, nutritional products and self-healing. Price: £89.00.

The harmonisation of water, food and nutritional products was one of the unanswered questions of science. It was thought to be theoretically possible and much research had been directed towards the solution but now that a working device has been constructed science should spend £89.00 and discover where it went wrong.

If you are a Reiki Master you make a bigger difference. Life gets better when you learn how to take care of the body! Some kind of treatment is required when the body 'gets sick'. Pain, illness, sickness only means the vibrational frequency of the body is low. As you may know: In Quantum physics the view is beyond the physical!

Quantum physicists spend a lot of their time measuring vibrational frequency and although their view is beyond the physical such results which make it back are tending to confirm that some kind of treatment is indeed required when the body 'gets sick'.

With a Laser Reiki healing you can raise the frequency of the body and it can get well.

Science has been slow to experiment on exactly how the frequency of the body can be raised but here we seem to have the answer. Laser Reiki healing. Science invented the laser and it has been aware of Reiki but it never thought to put the two together.

Health and wellness are natural to the body. When a person has neck pain it can be proven that the frequency of the neck is low. If you have a lower back pain that means that the energy flow in your lower back is low or partly blocked. An energy healing would help release the blockages so the body can heal itself again!

Science has been even slower to measure the frequency of the neck. Doctors who specialise in the lower back have had reports from patients that their energy flow in that area seems low yet have not motivated themselves to do much about it.

Practitioners who give Laser Reiki healings remove the energy blockages and then test again. After the energy healing the pain is gone! (Well, it is usually completely gone, but sometimes it may require several layers of energy healing to get through to the root cause of the problem or disease.)

This is a common feature across all forms of health practice. Ill-informed people suggest that energy healers are tricksters as they make people pay for many treatments rather than just one. However, the same people are perfectly willing to accept that radiotherapy must be given multiple times to be fully effective and that a whole course of antibiotics should be taken in order to completely eradicate a bacterial infection. When this is pointed out they change the subject.

The results of an energy healing -- The frequency is higher and into the range of health, joy, happiness!

Experts in frequency analysis are in agreement that the higher the frequency the better and many will also confirm that such frequency enhancement can extend into the range of joy.

Healing the body is a mechanical process. You may have energy blockages causing sickness? We can check it for you... Taylore Vance and partner Roi Halse, Reiki Masters.

Healthcare systems worldwide continue to rely on the old methods of drugs and surgery when an alternative is available. But the wheels of science spin slowly and such is the sclerotic nature of the scientific mind that scientists find it difficult to accept new ideas quickly and so repeat the mistakes of the past. Science is slowed by institutionalised inertia and should be making a greater effort to overcome the problem. It is fortunate that Taylore and Roi are available to offer modern treatment.

Reiki Blog at Reiki Energy Healing Remote Distance Healing where Taylore shares techniques. She and her partner have an energy healing centre at Reiki Ranch, Chehalis, WA.

Reiki Master - normally $350.

Take the whole Reiki 1, Reiki 2 and Reiki 3 training weekend for becoming a Reiki Master – Special only $200 plus books.

And while science prevaricates this is a very reasonable price when one considers the potential gains. Individual doctors should not wait for science to overcome its lethargy but should sign up for the course. Their patients expect the best, most modern therapy and their own conscience should dictate that they have a duty of care which relies on them becoming familiar with the most up to date treatments available.

Doctors take the Hippocratic Oath to ensure this but many seem to have forgotten exactly what it requires of them. In this case it clearly obliges doctors to independently embrace a new, ground-breaking, innovative technology. $200 is a small price to pay to keep allegiance to the Oath.

Which doctor could honestly claim to be holding true to his principles while at the same time not becoming a qualified Reiki Master?

Chapter 24: Solar Healing

Welcome to Solar Healing Centre: The solar healing centre is focused on helping humanity to develop a better understanding of how the sun can be used to heal the mind, body and spirit as demonstrated by Hira Ratan Manek, who, as a result of sungazing, has claimed better physical, mental, emotional and spiritual health. Hira Ratan Manek, HRM, among others, has proven that a person can live just on solar energy for very long periods without eating any food.

Given the world shortage of food which is expected to become worse in future years because of global warming and drought as well as an increase in world population, this is very good news.

It has come to be known as the HRM phenomenon. The method is used for curing all kinds of psychosomatic, mental and physical illnesses as well as increasing memory power and mental strength by using sunlight.

It is not too much of an exaggeration to suggest that confirmation of the HRM phenomenon should be one of science's next steps and should be undertaken with urgency.

One can get rid of any kind of psychological problems and develop confidence to face any problem in life and can overcome any kind of fear within 3 months after starting to practice this method.

Developing the confidence to face any problem in life has long been a goal of science though modern methods take longer than three months. Overcoming fear is a major project in phobia-research and HRM's methods could well find an application here.

As a result, one will be free from mental disturbances and fear, which will result in a perfect balance of mind. If one continues to apply the proper sungazing practice for 6 months, one will be free from physical illnesses. Furthermore, after 9 months, one can eventually win a victory over hunger, which disappears by itself thereafter.

Freedom from physical illnesses is of obvious importance to stressed health services globally while the victory over hunger and its benefits has been mentioned above. That it should disappear by itself, without further intervention, is another point in HRM's favour.

This is a straight-forward yet effective method based on solar energy which enables one to harmonize and recharge the body with life energy and also invoke the unlimited powers of the mind very easily. Additionally, it allows one to easily liberate from threefold sufferings of humanity such as mental illnesses, physical illnesses and spiritual ignorance.

Liberating humanity from the threefold sufferings would be a boost to people all over the world while invoking the unlimited powers of the mind could be a real step forward.

About Hira Ratan Manek: He was born on 12th of September 1937 in Bodhavad, India, and was raised in Calicut, Kerala, India, where he had his mechanical engineering degree from the University of Kerala. After graduation he joined the family shipping and spice trading business and continued working there until he retired in 1992.

After he retired he began to research and study the ancient practice of sungazing in which he had been interested since his childhood. This method was an old but forgotten method which had been practiced in the ancient times in many different parts of the world.

Science really should investigate the practices of ancient times. As noted elsewhere, ancient people often had a better grasp of things than is the case today.

After working on this method for 3 years he was able to re-discover the secrets of sungazing. During his study he was mainly inspired from the teachings of Lord Mahavir of Jains who was also practicing this method two thousand six hundred years ago. Other inspirations for sungazing came from ancient Egyptians, Greeks and Native Americans.

That Native Americans knew of this method as well as Ancient Egyptians and Greeks confirms just how widespread the technique was. As a side-project, perhaps science could investigate how such important knowledge came to be lost. Perhaps it had something to do with the rise of the so-called 'Age of Enlightenment'.

Since June 18th 1995 HRM has and continues to live only on sun energy and water. Occasionally, for hospitality and social purposes, he drinks tea, coffee and buttermilk. Until now he had three strict fastings, during which he had just sun energy and only water and was under the control and observation of various science and medical teams.

The first of these fasting lasted for 211 days during 1995-96 in Calicut, India directed by Dr. C.K. Ramachandran, a medical expert on allopathy and Ayurvedic medicine.

It is good that a doctor was involved in these preliminary experiments, not just to ensure the wellbeing of HRM but also to ensure that the sungazing project from its very beginning was monitored by a specialist. This will make it easier for modern-day scientists to pick up where he left off and will ensure that sungazing has little difficulty being accepted as a legitimate field of scientific research.

This was followed by a 411 day fast from 2000-2001 in Ahmedabad, India directed by an international team of 21 medical doctors and scientists led by Dr. Sudhir Shah.

Again, this shows that strict medical and scientific protocols were observed.

After the excitement of the findings at Ahmadabad, HRM was invited to Thomas Jefferson University and the University of Pennsylvania in Philadelphia where he underwent a 130 day observation period. This science/medical team wanted to observe and examine his retina, pineal gland and brain, therefore this observation team was led by Dr. Andrew B. Newberg, a leading authority on the brain and by Dr. George C. Brainard, the leading authority on the pineal gland. Initial results found that the grey cells in HRM's brain are regenerating.

700 photographs have been taken where the neurons were reported to be active and not dying. Furthermore, the pineal gland was expanding and not shrinking which is typically what happens after mid-fifties and its maximum average size is about 6 x 6 mm, however for HRM, it has been measured to be at 8 x 11 mm.

Size is important in this field as in others.

There have been many other sungazers who have achieved similar results and have volunteered to be tested, however due to lack of funding and other lifestyle restrictions the results have not been documented. The uniqueness of HRM is that he has surrendered his living body for observation and experiment to the scientific firmament for several extended periods of time.

Although scientists and doctors have agreed that hunger is being reduced if not eliminated, due to the complexity of the various brain functions they have not been able to explain how sungazing has such positive effects on the human mind or body, however more research is underway.

And this is how it should be, because this is how science makes progress.

From 2002 HRM has been travelling all over the world to preach about the practice of sungazing so that humanity can heal their problems without any cost. In 2009 HRM travelled to 80 countries in 210 days teaching this ancient solar methodology. There are now solar healing centres coming up all over the world and facilitating a global group of HRM phenomenon practitioners.

None of which is surprising given its obvious benefits.

Sun gazing process: We have a supercomputer in our bodies given to us by nature which is our brain. HRM Hira Ratan Manek calls it the brainutor. The brain is more powerful than the most advanced supercomputer. Each and every human being is gifted with innumerable talents and infinite inherent powers by nature. Individuals should never underestimate themselves.

Everyone is gifted. If we make use of these powers we can take ourselves to great levels. Unfortunately these infinite inherent powers are programmed in that part of the brain that is largely dormant and goes unused. Even medical science agrees we hardly make full use of the brain but only about 5-7%. The most brilliant of humans like Albert Einstein is reported to have used only about 32% of their brains.

The world record holder, at least in modern times, is indeed Einstein with 32%. This was achieved as he thought of the first bit of his theory of relativity.

If we can activate the human brain and awaken these infinite powers inherent in ourselves then we can raise ourselves to higher levels. We can achieve any results we want. In order to operate the brain effectively it needs to be activated. Being a holistic entity it needs a holistic power supply.

It is of course true that the brain is a holistic entity and no reputable scientist seriously questions it. Indeed there would be doubt as to his status as a scientist were he to do so. There can also be little doubt that in order to operate the brain effectively, it must be activated.

Science has been hunting for a holistic power supply to awaken the infinite inherent powers in ourselves but perhaps they have been pouring too much money into wind turbines and nuclear fission. These are not holistic, and there is a more appropriate power supply they have overlooked.

Sun energy is the source that powers the brain, which can enter and leave the human body or the brain only through one organ that is the human eye. Eyes are the sun energy's entry door to the human brain.

Opticians will confirm this fact.

Recent research has found out that the eye has many functions other than vision. And more information is continuing to be revealed about the functions of the eye. The eyes are complex organs and they have 5 billion parts much more than a spacecraft that has about 6-7 million parts. By this, you can see the immense capacity of the human eye.

Scientists are still counting but initial results suggest that the eye does have more parts than a spacecraft.

Since eyes are delicate parts of the body we have to use them in such a manner that they serve our purposes without getting damaged. Present day teachings and ideas such as don't look at the sun at all, you will damage your eyesight, never go out in the sun as you will get cancer, are causing needless hysteria and paranoia. The more you are away from nature, the more there is a cause for illness and you will automatically support global corporations.

These days it is recognised that we should get back to nature, to do otherwise is certainly a possible cause of illness. Science should make it a priority to confirm that by being more away from nature one will support global corporations. This sounds intuitively correct.

HRM phenomenon: Healing Powers of the Sun: Play this 1+ hour audio CD in your car while you're driving and listen to a live recorded lecture by Hira Ratan Manek describing the HRM phenomenon and how you can harness the sun's energy to heal illnesses within the mind and body through the process of safe Sungazing. $19.95.

This seems a very reasonable sum to pay for advice on *safe* sungazing.

HRM phenomenon: Healing Powers of the Sun: Watch this 2+ hour DVD and gain a deeper insight into the HRM phenomenon and how you can harness the sun's energy to heal illnesses within the mind and body through the process of safe sungazing through a 1 hour lecture by Hira Ratan Manek. Part 2 includes bonus footage for a 1 hour interactive question & answer forum with the audience. Additional bonus media clips are also included in part 3 on this DVD. $24.95.

Again, a small price to pay for knowing how to heal illness by staring at the sun.

To get energy for their body, native Indians take sunbaths by standing in the sun for two hours exposing maximum parts of their body and they don't need to eat food on those days. They sustain on micro food of the sunlight.

This is powerful and convincing evidence of non-standard sustenance.

The scientific technique practiced by Hira Ratan Manek is how he derives his energy from the solar energy of the sun. Mankind is also using solar energy for running solar cookers, solar heaters and solar cars. Similarly, what HRM does is to convert himself into a solar car by using sunlight.

This is very impressive.

There is a scientific explanation of what may be occurring during the exercise of this method. Since the brain is a powerful recipient and the retina and the pineal gland or third eye being equipped with photoreceptor cells, a kind of photo analysis could be taking place which provides a person with all kinds of energy and vitamins that are required for the proper functioning of his internal organs.

Additionally, the rays of the sun with seven colours have a cure for all kinds of diseases such as Alzheimer's, Parkinson's, obesity, arthritis, osteoporosis, cancer and others. If world peace is to be achieved everyone should have a perfect balance of mind. Thus, this method can also contribute to world peace.

Neuroscientists will confirm that the brain is indeed a powerful recipient. The role of the retina is well understood and the function of the pineal gland is becoming more so as research progresses. Certainly a kind of photo analysis could be occurring.

Nature has coded inside the brain infinite powers which are largely dormant but if invoked or awakened we heal by ourselves. That's why sungazing becomes very important. The brain becomes more powerful or energetic only through safe sungazing.

The brain has countless neurons and these are important for the efficiency of the brain. With safe sungazing the neurons increase in number and become healthier and that takes care of human health. The simple key to health is to see that we have not only more neurons but also totally healthy neurons. It is only through the eyes

that the light from the sun reaches the brain and creates beneficial effects in the brain.

Neuroscientists will confirm that the brain at our present state of knowledge does indeed have countless neurons but with scanning techniques advancing all the time it should be possible to count them at last. As sungazing increases the number science will have to keep up and scan brains more often in order to properly quantify the exact rate of increase. This will be important as the dormant powers start to surface.

Scientists will be interested in exactly which powers show themselves as the number of brain neurons multiply. It is obvious that light from the sun reaches the brain only through the eyes; the less trivial fact is that this sunlight also makes already existing neurons healthier. This fact alone promises to revolutionise neurosurgery. Brain operations could take place out of doors, for example, on a nice sunny day.

In general much of humanity is afraid of sun to some degree. We have believed that looking at the sun or doing sungazing harms the eyes but that is a wrong belief. Safe sun never harms the eyes and, on the contrary, it makes eyes healthier. The sungazing practices of the Egyptian, Mayan, and Incan priests and priestesses, as well as others, serve to remind us of this truth.

Unfortunately, we seem to have forgotten about it for some reason.

Safe sungazing is a must for better health and for alleviation of health problems. When safe sungazing is added with sun warming of the body, drinking sun charged water, one easily gets cost free health. One has to understand the practice and start practicing.

Warming the body via the sun has obvious health benefits but less well widespread is the practice of drinking sun charged water. Science should set up a research program to make sun charged water available to people in poor health and measure their improvement in relation to the volume of sun charged water they drink per day.

Many people experience a reduction in their appetite/less hunger and/or more joy in their lives after sungazing. Some are able to subsist more and more on light as they expand their practice of sungazing, from days, weeks, or even months at a time without eating regular food and/or no food.

With famine common in many parts of the world, sungazing will find immediate application. The ability to survive for months at a time without eating regular food would take the pressure off relief agencies and allow them more time to build up stocks of different types of food instead of rushing to send sacks of rice which, while nutritious, hardly makes for a varied diet.

Safe sungazing practice: You can break up the practice in three phases. 0 to 3 months, 3-6 months and 6- 9 months. The practice entails looking at the rising or setting sun one time per day only during the safe hours. No harm will come to your eyes during the morning and evening safe hours. The safe hours are anytime within 1-hour window after sunrise or anytime within the 1-hr window before sunset.

It is scientifically proven beyond a reasonable doubt that during these times one is free from UV and IR rays exposure which is harmful to your eyes. To determine the timings of sunrise or sunset you can check the local newspaper which also lists the UV index as 0 during these times. Both times are good for practice - it depends on individual's convenience.

Sungazing also has the added advantage of getting vitamin A and D during the 1-hour safe period window. Vitamin A is necessary for the health of the eye, the only vitamin that the eye requires. If you sun gaze the spectacles will go away and this will provide better eyesight without glasses.

And here we have yet another benefit of sungazing. Science will undoubtedly be able to check that Vitamin A is the only vitamin that the eye requires and that infra-red exposure falls to zero during the safe hours. As sungazing proceeds for even quite modest times, one can be fairly sure that spectacles will no longer be required.

For those who cannot initially sungaze during the safe periods sunbathing is an effective method for receiving the sun energy at a slower pace until one is able to sungaze. Also do not use sunscreen. When the body gets heated up you perspire and sweat is a waste product and needs to go out of the body. When you are painted or coated with lotions and creams they get degenerated and the chemicals enter your body. It is our malpractice, our wrong use, why we blame the sun for skin cancers.

One of Science's Next Steps must be to make this knowledge more widely spread. It is completely unfair that we blame the sun for causing skin cancers. As with so much of modern life it is us ourselves who are to blame for our ills and, if people cannot be persuaded from applying suncream, science should find ways of stopping the chemicals entering our bodies.

Manufactures of suncream will deny this happens but they would say that, wouldn't they?

0 - 3 months: First day, during the safe hours, look for a maximum of 10 seconds. Second day, look for 20 seconds at the rising sun adding ten seconds every succeeding day. So at the end of 10 continuous days of sun gazing you will be looking at the sun for 100 seconds i.e. 1 minute and 40 seconds.

This follows good practice. The body should acclimatise itself as is the case with any new demand put upon it. You would not attempt to climb Everest having only ever walked up a small hill, for example.

When you reach three months you will have gazed at the sun 15 minutes at a stretch. If you can watch TV for 3 hours, surely you can see the sun for that long.

And be more entertained!

What is happening as you go up to 15 minutes? The sun energy or the sunrays passing through the human eye are charging the hypothalamus tract, which is the pathway behind the retina leading to the human brain.

One of the failures of modern day medicine has been to neglect to attach a Coulombmeter to this tract in order to measure the exact amount of charge. This oversight should be corrected in the near future as sungazing research proceeds.

One of the software programs inherent in the brain will start running and we will begin to realize the changes since we will have no mental tension or worries. Moreover, we will become fearless since our psychosis will have disappeared, so will have all the ills of the mind. This is the first phase of the method and lasts around 3 months.

The disappearance of worries is a prime target of recent neurological research. That the use of sungazing should take the time of implementation of these beneficial effects down to around 3 months represents a real step forward in the field.

Besides, mental depression will go away. Psychiatrists are observing that sadness is caused by lack of sunlight. With the practice of sungazing you will not have depression in your whole lifetime. You will achieve a perfect balance of mind. Fear of death will go away. The state of mind is such that you will welcome death.

It is not difficult to believe that extended periods of staring at the sun could well induce this feeling.

What is to happen, you will be able to let it happen. There will be no worries. Everyone has some sort of mental disorder which is the biggest human problem, which can be removed by the proper use of sunlight.

This is good news on the psychiatric front.

3 - 6 months: Next, physical diseases will start being cured. 70 to 80% of the energy synthesized from food is taken by the brain and is used up in fuelling tensions and worries. With a lack of mental tension, the brain does not require the same amount of energy as before. As you proceed in sun gazing and as your tensions decrease the need for food intake will go down.

People know intuitively that tension and worry drain them of energy. So it stands to reason that when tension is removed they won't want to eat as much.

When you reach 30 minutes duration of continuously looking at Sun you will slowly be liberated from physical disease since by then all the colours of the sun will have reached the brain through the eye.

Brain regulates the flow of colour prana appropriately to the respective organs. All the internal organs get ample supply of the required colour prana. The vital organs are dependent on certain sun colour prana. Kidney red, heart yellow, liver green etc. Colours reach the organs and address any deficiencies. This is how colour therapies work, reiki and Pranic healing. There is a lot of information available on colour therapy. This is the process of getting liberated from physical ailments over a six-month period.

After 3-4 months you can become cured of your physical ailments with autosuggestion, which is imagining and visualizing healing your ailments while gazing at the sun. Scientific methods such as solariums, crystals, colour bottles, natural stones, gems, all utilize sun energy which is stored in these natural stones. You can keep natural colour stones in drinking water to further hasten healing.

There is so much science in this paragraph that I need to deal with it in shorter segments. After 30 minutes of sungazing all of the colours of the spectrum will have reached the brain – this is accurate because it is well known that different colours travel at different speeds through the same material. Science should confirm that the figure of 30 minutes is accurate.

That the brain regulates the flow of colour to the vital organs has long been suspected. The major organs which respond to the different colours are detailed above though minor organs such as the pancreas and the lungs need to have their shade of colour determined as well. This would be a sensible use of Science's time and research should be begin immediately.

There is indeed much information available concerning colour therapy and Scientists should begin researching the methodology of Reiki and Pranic healing as soon as possible. There are a number of websites which they will find invaluable as a starting point.

In solariums there is usually a platform at the height of 100 feet where each of the 7 glass cabinets is constructed for each of the r.o.y.g.b.i.v. colours. This platform revolves around the sun all day and according to the nature of the disease diagnosed, the patient is placed in the appropriate colour for healing. Similarly, glass drinking water bottles with different colours are kept in sun for 8 hours. The water gets solarised and water develops medicinal value and is used to treat different diseases.

Solarized water should be investigated more thoroughly by scientists. It may be possible to reduce the 8 hours presently required by adding water to liquids which are already coloured, e.g. orange squash or Ribena.

Photosynthesis, which we misunderstand, does not in fact need chlorophyll. Only the plant kingdom needs chlorophyll. Human body can do it with a different medium. Photosynthesis is transforming the sun energy into a usable energy format. This is how photovoltaic cells work and electricity is produced, similarly water is heated, food is cooked in solar cooker, and solar batteries run automobiles.

Many scientists have long suspected that we misunderstand photosynthesis. It seems obvious that the different kingdoms of biology will achieve the same aim using different means.

6 - 9 months: In 6 months time you will start to have the original form of micro food, which is our sun. Additionally, this can avoid the toxic waste that you take into your body while you eat regular food.

Palaeontologists have been leaning towards the idea that our original form of micro food was the sun and this is further proof. That toxic wastes can be introduced to the body whilst eating is an all too common experience of aficionados of 'fast-food' to take one obvious example.

Therefore, as you consume the original form of food, hunger goes down starting to disappear eventually. By eight months, you should see hunger almost gone. For a dull or weak student or with no belief, this time period may be 9 months. After that time, hunger

disappears forever. All mechanisms associated with hunger like aroma, cravings and hunger pangs also disappear. Moreover, energy levels are at a higher level. There is a judgment, having had this experience, that the brain is well activated with the sun energy.

This will dramatically reduce the strain on the Earth's farming resources.

An important gland in the brain's centre called the pineal gland or the third eye is activated. 25 years ago it was considered a useless gland. Now it has become an important gland for study and up to now about 18,000 papers have been published about it. It has always been known as the seat of the soul.

Religious leaders could assist with this functional identification by helping to interpret MRI scans.

Amygdala (another brain gland) for the last 2 years has been gaining importance in medical research. It's a nucleus of the sun or cosmic energy and plays an important role in the photosynthesis via the sunlight reaching the brain through the eye. These create a magnetic field and the body/brain recharges with the energy of the sun entering in you.

Science has not yet caught up with the array of new physics demonstrated here. Cosmic energy has not until now been thought to be important for energy transfer within the human body so this is an avenue which needs to be explored. The amygdala had not been supposed to be involved in photosynthesis but the situation has now changed.

The production of the resultant magnetic field seems to have been neglected by physics as a trawl through relevant research documents proves, so this work must also be high on science's priority list. That the body/brain recharges with the energy of the sun is in little doubt but science still needs to outline the exact mechanism by which it occurs.

Historically, a lot of people have remained without food. In 1922 the Imperial Medical College in London decreed that solar rays were the ideal food for humans.

So we see that scientists were well aware of the beneficial effects of sungazing almost 100 years ago. Given that, it is somewhat surprising that science has left it so long before investigating. It would be interesting to see why exactly this has been the case. Perhaps, and not for the first time, scientists have been influenced by the powerful political muscle of the agricultural industry. If so, it is time for science to apologise and put the record straight.

In the meantime it is not just HRM who is an advocate of sungazing. Here we have further evidence from Sheryl Walters.

Why sun gaze? We all know that the sun is an incredible source of energy. The world's top research is continually showing us that the sun is not this evil horrible monster that we need to constantly protect ourselves from. The sun is actually a source of incredible nutrients such as vitamin D. More sun generally means less depression and even less cancer.

Now that the sun has been excluded, science should turn to investigating melanomas more seriously to find out what does cause them.

Sungazing takes this idea and runs with it. With the aid of sunlight you can recharge each cell and atom in your body to its full potential. Sungazing can be done to tap into your full human potential. Apparently, sungazers who reach that 44 minute mark actually use 50% of their brain rather than the usual 10-12%.

Here Sheryl agrees with HRM. It is good scientific practice to have independent corroboration of the facts.

You may begin to desire healthy foods and junk food and toxic food will seem less appealing. Your overall sensitivity will increase.

Again, this is in broad agreement with HRM.

I am thinking we need to take this practice to some poor countries around the world. Imagine if we could actually teach people in Africa, for example, how to live on the sun which is blazing above them.

This is an admirable humanitarian proposal. People in Africa only need to be told how to look at the blazing sun properly. Their ancestors probably did this, or it is difficult to imagine how else they survived. If starving people learned how to sungaze there would be less need to send them emergency supplies, saving on transport costs and reducing the carbon footprint of the means of transport, particularly aircraft. In this scenario everyone wins.

About the author: Sheryl is a kinesiologist, nutritionist and holistic practitioner. Her website www.younglivingguide.com provides the latest research on preventing disease, looking naturally gorgeous, and feeling emotionally and physically fabulous. You can also find information on some of the most powerful super foods on the planet including raw chocolate, purple corn, and many others.

For context, I include a section on the pineal gland and its importance.

The pineal gland's location deep in the brain seems to intimate hidden importance. In the days before its function as a physical eye that could see beyond space-time was discovered, it was considered a mystery linked to superstition and mysticism.

As has been obvious to neurosurgeons for a long time structures and glands which are located deep within the brain do indeed seem to indicate hidden importance. But now, mysticism can be discarded and we can understand it as a physical eye which can see beyond space-time.

The pineal gland is activated by light and it controls the various bio-rhythms of the body. When it awakens one feels a pressure at the base of the brain.

Until recently it was not realised that a gland could become conscious.

When awakened, the third eye acts as a stargate that sees beyond space-time into time-space. To activate the third eye is to raise one's frequency and move into higher consciousness.

If Einstein had thought a little more deeply he may have invented time-space in the same way he invented space-time. Fortunately we can now see directly into it so his omission has been resolved.

Planetary vibration/frequency is accelerating exponentially allowing souls to peer into other realms far more easily than in the past. Frequency will continue to rise until consciousness evolves out of the physical in the next few years.

One of the major results of the Hubble Space Telescope has been a more complete understanding of planetary vibration. It is hoped that the successor to Hubble, the James Webb telescope will further our understanding in this important area.

To activate the third eye and perceive higher dimensions the pineal gland and the pituitary body must vibrate in unison. When a correct relationship is established, between personality, operating through the pituitary body, and the soul, operating through the pineal gland, a magnetic field is created. The negative and positive forces interact and become strong enough to create light in the head.

If this happened it would come as no surprise at all that a person would report feeling light in the head.

Beginning with the withdrawal of the senses and the physical consciousness, the consciousness is centred in the region of the pineal gland. The perceptive faculty and the point of realization are

centralized in the area between the middle of the forehead and the pineal gland. The trick is to visualize, very intently, the subtle body escaping through the trap door of the brain.

A popping sound may occur at the time of separation of the astral body in the area of the pineal gland. Visualization exercises are the first step in directing the energies in our inner systems to activate the third eye. The magnetic field is created around the pineal gland by focusing the mind on the midway point between the pineal gland and the pituitary body.

This all seems rather straightforward once you get the hang of it.

Finally in this section I include a plea from a well-wisher.

Why are folks like HRM Hira Ratan Manek not lecturing and teaching it to the starving in sun/sand rich areas that are lacking in food? Places like Somalia, Ethiopia and Central America? If these people can learn to thrive on abundant photons are these not the places to be teaching this concept? Could this be the way to end world hunger? Why aren't legions of these teachers, Yogis, and practitioners setting out into these places?

The answer, from his own website, is:

From 2002 HRM has been traveling all over the world to preach about the practice of sungazing so that humanity can heal their problems. In 2009 HRM travelled to 80 countries in 210 days preaching and teaching this ancient solar methodology. Each year for the last few years HRM gives on average approximately 300 lectures on sungazing in different languages. Several interviews have appeared in leading media and TV all over the world including the BBC World Service and there are several documentaries on the HRM method of sungazing exhibited on different worldwide channels to encourage the people's practice of sungazing.

He could hardly do more.

Chapter 25: Urine Therapy

You can get your average pH as follows: Saliva pH x 2 + Urine pH / 3

This is an equation which science has been working towards. Individual scientists have proved each part of the equation but this is the first time that a convincing synthesis has been achieved.

The product of your own metabolism: The first time I was exposed to the concept that the body makes its own medicine it was through a friend who was introduced to it through an Eastern Indian guru type of fellow. I thought it was interesting but a wacky kind of idea and I never thought about it again until I came across a book. The book is entitled Your Own Perfect Medicine by Martha Christy. She has an interesting story to tell.

As usual, scientists are loath to allow Eastern Indian guru types of fellows very little credit. However, given Martha Christy's story they should change their attitude.

Ms. Christy was sick. Very sick. For a very long time. Pelvic inflammatory disease, ulcerative colitis, Chron's disease, chronic fatigue syndrome, Hashimoto's disease, mononucleosis. She had severe kidney infections, two miscarriages, chronic cystitis, severe candida, endometriosis, adrenal insufficiency, serious chronic ear and sinus infections, food and chemical allergies.

And that wasn't the half of it. She had every conceivable medical test, her share of surgery and drugs - plenty of them. Then she tried all forms of alternative therapy. Homeopathy, herbs, mega-vitamins and live-cell treatments in Mexico. After traditional medicine failed to work she and her husband spent over $100,000 trying to get her well with alternative approaches. Nothing worked.

This is hardly surprising. Anyone who has tried to rid themselves of these ailments would quickly come to the conclusion that modern medical practice has little to offer.

And then one day, her husband brought home a little book that told of how individuals had been cured of even the worst diseases with a seemingly strange and little-known natural therapy. Soon afterwards she began the therapy herself. From the first day she began she received almost instantaneous relief from her incurable

constipation and fluid retention. **Within a week her severe abdominal and pelvic pain was gone.**

This period of recovery alone should have alerted scientists to the effectiveness of the therapy.

The chronic cystitis and yeast infections, internal and external, soon disappeared and her food allergies, exhaustion, and digestive problems all began to heal.

And the removal of yeast infections in particular should have indicated that here was something which required urgent investigation.

After a few more months her colds, flu, sore throats and on again off again viral symptoms disappeared. Her hair, which had fallen out by the handfuls after her fifth surgery, became thick and lustrous. Her weight normalized and her energy and strength came back. After nearly 30 years of non-stop illness Martha Christy was whole again.

But still science ignored the facts. Why is it that scientists so often doubt evidence when it is right there in front of their faces?

What was this therapy she had discovered? What was this therapy that helped seriously ill patients gain complete remission from their afflictions? What was it she actually did? Well, here it is. She orally and medicinally re-consumed her own urine.

It is surprising that science has missed this important cure but once more, international drug companies may have to bear the burden of blame. They are in the business of manufacturing drugs for us to take but what if those drugs were produced by our own body?

A nutrient rich powerhouse: In 1975 one of the founders of Miles laboratories, Dr. A. H. Free, published his book Urinalysis in Clinical Laboratory Practice in which he remarked that not only is urine a sterile body compound purer than distilled water, but that it is now recognized that urine contains literally thousands of compounds. Among the urine constituents mentioned in Dr. Free's treatise is a list of nutrients that will knock your socks off. Here's just a few...

Alanine, arginine, ascorbic acid, allantoin, amino acids, bicarbonate, biotin, calcium, creatinine, cystine, DHEA, dopamine, epinephrine, folic acid, glucose, glutamic acid, glycine, inositol, iodine, iron, lysine, magnesium, manganese, melatonin, methionine, nitrogen, ornithane, pantothenic acid, phenylalaline, phosphorus, potassium, proteins, riboflavin, tryptophan, tyrosine, urea, vitamin B6, vitamin B12 and zinc.

How much would it cost for a person to buy these compounds from a licensed manufacturer? But here they are, freely available in our own urine.

Despite what you may have been led to believe about urine, pharmaceutical companies have grossed billions of dollars from the sale of drugs made from urine constituents. Research is happening every day in labs attempting to isolate specific elements of urine so they can create new drugs and patent the substances.

For instance, Pergonal is a fertility drug made from human urine. 1992 sales of this drug were reported at $855 million while it costs a patient $1,400 a month to buy. Urea, medically proven to be one of the best moisturizers in the world, is packaged in expensive creams and lotions.

Scientists should make it a priority to disseminate the information that patients currently taking the medicines mentioned above could save themselves a considerable amount of money and gain exactly the same benefit simply by not flushing away their wee. But don't expect a rush on behalf of science to educate the public. Urine scientists should be investigated to see whether or not they are associated too closely with pharmaceutical companies and given the facts above I know where my suspicions lie.

More electrons for your body: Most of the time when a measurement of a urine sample's oxidation and reduction reading is given the Orp value is taken. It often shows the urine in a reduced state. Chemically this means that there are more electrons, the charge is reduced in the negative direction. We like to say that the flow of life moves on the flow of electrons.

This seems very plausible.

If there are a lot of electrons in the urine then where are they coming from? Yep, from you. So if your papa ever said, child, you're pissing your life away--might there be more to that statement then meets the ears?

It is almost universally true that colloquial sayings are accurate.

Free electrons can be given to free radicals to fight oxidation so perhaps urine therapy also works along these lines. Something to think about. Or maybe it is just that things are better the second time around, they've already been processed somewhat once.

This is certainly something for science to think about.

To take this isopathic idea a bit further let's come back to the theories of Guenther Enderlein and his biological perspective. As you'll recall, Enderlein developed remedies that were based on his observations that how we get sick is a function of the blood pH getting thrown off causing microbes in the blood to grow into pathogenicity.

His biological remedies involved re-introducing the small colloids of life, the protits, back into the body to mate with the pathogenic forms so they would de-evolve back into a non-pathogenic state.

Nothing that Darwin wrote ever denied the possibility of de-evolution, that is backwards evolution. The theory of Enderlein, improving upon Darwin's original idea, posits a testable fact.

Evolutionary Biologists have become stuck in a rut in recent times, content with what they call Neo-Darwinism in an attempt to pretend that they have come up with something new. Instead they have been marking time waiting for a truly new idea to emerge.

Well now they have one and they are trying to ignore it. Science often works like this though it shouldn't. Never underestimate the jealousy of scientists when it comes to original ideas.

Now let's take the idea in relation to urine therapy and expand it to meet this biological framework. Urine therapy has been able to cure seemingly incurable disease states. Why is that? Well here's an idea. The colloids in your blood that are the protits/somatids are very small particles. In fact they are so small as to be unfilterable by something like the kidneys.

Though researchers haven't been looking, perhaps one of the invisible constituents of urine are these colloid particles that naturally just get passed through the kidneys out of the blood. When one consumes their own urine through the process of urine therapy they are receiving a dose of these pure colloids.

These biologically proceed to support the immune function and possibly lead to taking the pathogenic microbes in the blood back down to apathogenic states. In this way urine truly becomes the body's own perfect medicine.

This is very specialised and I urge you to read it carefully or perhaps skip over it if you lack the necessary expertise. For those with the necessary technical knowledge the conclusion presented is highly plausible and the arguments underlying it deserve to investigated at the earliest possible opportunity.

Urine contains compounds that are very specific to the individual from which it comes. It is antibacterial, antifungal, antiviral, antineoplastic, anticancer, anticonvulsive and antispasmodic. It is totally non-toxic.

The antispasmodic industry alone would collapse if news of this got out.

With research that could fill volumes you may be wondering why you haven't heard of this before. Well, urine is abundant everyone has their own supply. It's free and it can't be patented. Reasons enough?

Indeed. But don't forget that politicians in the pay of drug companies might have something to do with it. When was the last time you saw a cabinet minister appear on television and tell you to drink your own wee?

How to do it: Very briefly here are two main ways that are suggested to do urine therapy. This is in no way a complete discussion of how to use the therapy but simply an introduction.

Use your own urine in a homeopathic fashion. First, collect midstream urine in a clean cup or container. This should be a clean catch, meaning the genital area, important for women in particular, has been cleaned beforehand. To 1/6 ounce of distilled water in a sterile bottle add one drop of fresh urine. Cap and shake 50 times. Take one drop of this mix and add to another 1/6 ounce of distilled water and shake 50 times. Take one drop of this mix and add to 1/6 oz. of 80 to 90 proof vodka which acts as a preservative.

One must be careful with the amounts here. There is a big difference between drinking piss and getting pissed.

Begin with oral drops then increase dosage as needed. Use fresh urine drops direct. For some cases sub-lingual drops work well. Should always use fresh urine immediately upon collection. You should not boil or dilute the urine in any way. You must use it in its natural form.

Start by taking 1-5 drops of morning urine on the first day. On the second day take 5-10 drops in the morning. On the third day take 5-10 drops in the morning and the same amount in the evening before you go to bed.

This might not seem much of a nightcap but it should be one of sciences next steps to prove that such a drink before bedtime can offer significant health benefits.

Once you feel accustomed to the therapy gradually increase the amount as needed for obtaining results for your condition. As you use the therapy you will learn to adjust the amount you need by observing your reactions. It may be that you'll work up to actually drinking an ounce or two at a time.

This may wake you up in the night because you want to go to the toilet but try to resist the urge. If not, take a small bottle with you.

Complete guide urine therapy: Urine consists of 95% water, 2.5% of urea and the remaining 2.5% is a mixture of minerals, salts, hormones and enzymes. Only urea, the substance after which urine is named, can be poisonous when present in the blood. However, this is irrelevant in the practice of drinking urine as urine is not immediately put back in the bloodstream. In small amounts urea gets back into the body, it is purifying, and clears up excess mucus.

Another helpful by-product of urine consumption. In the morning you probably won't need to blow your nose.

The kidney's most important function consists of balancing out all elements in the blood. They remove all superfluous vital substances from the blood and filter out a surplus of water. This water and the vital substances consequently form urine. In order to save energy and bring the blood into balance the kidneys remove unused enzymes from the blood. The same goes for hormones, minerals and other substances. It is clear that urine is full of vital elements which can hardly be called waste products.

It has long been known that soap-operas are widely used as a form of social engineering by having their characters espouse opinions intended to modify the behaviour of viewers by encouraging them to imitate the actions of their favourite soap stars. An small increase in the amount of wee drinking by these minor celebrities on screen could have an effect on the nation's health well out of proportion with the amount of money spent on persuading scriptwriters to include such scenes.

Indeed, with government intervention, television companies might well be obliged to have a certain amount of urine consumption written into popular shows as an integral part of the commissioning process.

And in the case of the BBC, a percentage of the licence fee could be made subject to the frequency at which urine is consumed in programmes across the network.

Human urine has strengthening and curative characteristics concerning many deficiencies. A mixture of potato and sulphur

powder mixed with heated old urine helps against hair loss. One should rub this mixture into the scalp; this slows down loss of hair.

More than one research establishment is attempting to discover the exact proportion of potato/sulphur powder which is most effective.

All kinds of throat inflammation can be helped by gargling with urine to which a bit of saffron has been added.

I thought everyone knew this already.

Trembling hands and knees can be helped by washing and rubbing one's own warm urine into the skin directly after one has urinated.

This was discovered by accident and is an example of empirical scientific research which can be reproduced by anyone. In this case, drunken men whose hands naturally trembled, urinated on themselves rather than getting it into the bowl. The next morning, after a good night's sleep, they found their symptoms had completely disappeared.

'A universal and excellent remedy for all distempers inward and outward. Drink your own water in the morning nine days together and it cures the scurvy, makes the body lightsome and cheerful.'

Science is in the process of confirming that wee is just as good a cure for scurvy as Vitamin C. In retrospect it is a shame the urine cure wasn't known in the days when mariners fell victim to this vitamin deficiency disease on long sea voyages. If they had been aware the cure would have been readily available and much hardship could have been avoided. It would also have led to a different nickname for English sailors as they would no longer have needed to take citrus fruits along with them in order to ward off the disease. Instead of becoming known as 'Limeys' they may have become known as something else. Pissheads, perhaps.

'It is good against the dropsy and jaundice, drunk as before. Wash your ears with it warm and it is good against deafness, noises and most other ailments in the ears. Wash your eyes with your own water and it cures soreness and clears and strengthens the sight.

'Wash and rub your hands with it and it takes away numbness, chaps and sores and makes the joints limber. Wash any green wound with it and it is an extraordinary good thing. Wash any part that itches and it takes the itch away. Wash the fundament and it is good against piles and other sores.'

This would explain why the source of urine is positioned so close to the fundament, the part of the body prone to piles. We may be seeing evolution in action.

One's own urine is the best medicine for the kidneys that we could imagine. It is clear for any holistic health practitioner that all cycles within the body are interconnected and this means that the healing of one of those cycles will have a positive effect on the others. Urine was often used at the front, for lack of other medication, and as a disinfectant for surgery instruments.

A Russian doctor treated many people from far and wide and was able to alleviate or completely cure illnesses with nothing else than urine therapy, while other methods up until then had failed.

There is a whole history to be written about Russian doctors and their seemingly infallible eye for novel cures using limited resources in the most pressing circumstances. Perhaps a cure for one of today's intractable diseases could be found by having a Russian doctor present with terminally ill patients in a poorly equipped ward bereft of nursing care. Science should make it a priority to try such an experiment. It appears to have succeeded so often in the past that it should not necessarily be considered a long shot.

Drinking urine is a good alternative wherever water is scarce. It not only satisfies the need for liquid but also actually keeps the body healthy. Some time ago there was an earthquake in Egypt. A survivor was pulled out of the rubble in Cairo after being trapped for three days. The man had kept himself alive by, among other things, drinking his own urine and he was in excellent condition.

Evidence from some time ago is still admissible as scientific evidence.

I heard another story about a man who kept himself alive with his own urine for a week in a collapsed mine. At the time of his rescue he looked fine and was in extraordinary health.

Hearing stories is also allowed.

I also recently read an article about an Italian athlete who was lost in the Sahara for ten days. Upon returning to the civilised world he told how he had drunk his own urine for lack of other liquids. He had kept himself alive by eating desert plants and insects and drinking his own urine.

However, reading articles is the best scientific evidence of all. Written articles go through a review process. This is science's way of ensuring that any mistakes which may slip through are detected and removed.

How many people do you know who have drunk enough urine to really know what it tastes like? Probably not too many. But taking urine into your mouth might be too big a step to begin with. Rubbing a drop into the skin and first smelling your own urine can help you to overcome part of the barrier. Really, it often does not smell bad at all. Many people even like its sometimes sweet odour.

This is wise advice. It also conforms to modern medical practice as in allergy testing. A small amount can gradually be built up into larger and larger amounts as the body becomes used to the increased dosage. Note: there is no evidence at all that your own urine is actually allergic for your own body, this is merely an example.

More extensive massaging of urine into your skin is also a good way to become accustomed to your life water.

It is also a means of ensuring other people become accustomed to the way you now smell.

Before a fast: Two days before the fast decrease the intake of protein-rich and heavy foods, especially fried and fatty foods. Fruit and raw vegetables are easily digestible and ensure that the intestines clean themselves so the actual fast can easily begin. In this period start drinking greater amounts of urine.

Heavy foods are implicated in the phenomena of unclean intestines so this makes sense.

Actual fast: In this period exclusively drink water and urine. It is best if you do not work during the fast. Although some exertion is possible rest and relaxation are important in order for the purifying process to take place undisturbed. In the beginning, stay with drinking the middle stream. Alternated with urine, pure, clean water can be drunk. Once the fast has been in progress for some time all the urine can be drunk. In this period you will urinate quite easily: Urinating every fifteen minutes is not unusual.

It is sound advice. Your co-workers may find your behaviour a little out of the ordinary.

A complete body massage every day with old, heated urine is highly recommended. Urine massage is good for blood circulation and massaging with old urine also ensures that you do not have heart palpitations. Furthermore it serves as a way of feeding the body through the skin, immediately into the muscle and lymph tissue.

Modern cardiac surgeons are increasingly becoming interested in stopping heart palpitations by having a supply of old urine close at hand.

Nutritionists, likewise, are coming around to the idea of feeding the body bypassing the traditional and increasingly outmoded digestive system.

Skin feeding of the body should become more widespread in health services throughout the world as the benefits would be immediate. One thinks of dental patients who cannot chew. Massaging urine into them offers many advantages.

During fasts urine enemas are highly recommended. Many illnesses begin in the intestines and it is very important to get rid of toxic waste products stored in them and to keep them clean.

During a chronic illness, the body often has a high level of toxicity, which means that poisonous substances can be found in the tissues. Administering an enema is a good way to remove poisonous substances from the body and especially from the intestines. Furthermore, a number of substances found in urine are better absorbed by the body in this way than by oral ingestion.

It is intuitive that an enema can remove poisonous substances from the body so this is good practice.

Urine should be kept in the mouth for twenty to thirty minutes or, when that seems too much, as long as possible.

Another reason for not attempting to go to work.

Vomiting particularly occurs when urine tastes and smells very strong and unpleasant, such as is the case with fever, jaundice and a number of other diseases. In some cases drinking urine can be extremely unpleasant. However, if you drink as much urine as possible the urine will quickly become thinner and taste more pleasant.

As doctors will confirm, the start of any new therapy can make you feel sick so this is no cause for concern. By increasing the amount of therapy the body quickly becomes accustomed.

During a cleansing process the body sometimes removes excess mucus from the lungs and bronchial tubes. If a great deal of mucus is released, reduce the amount of urine to be ingested or stop temporarily. Start inhaling urine through the nose as this clears the upper part of the bronchial tubes.

Science should research the type and size of straw necessary for optimum results.

Urine is free and always available for those who need it. In a world which money talks urine therapy can appear be threatening to those who earn their living by manufacturing or prescribing medicine.

And that, in a nutshell, is why the beneficial effects of drinking urine has been denied to the population for so long. It has very little to do with vomiting and much to do with the money orientated vested interests of modern medicine. Science is duty bound to expose this.

Many people believe we are healthier nowadays thanks to advances in medical science. This is partly true. The flip side of the coin is that we have had to give up a great deal of freedom and independence. The enormous efforts of medical science are partly based upon the failure to really cure illnesses. Fighting symptoms is considered to be crucial but this does not take care of the cause. Urine therapy, being a real nature cure, not only reduces the symptom but also deals with the cause of the illness.

It is difficult to think of one single modern medical advance which deals with the cause of illness. Modern medical practice should move decisively towards real nature cures such as urine therapy.

When you are convinced it works for you the reluctance to talk about it will gradually subside and maybe even disappear. Perhaps you will start to enjoy talking about it, as I do. You can be sure that you will be surrounded by plenty of laughter, which in fact can be rather pleasant.

The odour of urine is much less repulsive than most people think since most associate urine with public toilets. Actually, it often smells rather pleasant when applied fresh on the skin. I myself regularly use pure urine as aftershave and hair lotion.

Discussion of this could certainly break the ice when entertaining a new date, at an agreeable restaurant for example. One might well anticipate plenty of laughter as an enjoyable evening passes.

The majority of useful, vital substances is found in the morning urine. This is because at night, while you sleep, your body is totally relaxed. This deep relaxation gives the body the chance to carry out its recovery activities. The decomposition products partly end up in the urine and can be re-absorbed and used for new build-up processes. This process of filtering by the kidneys returns the so-called raw decomposition products to their original substances, which can subsequently be absorbed and reused by the body.

Indeed, restoration now appears to be more important than the actual filtering process itself. This new insight is being applied practically by manufacturers of the new generation of dialysis machines.

Certain hormones are also released during sleep, a number of which are intended to bring about the above mentioned deep

relaxation. Reabsorbing these hormones ensures that we are more rested during the waking hours. It saves the body energy because it does not have to manufacture these hormones again. Ingestion of the morning urine, which is full of hormones, regulates the entire hormonal process. These hormones have the particular function of maintaining hormonal balance.

Yet another reason for drinking your urine. As well as splashing it on yourself as a pleasant aftershave and applying it to your head as a natural hair lotion you can save your body the effort making hormones first thing in the morning when it can be rather hectic.

In theory, it is best to use exclusively your own urine, especially if internally applied. However, if you are in a state of shock and cannot urinate the urine from somebody else can safely be administered. If possible use the urine from somebody of the same sex. Different hormones can be found in the urine from a male than in that of a female.

For certain illnesses it seems to be beneficial to ingest the urine from children. The urine from a child is often very pure, especially if the child follows a healthy diet. In some cases the urine from a child can also be used in the external massage application for the seriously ill who cannot produce enough of their own.

Science should now stop spending so much time and effort searching for esoteric and probably undetectable gravitons and start advising the public that drinking children's wee is beneficial for them. With the backing of scientists opinions public misgivings should soon fade away. Making sure children have followed a healthy diet may prove more problematical. Science should develop tests which can be applied so that patients are not offered wee from children whose diet is largely pizza, for example.

Urine from different people usually does not differ much in its ingredients, which is why the urine from one person will also to some extent work for somebody else. However your own urine contains personal, characteristic substances and provides the particular information the body needs in order to carry out the healing process as effectively as possible.

That information can be contained in urine seems beyond doubt but science should now move towards establishing exactly what information is provided.

In general, it is no problem to combine urine therapy with any other form of natural treatment. This also applies to the use of nutritional supplements as long as they are natural. One should refrain from any chemically manufactured supplements.

There is a world of difference between a compound derived from natural sources and the same compound synthetically manufactured in a laboratory and it should not be beyond science to convincingly demonstrate the difference between the two.

Experience has shown that the use of vitamin supplements in combination with the practise of urine therapy can considerably cut down the amount of supplements you need to take because of the recycling effect. Many substances, such as vitamins and enzymes, act as carriers for the other.

The symbiosis between vitamins and enzymes is completely new and has either been overlooked or disregarded by scientists. This is a major error which should be corrected immediately.

Urine therapy can also be seen as a form of self-vaccination: Certain bodily substances which have been removed from the body, some of which may have been produced as a result of illness, are reintroduced into the body in small amounts. These substances are reabsorbed into the blood through either the intestines or the skin. According to this hypothesis, the immune system is then given the chance to react appropriately.

It should come as no surprise that urine contains substances which are capable of self-vaccination, after all substances produced as the result of illness should have been suspected all along to be present in the excretory fluids. How science could have missed this for so long is something of a mystery but now it has the chance to make up for its negligence and show once and for all that rubbing urine into one's skin is pretty much the same as being vaccinated.

In the early nineteenth century Dr. Charles Duncan conducted research into therapies with self-produced substances, including urine therapy. He demonstrated that patients suffering from gonorrhoeic urethritis infection of the urinary tube, as a result of the venereal disease gonorrhoea, produce their own medication in the form of their own discharge.

Science seems to have completely missed the importance of Dr. Duncan's work.

Auto-therapy was applied here by placing a drop of a patient's discharge directly on the tongue, in order to stimulate the body's natural powers. This method had a strong healing effect at every

stage of the illness: If applied at an early stage, it could cause the gonorrhoea to disappear.

Nothing appears more natural than placing the discharge from a gonorrhoeic patient directly onto their tongue yet somehow the benefit of the procedure seems to have been forgotten.

Auto-therapy is based on the principle that the body can use all fresh, self-produced, unaltered diseased tissue substances which originate from the micro-organisms causing the illness. Seen in this light, patients have their own medication in exactly the form constructed by nature to heal their condition.

Diseased tissue substances promise much in the decades ahead. However, it can be surmised that multi-national pharmaceutical companies, with their vested interests and heavily capitalised manufacturing plants, will stand in the way. How we can rid ourselves of such an outmoded paradigm of treatment is an open question.

One way forward might be for people with a specific illness to appear on television and openly avow to reject drugs created in laboratories. They should allow diseased tissue substances produced by their own body to be administered to themselves perhaps by rubbing the substances onto their skin or pipetting it onto their tongue. They could then come back a week later and show viewers how effective the treatment has been.

No modern drug-producing company could possibly compete with this sort of demonstration.

I have already referred to research on the effects of melatonin, a hormone found in urine which possibly has a calming effect. Melatonin also has a powerful anti-cancer effect.

Doctors should offer cancer patients the opportunity to drink their own urine without delay.

Some people recommend to use the first urine after sexual intercourse. During the process of sexual stimulation certain hormones are released in the higher endocrinal glands which have a regenerating effect on the body for men as well as women.

For those unable to wait for science to conduct its time-consuming tests there is no reason why they should not pre-empt the findings and go ahead with their own self-medication, taking a bottle to bed with them as well as their partner.

Another important aspect is the theory of structured water. The body consists for the biggest part of water and so does urine. Not all water is the same though. The molecular structure of water can be

less or more organized and in the latter case one speaks of structured water.

The more it is organized the better all kinds of enzymatic processes can do their job. These enzymatic processes, in their turn, are responsible and necessary for the digestion, absorption and transmutation of all nutrients. Urine is thus a crystalline-like substance containing a high amount of structured water.

Science has been lax in investigating structured water. Too often it has left water unstructured in its experiments and this is increasingly being seen as tantamount to scientific negligence.

A higher amount of structured water in the body system is correlated with better health and more energy. The fact that urine is a liquid crystal substance, particularly because of the various salts in it, implies that it contains crystalline vibrations completely in tune with the vibrational condition of the body. Reingestion might give the body valuable vibrational information. Healthy vibrations will strengthen the already existing, healthy body resonance.

It seems good common sense that drinking wee will give the body vibrational information.

It is known that disturbing sounds of any sort can be counteracted best by confronting it with the same sounds. The vibratory patterns of the body, both in the bones solid crystals and in the tissues and fluids liquid crystals, play an important role in the process of transmutation. The resonance field of a crystal can make a protein, for example, change its form into one that is more useful for the body, or easier adaptable by it.

Biologists have been slow to realise that proteins, under the influence of crystals, can change their form into one more useful to the body. This should now open up a fruitful new field of research.

Applying your own bodily substances in an attempt to heal can lead to a considerably broadened outlook on the intelligence and power of the body and can increase your appreciation and love for yourself as a physical and spiritual being. Instead of regarding excretions, really just a part of yourself, as enemies, you regard them as your helpers.

There is no intrinsic reason why beneficial excretions should stop at urine. Science should move towards investigating other bodily substances formerly thought of as waste in the same philosophical light and view them instead as helpers in the attempt to heal.

There is therefore every reason to suppose that the substances harvested from blowing ones nose could find application in treatment of rhino-specific ailments just as the application and/or ingestion of faeces could lead to the elimination of all sorts of alimentary canal disorders.

For example, discharges from the nose may well be found to be specific against the common cold, an ailment notoriously immune to traditional science's efforts at cure. In the same spirit poo might be internally reintroduced into the body to provide valuable vibrational information. The use of these formerly unheralded substances should be demystified by science as soon as possible.

This healthier way of seeing yourself might well have a powerful healing effect on your body. Urine therapy confronts us with a very concrete 'healer within' which works both on a mechanistic and on an energetic level. The latter implies that urine, as a holographic substance, can affect all levels of being from the physical, through the electromagnetic fields of the emotions and the mind up to the subtler genetic vibrational information of the soul.

While the soul appears out of reach for science as presently constituted (though we can hope for progress as the debate continues) the status of urine as a holographic substance seems well established. That it works at both a mechanistic and an energetic level is denied by no-one in the scientific community that I am aware of.

The electromagnetic fields of the mind are already the subject of research in some of the most well-funded laboratories in the world and the more advanced of these are moving quickly to establish the same electromagnetic basis for the emotions. Reports from individual research centres hint at exciting discoveries concerning subtler vibrational information at least some of which seems, at this early stage, to be almost certainly genetic in origin.

Given all this, the stage seems set for urine therapy to become one of the most important of Science's Next Steps.

Science's Next Steps

Is dedicated to scientists everywhere
Now and in the future
In the hope they will
Keep an open mind

Chapter 1: Aromatherapy

http://www.organic-aromatherapy.co.uk/

Chapter 2: Astrology

http://www.cafeastrology.com/celebrityastrology.html

http://www.exploreastrology.co.uk/basicscharts.html

Chapter 3: Bermuda Triangle

http://www.bermuda-triangle.org/html/vortex_kinesis.html

Chapter 4: Better Vision

http://www.bettervision.com/

http://www.i-see.org/bates_nutshell.html

Chapter 5: Bio-Energy

http://www.thunderbolts.info/forum/phpBB3/viewtopic.php?p=34960&sid=275528702d54f7998ed538d83e53907f

https://www.quantumbalancing.com/pwm45.htm

http://www.energetic-devices.com/

http://energy-medicine.org/

Chapter 6: Biology

http://biomedx.com/microscopes/rrintro/rr1.html

Chapter 7: Diode

http://www.quantec.eu/english/

Chapter 8: Dowsing

http://www.mystical-
www.co.uk/index.php?option=com_content&view=article&id=484&Itemid
=615

http://www.selfgrowth.com/print/584050

http://www.intuitivedowsing.com/

Chapter 9: Energy Healing

http://www.newpathwaytohealing.com/healing/ian-stone-founder-of-
metaphysical-institute-in-australia-metaphysician-energy-healer-light-
worker/

http://www.selfgrowth.com/articles/the_value_of_our_own_healing

Chapter 10: Feng Shui

http://www.selfgrowth.com/articles/feng-shui-and-electronics

http://www.selfgrowth.com/articles/using-feng-shui-to-help-illnesses

http://www.selfgrowth.com/articles/feng-shui-fun-crystals-to-the-rescue

http://www.tomorrowskey.com/FengShuiConsulting.html

Chapter 11: German New Medicine

http://www.learninggnm.com/home.html

Chapter 12: Golden Ratio

http://www.selfgrowth.com/articles/human_evolution_uses_golden_ratio_ne
ural_patterns

https://www.createspace.com/1796253

Chapter 13: Homeopathy

http://www.naturalnews.com/024670_home_homeopathy_homeopathic.html

http://www.naturalnews.com/029419_homeopathic_medicine_evidence.html

http://www.homeopathic.com/Homeopathic_Revolution/

Chapter 14: Medical Products

http://www.drclark.net/products-devices-a-techniques/zapper-basics/zapping

http://www.drclarkinfocenter.com/en/products_devices/devices/zappicator.php

http://www.orgoneenergy.org/orgone-reviews

http://ezinearticles.com/?Orgone-Energy-and-Schumann-Resonance-Metaphysical-Products-and-Their-Effects&id=1899686

http://schumannresonator.com/

https://www.natures-energies.com/health/the-orgone-geoclense

http://www.orgoneenergy.org/pendant-frequent-questions.html

Chapter 15: Medicine

http://abcnews.go.com/Health/story?id=6596366&page=1

http://www.appleyoga.com/tpo3/

http://www.diseaseproof.com/archives/healthy-food-how-asparagus-are-you.html

http://www.cancertutor.com/bob-beck/

http://www.thunderbolts.info/~thundes2/forum/phpBB3/viewtopic.php?p=25250&sid=db1ef91a825ff9fb8ea9005b0cabd6dd

Chapter 16: Mind Science

http://www.pillaicenter.com/whatismindscience.aspx

http://www.pillaicenter.com/Miracle-Stories.aspx

Chapter 17: Numerology

https://turntheperspicacityup.wordpress.com/2013/08/25/numerologysymbolism-the-study-of-the-occult-significance-of-numbersthe-use-of-symbols-to-express-things-by-steve-gamble-2/

Chapter 18: Orgone Energy and Tesla

http://www.thunderbolts.info/forum/phpBB3/viewtopic.php?p=37353&sid=5693c2d33b44f473062bef3f152709c0

http://peswiki.com/index.php/PowerPedia:Eric_Dollard

Chapter 19: ORMES

http://breakthru-technologies.com/sites/breakthru-technologies.com/files/Hydrosil%20Ormes.pdf

Chapter 20: Physics

http://www.thunderbolts.info/forum/phpBB3/viewtopic.php?f=6&t=457

http://www.thunderbolts.info/forum/phpBB3/viewtopic.php?f=8&t=1106

Chapter 21: Products

http://www.equilibrauk.com/products.shtml

Chapter 22: Radionics

http://www.telegraph.co.uk/lifestyle/wellbeing/5356013/Radionics-can-a-lock-of-hair-hold-the-key-to-health.html

http://homoeonic.com/product/evolution-mxd/

http://www.alibaba.com/product-detail/Radionic-Antenas-Homoeonic-Equipment_106873115.html

http://www.remedydevices.com/

Chapter 23: Reiki

http://www.equilibrauk.com/shop1.shtml

http://www.equilibrauk.com/importanceofwater.html

http://www.equilibrauk.com/illusion.shtml

http://www.toolsforenergy.com/article_stevegamble.asp

http://www.equilibrauk.com/illness.shtml

http://www.byregion.net/cgibin/users/profiles.pl?username=taylore

Chapter 24: Solar Healing

http://solarhealing.com/

http://www.naturalnews.com/024256_sun_gazing_food_life.html

http://www.healingenergytools.com/pineal-gland-2/

Chapter 25: Urine Therapy

http://www.biomedx.com/urine/

http://www.universal-tao.com/article/urine_therapy.html

www.ingramcontent.com/pod-product-compliance
Lightning Source LLC
Chambersburg PA
CBHW071409180526
45170CB00001B/25